T0140033

Theory and Applications of Computability

In cooperation with the association Computability in Europe

Books published in this series will be of interest to the research community and graduate students, with a unique focus on issues of computability. The perspective of the series is multidisciplinary, recapturing the spirit of Turing by linking theoretical and real-world concerns from computer science, mathematics, biology, physics, and the philosophy of science.

The series includes research monographs, advanced and graduate texts, and books that offer an original and informative view of computability and computational paradigms.

More information about this series at http://www.springer.com/series/8819

S. Barry Cooper • Mariya I. Soskova
Editors

The Incomputable

Journeys Beyond the Turing Barrier

Editors
S. Barry Cooper
School of Mathematics
University of Leeds
Leeds, United Kingdom

Mariya I. Soskova
Dept. of Mathematical Logic
& Applications
Sofia University
Sofia, Bulgaria

ISSN 2190-619X ISSN 2190-6203 (electronic)
Theory and Applications of Computability
ISBN 978-3-319-82881-7 ISBN 978-3-319-43669-2 (eBook)
DOI 10.1007/978-3-319-43669-2

Printed on acid-free paper

This Springer imprint is published by Springer Nature
The registered company is Springer International Publishing AG
The registered company address is: Gewerbestrasse 11, 6330 Cham, Switzerland

Preface

As coeditor, I would like to express my very great appreciation to a number of individuals and institutions who supported this project throughout its development. The event that gave rise to this book would not have come to be without the practical and financial support of the Isaac Newton Institute and their staff and the funding from the John Templeton Foundation. The Chicheley Hall International Centre and its Royal Society staff were magnificent in providing the right atmosphere for the workshop. And the remarkable participants were the ones who made the event such an exciting one.

I am grateful to Springer and the editorial board of the book series "Theory and Applications of Computability", who accepted our idea for this book with no hesitation and supported us throughout its development. Ronan Nugent's advice and encouragement is deeply appreciated and was especially needed during the final stages of this project. I thank all contributors to this book for their diligent work and patience; the numerous reviewers for their valuable suggestions, comments and questions; the Sofia University Science Foundation and the programme "Women in Science" for their financial support; and my husband Joe Miller and my mother Alexandra Soskova for their unwavering moral support. Most of all, I would like to thank Barry Cooper for including me in this great adventure.

Madison, WI, USA
28 March 2016

Mariya I. Soskova

Contents

Contributors

Ruth E. Baker Wolfson Centre for Mathematical Biology, Mathematical Institute, University of Oxford, Oxford, UK

J. Mark Bishop Goldsmiths, University of London, London, UK

Cristian S. Calude Department of Computer Science, University of Auckland, Auckland, New Zealand

Peter A. Cholak Department of Mathematics, University of Notre Dame, Notre Dame, IN, USA

Kate Clements Blackett Laboratory, Imperial College, London, UK

Bob Coecke University of Oxford, Oxford, UK

Fay Dowker Blackett Laboratory, Imperial College, London, UK

Perimeter Institute, Waterloo, ON, Canada

Institute for Quantum Computing, University of Waterloo, Waterloo, ON, Canada

Hristo Ganchev Faculty of Mathematics and Informatics, Sofia University, Sofia, Bulgaria

Antonina Kolokolova Department of Computer Science, Memorial University of Newfoundland, St. John's, NL, Canada

Seth Lloyd Mechanical Engineering, Massachusetts Institute of Technology, Cambridge, MA, USA

Philip K. Maini Wolfson Centre for Mathematical Biology, Mathematical Institute, University of Oxford, Oxford, UK

André Nies Department of Computer Science, University of Auckland, Auckland, New Zealand

Dimiter Skordev Faculty of Mathematics and Informatics, Sofia University, Sofia, Bulgaria

Aaron Sloman School of Computer Science, University of Birmingham, Birmingham, UK

Mariya I. Soskova Department of Mathematical Logic and Applications, Sofia University, Sofia, Bulgaria

Vlatko Vedral Department of Physics, Clarendon Laboratory, University of Oxford, Oxford, UK

Centre for Quantum Technologies, National University of Singapore, Singapore, Singapore

Department of Physics, National University of Singapore, Singapore, Singapore

Center for Quantum Information, Institute for Interdisciplinary Information Sciences, Tsinghua University, Beijing, China

K. Vela Velupillai Madras School of Economics, Chennai, India

Petros Wallden School of Informatics, University of Edinburgh, Edinburgh, UK

IPaQS, Heriot-Watt University, Edinburgh, UK

Thomas E. Woolley Wolfson Centre for Mathematical Biology, Mathematical Institute, University of Oxford, Oxford, UK

Introduction

Mariya I. Soskova

What is the relevance of computation to the physical universe? Our theories do deliver computational descriptions, but the gaps and discontinuities in our grasp suggest a need for continued discourse between researchers from different disciplines.
Barry Cooper, 2014

Abstract In this introduction we briefly describe the history and motivation that lead to the workshop "The Incomputable" and to the idea for a book reflecting the main theme of the workshop - the mathematical theory of incomputability and its relevance for the real world. We explain the structure and content of the resulting book and clarify the relationship between the different chapters.

The first recorded use of the term "incomputable" known to the Oxford English dictionary is from 1606 in a book bound in the skin of a Jesuit priest, Henry Garnet, which describes his trial and execution for complicity in the gunpowder plot. From context, it was almost certainly a misprint for "incompatible", though the word is recorded in the seventeenth century as being used to mean "too large to be reckoned". Since then the meaning of this term has been shifting from being "too large" to "too complex". By 1936, the term had been transformed into a robust mathematical notion through the work of Church [2] and Turing [11]. However, the implications of incomputability for the physical world presented a difficult problem, one that far too few attempted to battle. A gap was forming, separating classically trained abstract thinkers from the practical scientist searching for explanations of the workings of the world.

Barry Cooper was my advisor, mentor, and friend. Both he and I ventured into the scientific world through the abstract and rigorous land of mathematics in the form of classical computability theory. By the time I met him, however, he had become more and more discontented by the isolation of this field from practical, as he called it "real world", science. For many years he devoted his energy to changing this.

M.I. Soskova (✉)
Department of Mathematical Logic and Applications, Sofia University, Sofia, Bulgaria
e-mail: msoskova@fmi.uni-sofia.bg

S.B. Cooper, M.I. Soskova (eds.), *The Incomputable*, Theory and Applications of Computability, DOI 10.1007/978-3-319-43669-2_1

The conference series Computability in Europe is one example of the success of his efforts. It started in 2005 with its first event held in Amsterdam. It was co-chaired by Barry and it was the first conference that I attended in my scientific carrier. The motto was "New Computational Paradigms", explained by the organizers as follows:

> These include prominently connections between computation and physical systems but also higher mathematical models of computation. The researchers from the different communities will exchange ideas, approaches and techniques in their respective work, thereby generating a wider community for work on computational issues that allows uniform approaches to diverse areas, the transformation of theoretical ideas into applicable projects, and general cross-fertilization transcending disciplinary borders.

Since then, the conference CiE has been held annually all over Europe and has been breaking the walls of the rigid classical use of the term "computability" and extending it to a much wider area of research. The list of topics included in the conference grew with every year: classical computability, proof theory, computable structures, proof complexity, constructive mathematics, algorithms, neural networks, quantum computation, natural computation, molecular computation, computational learning, and bioinformatics. The community grew into an association with a scientific journal and a book series, to which this book belongs.

The year 2012 marked a special milestone in the development of the community of computability theorists in Europe: we celebrated the centenary of the birth of Alan Turing, the man who showed through his work that a researcher need not be constrained only to the world of the abstract or the world of the applied. The year quickly came to be known as the "Alan Turing Year". Barry Cooper was the chair of Turing Centenary Advisory Committee, coordinating the year-long celebration. From all over, researchers whose work related to that of Alan Turing gathered in the United Kingdom, and Cambridge in particular. The Isaac Newton Institute for mathematical sciences hosted the six-month program "Syntax and Semantics: A Legacy of Alan Turing". Cambridge was the venue for the annual conference Computability in Europe, which was co-chaired by Barry, and with more than 400 participants was by far the largest in the series.

One of the special events envisioned by Barry Cooper for this year was the workshop "The Incomputable", celebrating Turing's unique impact on mathematics, computing, computer science, informatics, morphogenesis, philosophy, and the wider scientific world. The workshop was unique in its focus on the mathematical theory of incomputability and its relevance for the real world. Barry considered this to be a core aspect of Turing's scientific legacy, and this meeting attempted for the first time to reunite (in)computability theory and "big science". Barry had a vision for this workshop and I was thrilled when he shared it with me and asked me to join him as an organizer. The aim was to bring together mathematicians working on the abstract theory of incomputability and randomness with other scientists, philosophers, and computer scientists dealing with the incomputability phenomenon in the real world in order to establish new collaborations around a number of "big questions":

- *Challenging Turing: Extended Models of Computation.* There is growing evidence from computer science, physics, biology, and the humanities that not all features of the real world are captured by classical models of computability. Quantum physics, in particular, presents challenges even to the classical logic we use to reason about the world. To what extent is the classical model flexible, and can it be adapted to different settings? To what extent can we develop a new form of logic and new computational paradigms based on and meant to describe natural processes?

- *The Search for "Natural" Examples of Incomputable Objects.* In 1970, Yuri Matiyasevich [6] put the finishing touches on a remarkable proof [3], many years in the making, that the sort of arithmetical questions a school student might ask lead directly to a rich diversity of incomputable sets. The negative solution to Hilbert's Tenth Problem, collectively due to Martin Davis, Matiyasevich, Hilary Putnam, and Julia Robinson, demonstrated that the incomputable sets which can be abstractly enumerated by recursion-theoretic techniques are not artificial at all—they arise naturally as solution sets of Diophantine equations. Can one discover a specific natural set, ideally a computably enumerable set, that is incomputable but not computably equivalent to the halting set of a universal Turing machine? How does this play out in the physical world? Can we find new mathematical techniques for identifying incomputability, or discover new results explaining the obstacles we encounter?
- *Mind, Matter and Computation.* Turing's seminal role in artificial intelligence [12], for instance, his formulation of the "imitation game", later known as the "Turing test", has given rise to many fundamental questions in philosophy that remain unanswered. To what extent can machines simulate the human brain, to what extent are our decisions and actions automated? What is the relevance of the algorithmic world to the real world and how is the notion of computability reflected in modern philosophy? What are the consequences of quantum phenomena for the construction of an artificial brain?
- *The Nature of Information: Complexity and Randomness.* There are two ways related to incomputability in which information can be difficult to extract and process. In complexity theory, the focus is more on what is realistically possible in the way of information processing. The area has its origins in cryptology, the workings of which were blatantly present in Turing's mind from his time in Bletchley Park, working on the Bombe. The area has since developed into a rich and diverse theory; still there are basic issues that remain a mystery. For example, is there a difference between the efficiency of non-deterministic and deterministic computation? In the mathematical theory of effective randomness, the focus is on incompressibility and unpredictability. Randomness plays well both with the Turing world of incomputability and the practical world of complexity. What is the significance of the new definable classes emerging from computability theory and the study of effective randomness? What does effective randomness tell us about quantum randomness? Is quantum randomness really random, or merely incomputable?
- *The Mathematics of Emergence and Morphogenesis.* One of the less familiar, but most innovative, of Turing's contributions was his successful mathematical modeling of the emergence of various patterns in nature using mathematical

equations [13]. This key innovation now forms the basis of an active and important research field for biologists. There are still general questions concerning both the computational content of emergence, and the nature of good mathematical models applicable in different environments. Does emergence exist as something new in nature, and is it something which transcends classical models of computation?

There are many more important questions, of course, but we chose these five themes as guiding, when we designed the program of the workshop. And soon it was June 2012 and all of our planning triumphed into one of the most exciting and intellectually challenging event that I had ever attended, fully embodying Barry Cooper's vision. It was during that time, that a new goal emerged for us: to collect the numerous exciting ideas that were being discussed and capture the essence of this gathering in a book of chapters—the idea for this book was born. Since then we have been working to make this idea a reality. The chapters that we collected draw on the spectrum of invited speakers from the workshop (Fig. 1). They are organized into five parts, corresponding to the five "big question" themes that we started out with.

While this book was taking its shape, we were sad to lose one of the members in our community, my father Ivan Soskov. It was Barry's idea to ask my father's advisor and most recent student to collaborate on a tribute to his work. I am grateful to him and the authors Hristo Ganchev and Dimiter Skordev for this contribution. Ivan Soskov would have been a perfect fit for contributing an article on the abstract alternatives to Turing's model of computation, meant to bring the notion of computability to abstract structures, not restricted to the domain of the natural numbers, as this is the area that motivated most of his work. We begin

Fig. 1 The Incomputable, June 12–15 2012, Chicheley Hall

Part I, *Challenging Turing: Extended Models of Computation*, with the story of Ivan Soskov's work and life in computability which still conveys these ideas, even though it is related by a different storyteller. Next we explore the possibilities of a new logic system meant to capture the workings of the quantum world in a more accurate way. Kate Clements, Fay Dowker and Petros Wallden provide us with one such possible system and investigate the conditions under which it coincides with its classical predecessor. Finally, Bob Coecke argues for a paradigmatic shift from reductionism to "togetherness", exploring the relationship between the quantum physics and natural language that this gives rise to.

In Part II, *The Search for "Natural" Examples of Incomputable Objects*, we are brought up to speed with the current state of the art of an area that originated from Post's program [7]. The goal of this program was to find a natural combinatorial property of sets that isolates an incomputable computably enumerable set that is not equivalent to the halting problem. Post's program was a driving force in the development of classical computability theory, motivating the invention of some of the fundamental methods used today. It gave rise to the study of the lattice of c.e. sets under inclusion. Peter Cholak describes the main directions in this field that have been the focus of investigations in recent years. Seth Lloyd moves is swiftly to the real world setting, where he explores incomputable problems that arise when one considers physical law. He finds incomputability in the form of the energy spectrum of certain quantum systems, in the system Hamiltonian with respect to a particular sector and in general in the problem of determining the best expression of physical laws. Vela Velupillai explores the natural occurrence of incomputability in yet a different setting: economics. Computable economics is a growing field of research to which he has major scientific contribution. Particularly noteworthy is his attempt to construct a mathematical model of economic theory (both micro and macro theory) using methods from computability theory and constructive mathematics. This leads to the reappearance of incomputability and we are given many examples: the excess demand functions, the rational expectations equilibria, all explained and discussed in detail.

Part III, *Mind, Matter and Computation*, is devoted to understanding the right model of the universe and human life. Vlatko Vedral explores the fundamental role of quantum physics in living systems: in describing the state of every atom on earth, tin explaining DNA replication, and even in photosynthesis. The chapter concludes with an exploration of the possibility that quantum computers can be used to simulate life in a way that goes beyond the classical approach. In the next chapter Mark Bishop revisits "digital philosophy"—the belief in the algorithmic nature of the universe, governed my deterministic laws. This philosophy has been essential to the works of Konrad Zuse [15], Stephen Wolfram [14], Jürgen Schmidhuber [8] and Nobel laureate Gerard 't Hooft [10]. These authors see the universe is a digital computer and insist that the probabilistic nature of quantum physics does not contradict its algorithmic nature. Mark Bishop argues that "digital philosophy" is not the correct view of the world as it leads to panpsychism, the view that consciousness is a universal feature of all things.

Antonina Kolokolova is the first contributor to Part IV, *The Nature of Information: Complexity and Randomness*. She gives us an account of the many years of work that has been put into solving the famous P vs NP problem in complexity theory and describes a rich structure of complexity classes that has been discovered through this work. She explores the possibility of an intrinsic reason for our failure to overcome this long-standing barrier. Cristian Calude explains the connection between complexity and randomness, in particular with relation to cryptography. He discusses the practical disadvantages that we have by being limited to pseudo-randomness: strings generated by fully deterministic programs, that appear to have the statistical properties of true random strings. A possible solution to this is suggested by quantum randomness, and Calude explores the extent to which quantum randomness can be proven to be better than pseudo-randomness. The final chapter in this part is devoted to the mathematical concept of algorithmic complexity and randomness. Algorithmic randomness is a topic that goes back to works of Andrey Kolmogorov [4], Per Martin-Löf [5], Claus-Peter Schnorr [9], Gregory Chaitin [1] from the late 1960s and early 1070s. The reinvoked interest in this field comes from the nontrivial interplay between randomness and degree theory. Methods from algorithmic randomness lead to the discovery of a rich world of classes of Turing degrees, defined through their combinatorial properties. Of particular interest is a class that suggests a connection between information content and computational power: the degrees of K-trivial sets. A K-trivial set can be described as a set that is weak as an oracle, close to being computable, or equivalently as a set whose initial segments are easy to compress, i.e. a set that is as far from random as possible. Andre Nies explores a new method for characterizing sets that are close to computable via a new mathematical theory of cost functions. His work gives us new tools to characterize and explore the intriguing class of the K-trivials.

In Part V, *The Mathematics of Emergence and Morphogenesis*, we learn about Turing's original work on morphogenesis, how it has been developed by his academic successors and what are the future goals in the area through a detailed account provided to us by Thomas Woolley, Ruth Baker and Philip Maini. Morphogenesis is the fundamental biological process that causes an organism to develop its shape. It is the process which determines the spots on the body of a cow and the shape and distribution of the leaves on a daisy. Turing's ideas on morphogenesis are perhaps the ones that most clearly illuminate the fact that computation and mathematics are present in natural processes. Barry Cooper was deeply impressed by these ideas. In a description of the book "Alan Turing: His Work and Impact", of which he was an editor, he remembers:

> When I arriving in Manchester as a research student in 1968, I found Turing's only PhD student, Professor Robin Gandy, giving lectures on Turing's work on the emergence of Fibonacci sequences in sunflowers. I was mystified. Now, when so many people are seeking out form beyond form, computation beyond computation, and structure beyond chaos—investigating randomness, human creativity, emergent patterns in nature—we can only wonder that Turing was there before us. So much of what Alan Turing pondered and brought

into the light of day—techniques, concepts, speculations—is still valid in our everyday investigations.

The final chapter in this book is a special one. Aaron Sloman reports on his "Meta-Morphogenesis project". This project takes the ideas from Turing's original paper and transforms them to a whole new plane of topics: the evolution of biological and human intelligence. The idea for this project was born when Barry Cooper asked Aaron Sloman to contribute to the book "Alan Turing: His Work and Impact" with a chapter related to Turing's work on morphogenesis. Aaron Sloman, whose prior work was most significantly in artificial intelligence, responded to Barry's challenge with this novel idea and has been working on it ever since, motivated by his intuition that this project can lead to answers to fundamental questions: about the nature of mathematics, language, mind, science, life and on how to overcome current limitations of artificial intelligence.

Barry's 72nd birthday was blackened by devastating news: he was told that he is terminally ill with only a very short time to live. He generously devoted all this time to his family and friends, to his students, to his numerous projects, to mathematics. The disease was unexpected and unexpectedly treacherous. Less than 3 weeks later Barry Cooper passed away and we lost a great thinker. One of the very last emails that Barry and I exchanged was about this book. We had reached the stage when we had to put the final touches on our project: write an introduction and decide on an image for the cover. Unfortunately, these two final tasks were left to me and though my attempt to honor Barry's vision must prove inadequate, I hope I have conveyed some semblance of it. I will miss him for the rest of my life.

References

1. G.J. Chaitin, Information-theoretic characterizations of recursive infinite strings. Theor. Comput. Sci. **2**(1), 45–48 (1976)
2. A. Church, An unsolvable problem of elementary number theory. Am. J. Math. **58**(2), 345–363 (1936)
3. M. Davis, H. Putnam, J. Robinson, The decision problem for exponential diophantine equations. Ann. Math. (2) **74**, 425–436 (1961)
4. A.N. Kolmogorov, Three approaches to the quantitative definition of information. Int. J. Comput. Math. **2**, 157–168 (1968)
5. P. Martin-Löf, The definition of random sequences. Inf. Control **9**, 602–619 (1966)
6. J.V. Matijasevič, The Diophantineness of enumerable sets. Dokl. Akad. Nauk SSSR **191**, 279–282 (1970)
7. E.L. Post, Recursively enumerable sets of positive integers and their decision problems. Bull. Am. Math. Soc. **50**, 284–316 (1944)
8. J. Schmidhuber, Algorithmic theory of everything. Report IDSIA-20-00, quant-ph/0011122 (IDSIA, Manno/Lugano, 2000)
9. C.-P. Schnorr, A unified approach to the definition of random sequences. Math. Syst. Theory **5**, 246–258 (1971)
10. G. 't Hooft, Quantum gravity as a dissipative deterministic system. Classical Quantum Gravity **16**(10), 3263–3279 (1999)

11. A.M. Turing, On computable numbers, with an application to the Entscheidungsproblem. Proc. Lond. Math. Soc. **S2–42**(1), 230 (1936)
12. A.M. Turing, Computing machinery and intelligence. Mind **59**, 433–460 (1950)
13. A.M. Turing, Morphogenesis, in *Collected Works of A.M. Turing* (North-Holland Publishing Co., Amsterdam, 1992). With a preface by P.N. Furbank, ed. by P.T. Saunders
14. S. Wolfram, *A New Kind of Science* (Wolfram Media, Champaign, IL, 2002)
15. K. Zuse, *Rechnender Raum* (Friedrich Vieweg & Sohn, Braunschweig, 1969)

Part I
Challenging Turing: Extended Models of Computation

Ivan Soskov: A Life in Computability

Hristo Ganchev and Dimiter Skordev

Abstract On May 5, 2013, the Bulgarian logic community lost one of its prominent members—Ivan Nikolaev Soskov. In this paper we shall give a glimpse of his scientific achievements.

1 Introduction

Ivan Soskov (1954–2013) was born in the town of Stara Zagora, Bulgaria. After graduating from the Bulgarian National High School for Mathematics, Ivan Soskov continued his education at the Mathematical Faculty of Sofia University. In 1979 he obtained his M.Sc. degree and 4 years later his Ph.D. degree under the supervision

H. Ganchev (✉) • D. Skordev
Faculty of Mathematics and Informatics, Sofia University, 5 James Bourchier Blvd., 1164 Sofia, Bulgaria
e-mail: ganchev@fmi.uni-sofia.bg; skordev@fmi.uni-sofia.bg

S.B. Cooper, M.I. Soskova (eds.), *The Incomputable*, Theory and Applications of Computability, DOI 10.1007/978-3-319-43669-2_2

of Prof. Dimiter Skordev.[1] In 2001 Ivan Soskov defended a habilitation thesis and became a Doctor of Mathematical Sciences. In 1984 he became Assistant Professor and in 1991 he was promoted to Associate Professor at the Department of Mathematical Logic. From 1991 to 1993 he was an Adjunct Professor at UCLA in the programme PIC (Programming in Computing). He became full professor in 2004. From 2000 to 2007 he was head of the Department of Mathematical Logic and from 2007 until his last days he was the Dean of the Faculty of Mathematics and Informatics.

Ivan Soskov was an established scientist of international stature. He trained and educated a large number of Ph.D. and M.Sc. students and established himself as the undisputed leader of the Bulgarian school in the field of Abstract Computability. He has left us a great scientific legacy which will continue to inspire and guide us in the future.

Soskov's scientific interests lay in the field of classical computability theory. He started his career investigating the models of computation in abstract structures. The main problem he was interested in was the clarification of the connections between two basic approaches to abstract computability: the internal approach, based on specific models of computation, and the external approach, which defines the computable functions through invariance relative to all enumerations of a structure. Towards the end of the last century I. Soskov started his work in degree theory. His main motivation came from the ideological connections between one of the models of abstract computability, called search computability, and enumeration reducibility. In his works Soskov and his students developed the theory of regular enumerations and applied it to the enumeration degrees, obtaining a series of new results, mainly in relation to the enumeration jump.

In 2006 Soskov initiated the study of uniform reducibility between sequences of sets and the induced structure of the ω-enumeration degrees. With his students he obtained many results, providing substantial proof that the structure of the ω-degrees is a natural extension of the structure of the enumeration degrees, where the jump operator has interesting properties and where new degrees appear, which turn out to be extremely useful for the characterization of certain classes of enumeration degrees. The connection between enumeration degrees and abstract models of computability led to a new direction in the field of computable model theory. In the last few years of his life Soskov obtained various results concerning the properties of the spectra of structures as sets of enumeration degrees, among them two jump inversion theorems.

[1] He wrote: "In the academic year 1978/1979, as a supervisor of Ivan Soskov's master thesis, I had the chance to be a witness of his first research steps in the theory of computability. My role was rather easy thanks to Ivan's great ingenuity and strong intuition. Several years later, in 1983, Ivan Soskov defended his remarkable PhD dissertation on computability in partial algebraic systems. Although I was indicated as his supervisor again, actually no help of mine was needed at all in the creation of this dissertation. Everything in it, including also the choice of its subject and the form of the presentation, was a deed of Ivan only."

2 Ivan Soskov's Work on Computability on First-Order Structures

As it is known, classical computability theory concerns mainly algorithmic computability of functions and predicates in the set ω of the natural numbers. No serious doubts exist today that the partial recursiveness notion gives a proper mathematical description of the class of algorithmically computable functions and predicates in ω (the predicates can be regarded as functions whose values belong to the two-element set $\{0,1\}$). One can regard as clarified also the relative computability of functions and predicates in ω with respect to some given everywhere defined functions that are not required to be algorithmically computable—the mathematical description in this case is done by the notion of relative partial recursiveness. The situation is similar with the computability of functions and predicates in certain other commonly used sets of constructive objects, such as the set of words over a finite alphabet.

Even in ω, however, the situation is not so simple when one considers computability in some given partial functions—the notion of relative computability then splits into two, corresponding to computability respectively by using deterministic computational procedures and by using arbitrary ones (if the given functions are total then the two notions coincide, thanks to the possibility of an algorithmic sequential search in ω). The first of these two notions can be mathematically described as relative μ-recursiveness. Namely, a partial function φ is called *μ-recursive* in some given partial functions $\theta_1,\ldots,\theta_n$ if φ can be obtained from $\theta_1,\ldots,\theta_n$ and basic primitive recursive functions by means of finitely many applications of substitution, primitive recursion and the minimization operation μ. The other notion can be described as enumeration reducibility of the graph of φ to the graphs of $\theta_1,\ldots,\theta_n$ (we usually omit "the graph of" and "the graphs of"). The reducibility in question is usually described in the following intuitive way: a set A is said to be enumeration reducible (or more briefly e-reducible) to the n-tuple of sets $B_1,\ldots,B_n$, where $A\subseteq\omega^i$, $B_1\subseteq\omega^{j_1},\ldots,B_n\subseteq\omega^{j_n}$, if there is an algorithm transforming in arbitrary n-tuple of enumerations of $B_1,\ldots,B_n$ into an enumeration of A. The precise definition says that A is *enumeration reducible* to $B_1,\ldots,B_n$ if there exists a recursively enumerable subset W of ω^{i+n}, such that A consists of all $(x_1,\ldots,x_i)$ satisfying the condition

$$\exists u_1\ldots\exists u_n\left((x_1,\ldots,x_i,u_1,\ldots,u_n)\in W\ \&\ D^{(j_1)}_{u_1}\subseteq B_1\ \&\ \ldots\ \&\ D^{(j_n)}_{u_n}\subseteq B_n\right),$$

where $D^{(j)}_u$ is the uth finite subset of ω^j in an effective enumeration of the family of all finite subsets of ω^j.

The situation is even more complicated when computability on an arbitrary first-order structure $\mathfrak{A}$ is considered. If its universe is denumerable then the so-called external approach to computability on $\mathfrak{A}$ is possible. Namely, one could try to reduce it to computability in ω by means of using enumerations of the universe. In the other approach, which does not need enumerability of the universe, computability on $\mathfrak{A}$ is regarded as one which is relative with respect to the primitive functions and predicates of $\mathfrak{A}$, i.e. achievable by computational procedures using them (this approach is named internal). Now, however, even with everywhere defined primitive functions and predicates, the execution of non-deterministic computational procedures is not necessarily always reducible to the execution of deterministic ones. In addition, also the admissible degree of indeterminism may be different—for example, choosing an arbitrary element of the universe of $\mathfrak{A}$ could be, in general, not reducible to choices of natural numbers. And also there are no reasons to necessarily want the equality predicate to be computable.

Since computability on first-order structures is a notion, which is important from the point of view of theoretical computer science, research aiming at clarifying it has been done for more than half a century. Nevertheless, there are still too many questions to be answered. Ivan Soskov was one of the active and successful researchers in this area.

2.1 Ivan Soskov's Research on First-Order Computability Done in the Frame of the Internal Approach

A substantial part of Soskov's research of this kind concerns the notion of absolute prime computability introduced by Y.N. Moschovakis in [22]. If B is a set then any partial function or predicate of several arguments on B can be regarded as a partial unary function in the extension B^* of B defined by Moschovakis as follows: B^* is the closure of the set $B^0 = B \cup \{0\}$ under the formation of ordered pairs assuming that 0 is an object not in B and the set-theoretic operation representing ordered pairs is chosen so that no element of B^0 is an ordered pair. The natural numbers $0, 1, 2, \ldots$ can be identified with the elements $0, (0, 0), (0, (0, 0)), \ldots$ of B^*,[2] and then the predicates should be regarded as functions with values in the two-element set $\{0, (0, 0)\}$.[3] For any partial multivalued (p.m.v.) functions $\varphi_1, \ldots, \varphi_l, \psi$ in B^*, Moschovakis defines what it means for ψ to be absolutely prime computable in $\varphi_1, \ldots, \varphi_l$. This notion is a particular instance of a slightly more general one whose definition in [22] uses an encoding of certain p.m.v. functions in B^* by means of 11 schemata. The following simpler characterization of absolute prime computability was given later in [32, 33]: ψ is absolutely prime computable in $\varphi_1, \ldots, \varphi_l$ iff ψ can be generated from $\varphi_1, \ldots, \varphi_l, \pi, \delta$ by means of the operations composition, combination and iteration, where π and δ are the total single-valued functions in B^* defined as in [22] (namely $\pi(v) = t$ and $\delta(v) = u$ if $v = (t, u)$, $\pi(0) = \delta(0) = 0$, $\pi(v) = \delta(v) = (0, 0)$ if $v \in B$), and, for any p.m.v. functions θ and χ in B^*,

[2] Or with the elements $0, (0, 0), ((0, 0), 0), \ldots$, as it is in [22].

[3] There are situations when the identification in question could cause problems. This can happen when the authentic natural numbers or at least some of them belong to B. In such situations, it would be appropriate, for instance, to denote the elements $0, (0, 0), (0, (0, 0)), \ldots$ and their set by $\bar{0}, \bar{1}, \bar{2}, \ldots$ and $\overline{\omega}$, respectively.

(a) the composition of θ and χ is the p.m.v. function τ defined by the equivalence

$$v \in \tau(s) \Leftrightarrow \exists u(u \in \chi(s) \;\&\; v \in \theta(u)),$$

(b) the combination of θ and χ is the p.m.v. function τ defined by the equivalence

$$v \in \tau(s) \Leftrightarrow \exists t \exists u(t \in \theta(s) \;\&\; u \in \chi(s) \;\&\; v = (t, u)),$$

(c) the iteration of θ controlled by χ is the p.m.v. function τ defined by the equivalence

$$v \in \tau(s) \Leftrightarrow \exists t_0, t_1, \ldots, t_m(t_0 = s \;\&\; t_m = v$$
$$\& \forall i(i < m \Rightarrow t_i \in \chi^{-1}(B^* \setminus B^0) \;\&\; t_{i+1} \in \theta(t_i)) \;\&\; t_m \in \chi^{-1}(B^0))).$$

Obviously, the composition and the combination of any two single-valued, functions is single-valued again, and it is easy to check that the same is true also for their iteration. Of course, if θ and χ are single-valued, then all $\in$-symbols in (a), (b) and (c) except for the ones in front of χ^{-1} can be replaced with equality symbols.[4]

Ivan Soskov took his first research steps in his Master's thesis [35]. Suppose a partial structure $\mathfrak{A}$ with universe B is given. Let us call *prime computable in* $\mathfrak{A}$ the partial functions in B which are absolutely prime computable in the primitive functions and predicates of $\mathfrak{A}$ (intuitively, the functions prime computable in $\mathfrak{A}$ are those which can be computed by deterministic programs using the primitive functions and predicates of $\mathfrak{A}$). Soskov indicates that any equivalence relation $\sim$ on the set B can be extended to an equivalence relation $\approx$ on B^* in a natural way. Under certain natural assumptions about concordance of $\sim$ with the primitive functions and predicates of $\mathfrak{A}$, Soskov proves that any partial function on B prime computable in $\mathfrak{A}$ is representable in $\mathfrak{A}$ by some term on any equivalence class under $\approx$ which contains at least an element of the function's domain. By applying this (with the universal relation in B as $\sim$) to the particular case when $\mathfrak{A}$ has no primitive predicates, Soskov gives a rigorous proof of the following intuitively plausible statement: if $\mathfrak{A}$ has no primitive predicates, then any partial function on B that is prime computable in $\mathfrak{A}$ and has a non-empty domain can be represented in $\mathfrak{A}$ by some term.

Again in [35], a characterization of the prime computable functions is given for any of the structures with universe ω which have no primitive functions and predicates besides the successor function S, the predecessor function P and the equality to 0 predicate Z. A smart diagonalization is used to define a recursive function $\varphi : \omega \to \{0, 1\}$ such that all partial recursive functions are prime computable in

[4]The iteration defined in the way above is a particular instance of the iteration in so-called iterative combinatory spaces from [32, 33], where an algebraic generalization of a part of the theory of computability is given. The absolute prime computability is actually the instance of the considered generalized notion of computability in a specific kind of iterative combinatory space.

the partial structure $(\omega; P, S \upharpoonright \varphi^{-1}(0), S \upharpoonright \varphi^{-1}(1); Z)$, but the function S is not computable in it by means of a standard program (this example is published in [37]).

Several years later, in his Ph.D. thesis [40], Ivan Soskov proves that the functions prime computable in a partial structure are exactly the ones computable in it by means of standard programs with counters and a stack, the stack being superfluous in the case when the primitive functions of the structure are unary. Hence the statement in the above-mentioned example cannot be strengthened by replacing "a standard program" with "a standard program with counters".

In [36], a normal form theorem using the μ-operation is proved by Soskov for the functions prime computable in a total first-order structure. This result is applied to show the equivalence of prime computability in total structures with computability by means of Ershov's recursively enumerable determinants [4]—a notion similar to computability by means of Friedman's effective definitional schemes [6] and equivalent to it over total structures (the term "computability in the sense of Friedman-Ershov" is used in [36]).[5]

In Soskov's papers [38, 39, 42] and in his Ph.D. thesis [40], partial structures are considered and additional arguments ranging over ω are admitted in the partial functions in B (assuming that ω is identified with the set $\{0, (0, 0), (0, (0, 0)), \ldots\}$). The set of all partial functions from $\omega^r \times B^j$ to B is denoted by $\mathcal{F}_{r,j}$. Of course, all functions from $\mathcal{F}_{r,j}$ can be regarded as partial unary functions in B^*. The normal form theorem from [36] is generalized for them. For functions belonging to $\mathcal{F}_{0,j}$, the notion of computability by means of Ershov's recursively enumerable determinants makes sense also in the case of partial structures, and this computability is equivalent to computability by means of Shepherdson's recursively enumerable definitional schemes (REDS) introduced in [31] (they are a slight generalization of the effectively definable schemes of Friedman [6]). Prime computability of such a function ψ in a partial structure $\mathfrak{A}$ implies the REDS-computability of ψ in $\mathfrak{A}$, but, in general, not vice versa. Soskov gives a characterization of the REDS-computability by introducing another notion of computability in $\mathfrak{A}$. A function $\varphi \in \mathcal{F}_{r,j}$ is called *computable in* $\mathfrak{A}$ if a function $\psi \in \mathcal{F}_{r+1,j}$ exists such that ψ is prime computable

[5]The following intuitive description of Friedman's schemes (applicable also to Ershov's determinants) is given by Shepherdson in [31]: "Friedman's effective definitional schemes are essentially definitions by infinitely many cases, where the cases are given by atomic formulae and their negations and the definition has r.e. structure". As indicated there on p. 458, a result of Gordon (announced in an abstract of his) shows that, over a total structure, the computability by means of such schemes is equivalent to absolute prime computability (a detailed presentation of the result in question can be found in [15]). Unfortunately, Shepherdson's and Gordon's papers remained unnoticed by the logicians in Sofia quite a long time (the first publication of Soskov containing a reference to Shepherdson's paper is [42]; the first reference to Gordon's result is also in [42], but only his abstract is indicated there). This great delay was due to insufficient international contacts of the Sofia logic group at that time (especially before the several summer schools in mathematical logic organized in Bulgaria from 1983 on). An additional reason for missing these papers is that, at the time in question, the interests of the supervisor of Ivan Soskov's Ph.D. thesis were mainly in generalization of the theory of computability and in examples of this generalization different from the ones corresponding to the situation from [22].

in $\mathfrak{A}$, and for every $x_1, \dots, x_r \in \omega$ and every $s_1, \dots, s_j, t \in B$, the following equivalence holds:

$$\varphi(x_1, \dots, x_r, s_1, \dots, s_j) = t \Leftrightarrow \exists z(\psi(z, x_1, \dots, x_r, s_1, \dots, s_j) = t)$$

(intuitively, this corresponds to computability by means of parallel procedures). Clearly, prime computability in $\mathfrak{A}$ implies computability in $\mathfrak{A}$. The converse implication is not generally true, but it holds at least in the case when $\mathfrak{A}$ is total. The functions computable in $\mathfrak{A}$ are shown to be exactly the ones which are absolutely prime computable in the primitive functions and predicates of $\mathfrak{A}$ and the p.m.v. function whose graph is ω^2. The functions from $\mathcal{F}_{0,j}$ which are computable in $\mathfrak{A}$ are shown to be exactly the ones computable in $\mathfrak{A}$ by means of REDS (as indicated in [42], the functions from $\mathcal{F}_{0,j}$ which are prime computable in $\mathfrak{A}$ are exactly the ones computable in $\mathfrak{A}$ by means of tree-like REDS in the sense of [31]).

A function from $\mathcal{F}_{r,j}$ is called *search computable in* $\mathfrak{A}$ if it is absolutely search computable in the sense of [22] in the primitive functions and predicates of $\mathfrak{A}$ (this is equivalent to absolute prime computability of the function in the primitive functions and predicates of $\mathfrak{A}$ and the multiple-valued function whose graph is $(B^*)^2$). The functions computable in $\mathfrak{A}$ are search computable in $\mathfrak{A}$, but the converse is not generally true. In [40, 42] the following example of a structure $\mathfrak{A}_o$ is given in which not every search computable function from $\mathcal{F}_{0,1}$ is computable and not every computable function $\mathcal{F}_{0,1}$ is prime computable: $\mathfrak{A}_o = (\omega; P; T_1, T_2)$, where P is the predecessor function, $\mathrm{dom}(T_1) = \{0\}$, $\mathrm{dom}(T_2) = \{1\}$, $T_1(0) = T_2(1) = \mathbf{true}$. The function S is search computable in $\mathfrak{A}_o$ without being computable in $\mathfrak{A}_o$, and the restriction of the identity function in ω to the set of the even numbers is computable in $\mathfrak{A}_o$ without being prime computable in $\mathfrak{A}_o$.

In [39, 40], a partial structure with universe ω is called *complete* if any partial recursive function is computable in this structure. A necessary and sufficient condition for completeness is given in the spirit of [25]. A simplification of this condition is given for the case of partial recursiveness of the primitive functions and predicates. A partial structure with universe ω is called *complete with respect to prime computability* if any partial recursive function is prime computable in this structure. Completeness with respect to other kinds of computability is defined in a similar way. The following three examples are announced in [39] and presented in detail in [40]: $\mathfrak{A}_1 = (\omega; S, P; T_1, T_2)$, $\mathfrak{A}_2 = (\omega; \varphi; E)$, $\mathfrak{A}_3 = (\omega; \theta_1, \theta_2; Z)$, where T_1 and T_2 are as in the partial structure $\mathfrak{A}_o$ from the previous paragraph, E is the equality predicate,

$$\varphi(s,t) = \begin{cases} s+1 \text{ if } t = (s+1) - 2^{[\log_2(s+1)]}, \\ 0 \quad \text{otherwise}, \end{cases}$$

$$\theta_1(s) = \begin{cases} s \dot{-} 1 & \text{if } s \in A, \\ s+1 & \text{otherwise}, \end{cases} \qquad \theta_2(s) = \begin{cases} s+1 & \text{if } s \in A, \\ s \dot{-} 1 & \text{otherwise}, \end{cases}$$

where

$$A = \{s \mid \exists t\, (2^{2t} \leq s < 2^{2t+1})\} \text{ and } s \dot{-} 1 = \begin{cases} 0 & \text{if } s = 0, \\ s - 1 & \text{otherwise} \end{cases}.$$

It is shown that:

(a) $\mathfrak{A}_1$ is complete without being complete with respect to prime computability;
(b) $\mathfrak{A}_2$ is complete with respect to prime computability without being complete with respect to computability by means of standard programs with counters;
(c) $\mathfrak{A}_3$ is complete with respect to computability by means of standard programs with counters without being complete with respect to computability by means of standard programs.[6]

(The partial structure $\mathfrak{A}_0$ could be used as an example showing that, in general, completeness with respect to search computability does not imply completeness. Soskov does not mention this, maybe since there are much simpler examples for the same thing—for instance, the structure without primitive functions which has as its only primitive predicate the total binary one whose truth set is $\{(s,t) \mid t = s \dot{-} 1\}$.)

The paper [41] makes use of the notions of iterative combinatory space and of recursiveness in such spaces from [32] (cf. also [33]). So-called functional combinatory spaces are considered in the paper (they are one of the simple examples for the notion of iterative combinatory space). Let M be an infinite set and $\mathcal{F}$ be the partially ordered semigroup of all partial mappings of M into M with multiplication defined as composition (namely $\varphi\psi = \lambda s.\varphi(\psi(s))$) and the extension relation as partial ordering. Let $a, b \in M$, $a \neq b$, $J : M^2 \to M$, $D, L, R \in \mathcal{F}$, $\mathrm{rng}(D) = \{a, b\}$, $D(a) = a$, $D(b) = b$, $L(J(s,t)) = s$ and $R(J(s,t)) = t$ for all $s, t \in M$.[7] Let $I = \mathrm{id}_M$, $\mathcal{C}$ be the set of all total constant mappings of M into M, T and F be the elements of $\mathcal{C}$ with values a and b, respectively; and let $\Pi : \mathcal{F}^2 \to \mathcal{F}$, $\Sigma : \mathcal{F}^3 \to \mathcal{F}$ be defined in the following way:

$$\Pi(\varphi,\psi)(s) = J(\varphi(s),\psi(s)), \quad \Sigma(\chi,\varphi,\psi)(s) = \begin{cases} \varphi(s) & \text{if } D(\chi(s)) = a, \\ \psi(s) & \text{if } D(\chi(s)) = b. \end{cases}$$

Then the 9-tuple

$$(\mathcal{F}, I, \mathcal{C}, \Pi, L, R, \Sigma, T, F) \tag{1}$$

[6]Since computability by means of standard programs with counters is equivalent to prime computability in the case of unary primitive functions, the conclusion from the third of these examples can be made also from the example in [35, 37]. An improvement, however, is the fact that $\mathfrak{A}_3$ is a total structure, whereas the example in [35, 37] makes use of a partial structure.

[7]For instance, M can be the Moschovakis' extension B^* from [22] of a set B, and then we can take $a = 0$, $b = (0,0)$, $J(s,t) = (s,t)$, $D = \mathrm{id}_{\{a,b\}}$, $L = \pi$, $R = \delta$.

is an iterative combinatory space. According to the general definition from [32, 33], if $\varphi, \chi \in \mathcal{F}$ then *the iteration of* φ *controlled by* χ in (1) is the least fixed point in $\mathcal{F}$ of the mapping $\theta \mapsto \Sigma(\chi, I, \theta\varphi)$. If $\theta_1, \ldots, \theta_n \in \mathcal{F}$ then the elements of $\mathcal{F}$ *recursive* in $\{\theta_1, \ldots, \theta_n\}$ are those which can be obtained from elements of $\{\theta_1, \ldots, \theta_n, L, R, T, F\}$ by finitely many applications of multiplication, Π and iteration. The main result in [41] reads as follows: an element of $\mathcal{F}$ is recursive in $\{\theta_1, \ldots, \theta_n\}$ iff it is prime computable in the partial structure

$$(M; \theta_1, \ldots, \theta_n, J, L, R, T, F; \overline{D}),$$

where

$$\overline{D}(s) = \begin{cases} \textbf{true} \text{ if } D(s) = a, \\ \textbf{false} \text{ if } D(s) = b. \end{cases}$$

An extension of the notion of a Horn program is given in the paper [46], namely predicate symbols can be used as parameters which correspond to arbitrary subsets of the corresponding Cartesian degrees of the domain of a structure. Intuitively, these subsets are treated as effectively enumerable rather than semi-computable ones (although effective enumerability and semi-computability are equivalent in the case of computability on the natural numbers, the semi-computable sets are a proper subclass of the effectively enumerable ones in the general case). For the operators defined by means of such programs, a version of the first recursion theorem is proved.

2.2 *Ivan Soskov's Results on the External Approach to Computability on First-Order Structures*

In the computational practice, the operations with arbitrary objects are usually performed by using their representations or possibly approximations by some appropriate kinds of constructive objects. The description of computability on denumerable first-order structures by using a constructive representation of their elements can be considered as the main intuitive aspect of Soskov's papers [43, 44, 49] (as well as of the abstract [48] of the last of them), of the joint paper [65] with Alexandra Soskova and of Soskov's Dr. Hab. thesis [55]. It plays an essential role also in the technical report [45] and in the paper [47].

In theoretical studies, one always could use natural numbers as representing constructive objects. This leads to studying computability on a first-order structure whose elements are enumerated in some way (with the reasonable requirement that every element of the structure has at least one number and different elements have different numbers, and this, of course, sets the limitation for the universe of the structure to be a denumerable set). Such an enumeration approach to computability for objects other than the natural numbers has been used for quite a long time (for instance, computability on the set of the words over a finite alphabet is often reduced in such a way to computability on ω). Let α be a partial mapping of ω onto a denumerable set B. An m-argument partial function φ on ω will be called an *α-quasiassociate* of an m-argument partial function θ on B if

$$\theta(\alpha(x_1), \ldots, \alpha(x_m)) = \alpha(\varphi(x_1, \ldots, x_m))$$

for all $x_1, \ldots, x_m \in \text{dom}(\alpha)$ (where the equality symbol means full equality, i.e. if one of the expressions on the two sides of it makes sense, then the other one also does, and they have one and the same value); the function φ will be called an *α-associate* of the function θ if φ is an α-quasiassociate of θ and $dom(\alpha)$ is closed with respect to φ. An m-argument partial predicate χ on ω will be called an *α-associate* of an m-argument partial predicate π on B if

$$\pi(\alpha(x_1), \ldots, \alpha(x_m)) = \chi(x_1, \ldots, x_m)$$

for all $x_1, \ldots, x_m \in \text{dom}(\alpha)$ (under the same interpretation of the equality symbol as above). A partial function or predicate on B will be called *computable with respect to α* if there exists a partial recursive α-associate of it. Suppose now that B is the universe of some partial first-order structure

$$\mathfrak{A} = (B; \theta_1, \ldots, \theta_n; \pi_1, \ldots, \pi_k), \tag{2}$$

where $\theta_1, \ldots, \theta_n$ are partial functions in B, and $\pi_1, \ldots, \pi_k$ are partial predicates on B. If we are interested in computability on $\mathfrak{A}$, then it seems natural to require that $\theta_1, \ldots, \theta_n, \pi_1, \ldots, \pi_k$ be computable with respect to α. Let us say that α is an *effective enumeration* of $\mathfrak{A}$ if this requirement is satisfied. Unfortunately, it may happen that the structure $\mathfrak{A}$ admits no effective enumeration,[8] and, on the other hand, if $\mathfrak{A}$ admits effective enumerations, then the considered computability of functions and predicates on B may depend on our choice among these enumerations. The second of these problems can be overcome by calling a function or predicate on B computable in $\mathfrak{A}$ if it is computable with respect to any effective enumeration of $\mathfrak{A}$ (such a computability notion was essentially suggested by the second author in 1977 under the name admissibility). There are cases when this notion deserves attention, but this is not so in general—for instance, if $\mathfrak{A}$ admits no effective enumeration, then all functions and predicates on B trivially turn out computable in $\mathfrak{A}$.

A much more promising enumeration approach to computability on first-order structures originates from Lacombe's paper [18]. The case considered there, up to minor details, is the one when $\alpha, \theta_1, \ldots, \theta_n; \pi_1, \ldots, \pi_k$ are total and α is injective. In this case, any function and any predicate on B has exactly one α-associate. A total function or predicate on B is called *$\forall$-recursive in $\mathfrak{A}$* if its α-associate is recursive in

[8] In the paper [65] by Alexandra Soskova, and Ivan Soskov a characterization is given of the structures which have effective enumerations.

the α-associates of $\theta_1, \dots, \theta_n, \pi_1, \dots, \pi_k$ whenever α is an injective mapping of ω onto B. By a result of Moschovakis proved in [23], $\forall$-recursiveness in $\mathfrak{A}$ turns out equivalent to search computability in $\theta_1, \dots, \theta_n, \pi_1, \dots, \pi_k$, the equality predicate on B and constants from B.

First of all, let us note that a total enumeration α of B is injective iff the equality predicate on ω is an α-associate for the equality predicate on B. Therefore it is natural to attempt avoiding the presence of the equality predicate on B in the above-mentioned Moschovakis's result by omitting the requirement about injectivity of α in the definition of $\forall$-recursiveness. Unfortunately, some complications arise without this requirement, since the α-associate of a function on B can be no longer unique. One could try to solve this problem by considering enumerations α of B together with chosen α-associates for $\theta_1, \dots, \theta_n$. Soskov makes this step, and he additionally allows α to be partial. Then the α-associates of the predicates on B can also be no longer unique, and therefore α-associates of $\pi_1, \dots, \pi_k$ also must be assumed to be chosen. In [48–50, 55, 65], a notion of enumeration is used whose definition can be formulated as follows.

Definition 2.1 An *enumeration* of the partial structure (2) is any ordered pair $(\alpha, \mathfrak{B})$, where α is a partial mapping of ω onto B, and $\mathfrak{B} = (\omega; \varphi_1, \dots, \varphi_n; \chi_1, \dots, \chi_k)$ is a partial structure such that:

(a) φ_i is an α-associate of θ_i for $i = 1, \dots, n$;
(b) χ_j is an α-associate of π_j for $j = 1, \dots, k$.

The enumeration $(\alpha, \mathfrak{B})$ is called *total* if $\mathrm{dom}(\alpha) = \omega$.

Remark 2.2 The enumeration notion considered in [43] is somewhat different: we can get it from the above one by replacing "φ_i is an α-associate of θ_i" with "φ_i is an α-quasiassociate of θ_i" and adding the requirement that $\mathrm{dom}(\alpha)$ is enumeration reducible to the graphs of $\varphi_1, \dots, \varphi_n, \chi_1, \dots, \chi_k$; still, the main result of [43], as seen from its proof, remains valid if adding the enumeration reducibility requirement is the only change. Making only this change gives the definition used in [47]. As to the definition in [44, 45], it can be obtained from the above-formulated one by adding the requirement the primitive functions and predicates of $\mathfrak{B}$ are always defined when some of their arguments do not belong to $\mathrm{dom}(\alpha)$.

A p.m.v. function θ in B is called *admissible* in an enumeration $(\alpha, \mathfrak{B})$ of $\mathfrak{A}$ if θ has an α-associate which is enumeration reducible to the primitive functions and predicates of $\mathfrak{B}$ (under a natural extension of the notion of α-associate to multiple-valued functions). The main result in [43] is that a p.m.v. function θ in B is search computable in $\theta_1, \dots, \theta_n, \pi_1, \dots, \pi_k$ and constants from B iff θ is admissible in any enumeration of $\mathfrak{A}$ in the sense of [43]. The main result in [45, 48, 49] concerns computability by means of recursively enumerable definitional schemes (REDS) of Shepherdson [31]. Namely, it is shown that a p.m.v. function θ in B is REDS computable in $\theta_1, \dots, \theta_n, \pi_1, \dots, \pi_k$ and constants from B iff θ is admissible in any enumeration of $\mathfrak{A}$ (of course, the sense of this statement depends on the notion of enumeration—since the definition of this notion in [45] contains an additional requirement, the "if"part of the statement in [45] is actually somewhat stronger than the one in [48, 49]).

An analog of the results from [43] and [49] which concerns operators is proved in [50]. An operator Γ transforming p.m.v. functions in the universe of $\mathfrak{A}$ into such ones is called *admissible* in an enumeration $(\alpha, \mathfrak{B})$ of $\mathfrak{A}$ if there exists an enumeration operator Δ such that, whenever $\theta \in \mathrm{dom}(\Gamma)$ and φ is an α-associate of θ, the application of Δ to φ and the primitive functions and predicates of $\mathfrak{B}$ produces an α-associate of $\Gamma(\theta)$. Under the natural definitions of the notions of search computable operator in $\mathfrak{A}$ and REDS-computable operator in $\mathfrak{A}$, it is proved that:

(a) an operator in $\mathfrak{A}$ is search computable iff it is admissible in any enumeration $(\alpha, \mathfrak{B})$ of $\mathfrak{A}$ such that $\mathrm{dom}(\alpha)$ is enumeration reducible to the graphs of the primitive functions and predicates of $\mathfrak{B}$;
(b) an operator in $\mathfrak{A}$ is REDS-computable iff it is admissible in any enumeration of $\mathfrak{A}$.

In the papers [43, 47], certain sets are called *weak admissible* in an enumeration $(\alpha, \mathfrak{B})$ of a partial structure $\mathfrak{A}$. These are the sets of the form

$$\{(\alpha(x_1), \ldots, \alpha(x_n)) \mid (x_1, \ldots, x_n) \in W \cap \mathrm{dom}^n(\alpha)\},$$

where $n \in \omega$ and W is a subset of ω^n enumeration reducible to the graphs of the primitive functions and predicates of $\mathfrak{B}$. In the case when the primitive predicates of $\mathfrak{A}$ admit only the value **true**, the sets definable by means of logic programs in $\mathfrak{A}$ are studied in [47], and it is proved that they coincide with the sets which are weak admissible in any enumeration of $\mathfrak{A}$ in the sense of [47].

In the dissertation [55], another characterization is given of the search computability in $\theta_1, \ldots, \theta_n, \pi_1, \ldots, \pi_k$ and constants from B; namely, it is shown to be equivalent to admissibility in any total enumeration of $\mathfrak{A}$ (thus admissibility in any total enumeration of $\mathfrak{A}$ turns out to be equivalent to admissibility in any enumeration of $\mathfrak{A}$ in the sense of [43]). This is actually shown for a wider class of p.m.v. functions; namely, additional arguments ranging over ω and function values in ω are also allowed, the notion of α-associate being generalized in a natural way (a generalization in this direction is given also for the characterization of REDS-computability from [48, 49]). Similarly, the sets definable by means of logic programs in $\mathfrak{A}$ are shown to be the ones which are weak admissible in any total enumeration of $\mathfrak{A}$. A characterization in a similar spirit is given in the dissertation also of the prime computability in $\mathfrak{A}$. Let a p.m.v. function θ be called *μ-admissible* in an enumeration $(\alpha, \mathfrak{B})$ of $\mathfrak{A}$ if θ has an α-associate which is μ-recursive in the primitive functions and predicates of $\mathfrak{B}$. Soskov proves that θ is prime computable in $\theta_1, \ldots, \theta_n, \pi_1, \ldots, \pi_k$ and constants from B iff θ is μ-admissible in any enumeration of $\mathfrak{A}$ (the result is published in [44], where the "if" part of the statement is somewhat stronger due to the additional requirement in the definition of enumeration). He introduces also a new computability notion called *sequential definability in* $\mathfrak{A}$ and proves that a function is μ-admissible in any total enumeration of $\mathfrak{A}$ iff it is single-valued and sequentially definable in $\mathfrak{A}$.

The difficult part of the proofs of these results is the direction from admissibility or μ-admissibility to the corresponding kind of computability. It uses rather complicated Cohen-type forcing constructions and goes through certain normal form theorems of the admissible functions and sets. These constructions find their applications also in the proofs of other results of Soskov.

Some applications of the results to other problems, e.g. to completeness of partial structures with universe ω, are given in [55, Chap. 6].

In [45, 51] and [55, Chap. 8], applications are given to certain classifications of the concepts of computability on abstract first-order structures and to programming languages with maximal expressive power within certain classes of such ones. For instance, if $\mathcal{A}$ is a class of structures which is closed under homomorphic preimages, then, after the necessary definitions are formulated, a universality is shown of search computability for the class of the effective computabilities over $\mathcal{A}$ which are invariant under homomorphisms. Further, a programming language whose programs are codes of recursively enumerable sets of existential conditional expressions is shown to be universal for the class of the effective programming languages over $\mathcal{A}$ which are invariant under homomorphisms.

In [52] and [55, Chap. 9], an external characterization of the inductive sets on abstract structures is presented. The main result is an abstract version of the classical Suslin-Kleene characterization of the hyperarithmetical sets. Let $\mathfrak{A}$ be a structure with a countable universe B and with primitive predicates only. A subset A of B^n is called *relatively intrinsically* Π^1_1 on $\mathfrak{A}$ if, for any bijection $\alpha : \omega \to B$, the set

$$\{(x_1, \ldots, x_n) \in \omega^n \mid (\alpha(x_1), \ldots, \alpha(x_n)) \in A\} \tag{3}$$

is Π^1_1 relative to the diagram of the structure $\alpha^{-1}(\mathfrak{A})$. Soskov proves that a subset of B^n is relatively intrinsically Π^1_1 on the structure $\mathfrak{A}$ iff it is inductively definable on the least acceptable extension $\mathfrak{A}^*$ of $\mathfrak{A}$. This external approach to the definition of the inductive sets leads very fast to some of the central results presented in [24] and allows us to transfer some results of the classical recursion theory to the abstract case. In particular, a hierarchy for the hyperelementary sets is obtained which is similar to the classical Suslin-Kleene hierarchy.

In [53] and [55, Chap. 10], an effective version of the external approach is considered. Let again $\mathfrak{A}$ be a structure with a countable universe B and with primitive predicates only. A subset A of B^n is called *relatively intrinsically hyperarithmetical* on $\mathfrak{A}$ if, for any bijection $\alpha : \omega \to B$, the set (3) is hyperarithmetical relative to the diagram of the structure $\alpha^{-1}(\mathfrak{A})$. In the case when a bijection $\alpha : \omega \to B$ exists such that the diagram of $\alpha^{-1}(\mathfrak{A})$ is a recursive set, a subset A of B^n is called *intrinsically hyperarithmetical on* $\mathfrak{A}$ if the set (3) is a hyperarithmetical set for any such α. Soskov proves that the intrinsically hyperarithmetical sets coincide with the relatively intrinsically hyperarithmetical sets in this case.[9] As a side effect of the proof, an effective version is obtained of Kueker's theorem on definability by means of infinitary formulas.

Let us note that results and ideas from [52, 53] are substantially used in the paper [14] of Goncharov, Harizanov, Knight and Shore, where several additions are made to results from [52]. For instance, an analog of intrinsically Π^1_1 sets is proved there for the coincidence of intrinsical hyperarithmeticity and relative intrinsical hyperarithmeticity on a recursive structure.

The technique developed in [52] is applied in [54] also to a problem of degree theory. An essential strengthening is achieved there of the result of McEvoy and Cooper from [20] about the existence of a pair of sets which is minimal both with respect to Turing reducibility and with respect to enumeration reducibility. Namely, Soskov proves that, for any set A of natural numbers which is not Π^1_1, a set B of natural numbers exists such that A and B form a pair which is minimal with respect to each of the following four reducibilities: Turing reducibility, enumeration reducibility, hyperarithmetical reducibility and the hyperenumeration reducibility introduced and studied by Sanchis in [28, 29].

3 Enumeration Degrees

The enumeration reducibility is a positive reducibility between sets of natural numbers first introduced formally by Friedberg and Rogers. Intuitively, a set A is said to be enumeration reducible (or more briefly e-reducible) to the set B, denoted by $A \leq_e B$, if there is an algorithm transforming every enumeration of B into an enumeration of A. The precise definition (as already discussed in Sect. 2) says that $A \leq_e B$ *iff* there is a c.e. set W such that $A = W(B)$, where $W(B)$ denotes the set $\{x \mid \exists u(\langle x, u\rangle \in W \ \&\ D_u \subseteq B)\}$. The relation $\leq_e$ is a preorder that gives rise to a nontrivial equivalence relation, whose equivalence classes are called enumeration degrees. After the factorization, the preorder $\leq_e$ transforms into a partial order on the set of enumeration degrees. The respective structure is denoted by $\mathcal{D}_e$.

Enumeration reducibility is connected to Turing reducibility by the following equivalences:

$$A \text{ is c.e. in } B \iff A \leq_e B \oplus \overline{B} \tag{4}$$

$$A \leq_T B \iff A \oplus \overline{A} \leq_e B \oplus \overline{B}. \tag{5}$$

[9]It is known that a replacement of "hyperarithmetical" with "recursively enumerable" in this statement makes it refutable by counter-examples (enumeration reducibility and recursive enumerability being the corresponding replacements of relative hyperarithmeticity and hyperarithmeticity, respectively, in the above definitions).

The second equivalence means that the mapping $\iota : \mathcal{D}_T \longrightarrow \mathcal{D}_e$ acting by the rule $\iota(\mathbf{d}_T(A)) = \mathbf{d}_e(A \oplus \overline{A})$ is an embedding of the structure of the Turing degrees into the structure of the enumeration degrees. A set, say X, is said to be total if $\overline{X} \leq_e X$. Note that if a set is total, then it is enumeration-equivalent to the graph of its characteristic function. Examples of total sets are the graphs of total functions as well as the sets of the form $A \oplus \overline{A}$ for an arbitrary set A. Thus the range of ι coincides with the enumeration degrees of the total sets.

The enumeration jump operator J_e is defined by Cooper and McEvoy [3, 19] by setting $J_e(A) = E_A \oplus \overline{E_A}$, where $E_A = \{\langle x, i\rangle \mid x \in W_i(A)\}$.[10] The jump operator is monotone with respect to e-reducibility and hence it gives rise to a jump operation on enumeration degrees. Clearly, $J_e(A)$ is a total set and hence its degree is in the range of ι. Moreover, $J_e(A) \equiv_e A' \oplus \overline{A'}$ (where A' denotes the halting set relative to A), so the jump operation on e-degrees agrees with the jump operation on T-degrees and the embedding ι.

One of the most natural questions about the jump operation is about the range of the jump operation. It is well known (Friedberg [5]) that the range of the jump operation on the Turing degrees is the upper cone with least element $0_{\mathbf{T}}'$. Transferring this to the enumeration degrees we obtain that every total degree greater than or equal to $0_e'$ is the jump of a total degree. Further, a result by McEvoy shows that each total degree above $0_e'$ is the jump of a nontotal degree, so that total and nontotal degrees play similar roles with respect to the jump operation. However, neither of these results settles the problem about what the range is of the jump operator restricted to an upper cone of enumeration degrees (note that no result about Turing degrees can be directly applied to settle the problem since it would not be applicable to the upper cones that have a nontotal degree for least element). In a series of papers, Soskov and his students Baleva and Kovachev investigate the behaviour of the jump operation on arbitrary upper cones of enumeration degrees.

In order to illustrate the obtained results, let us consider the following problem: *Let* $\mathbf{q}$ *be a total enumeration degree in the upper cone having* $\mathbf{a}^{(n)}$ *as least element for some degree* $\mathbf{a}$.[11] *Does there exist an enumeration degree* $\mathbf{f}$ *such that* $\mathbf{a} \leq \mathbf{f}$ *and* $\mathbf{f}^{(n)} = \mathbf{q}$*?* In order to give an affirmative solution, it is enough to prove that for an arbitrary set Y and a total set Z,

$$J_e(Y) \leq_e Z \Longrightarrow \exists X(Y \leq_e X \;\&\; J_e(X) \equiv_e Z \;\&\; X \text{ is total}). \tag{6}$$

Indeed, let us suppose that $\mathbf{q}$ is greater than or equal to $\mathbf{a}''$ for some degree $\mathbf{a}$ (the assumption $n = 2$ does not affect the generality of the reasoning). Then according to (6) there is a degree $\mathbf{x} \geq \mathbf{a}'$, such that $\mathbf{x}' = \mathbf{q}$. Applying (6) once more we obtain a degree $\mathbf{f} \geq \mathbf{a}$ such that $\mathbf{f}' = \mathbf{x}$, so that $\mathbf{f}'' = \mathbf{q}$ as desired.

---

[10] In the original definition, $J_e(A) = K_A \oplus \overline{K_A}$, where $K_A = \{x \mid x \in W_x(A)\}$. This however is enumeration equivalent to the definition we use here.

[11] As usual for arbitrary $\mathbf{x}$, $\mathbf{x}^{(n)}$ denotes the result of the nth iteration of the jump operation on $\mathbf{x}$.

The formula (6) can be proven using the following simple forcing argument. Let us fix a set Y and a total set Z such that $J_e(Y) \leq_e Z$. We shall build X as the graph of a function $f : \omega \to \omega$ such that $f[2\omega + 1] = Y$. This directly implies that X is a total set and $Y \leq_e X$. Let us call a finite string of natural numbers, say τ, Y-regular if τ has even length and for every odd $x < \mathrm{lh}(\tau)$, $\tau(x) \in Y$. In order to satisfy the main claim of the theorem, it is enough to consider the monotone sequence $\tau_0 \subset \tau_1 \subset \cdots \subset \tau_k \subset \ldots$ of Y-regular finite parts such that for every natural s, τ_{s+1} is the least Y-regular finite part τ extending τ_s such that

(1) $\tau(\mathrm{lh}(\tau_s))$ and $\tau(\mathrm{lh}(\tau_s) + 1)$ are respectively the sth elements of Z and Y,
(2) either $\tau \in W_s(Y)$ or there is no Y-regular extension of τ in $W_s(Y)$.

Since $\{\tau \in \omega^{<\omega} \mid \tau \text{ is } Y\text{-regular}\} \leq_e Y$, the above sequence is computable in Z. Setting X to be the graph of the function $f = \bigcup_s \tau_s$, we have $J_e(Y) \oplus X \equiv_e J_e(X) \leq_e Z$. The reducibility $Z \leq_e J_e(X)$ follows from the fact that the sequence $\{\tau_s\}_{s<\omega}$ can be computed in $J_e(Y) \oplus X$.

The existence of jump inverts gives rise to the question about their distribution. For example, we might consider whether, if given a degree $\mathbf{a}$ and a total degree $\mathbf{q} \geq \mathbf{a}'$, there is a least $\mathbf{f} \geq \mathbf{a}$ such that $\mathbf{f}' = \mathbf{q}$, i.e. is there a least jump invert of $\mathbf{q}$ above $\mathbf{a}$. The answer to this question is negative, since for arbitrary sets Y, C and a total set Z such that $C \oplus \overline{C} \leq_e Z$ we have

$$J_e(Y) \leq_e Z \;\&\; C \not\leq_e Y \Longrightarrow \exists X(Y \leq_e X \;\&\; C \not\leq_e X \;\&\; J_e(X) \equiv_e Z \;\&\; X \text{ is total}). \tag{7}$$

In order to prove (7) we modify the proof of (6) in the following way. We consider the sequence $\tau_0 \subset \tau_1 \subset \cdots \subset \tau_k \subset$ of Y-regular finite parts such that for every natural s,

(1) τ_{2s+1} is the least Y-regular finite part τ extending τ_{2s}, such that
 (a) $\tau(\mathrm{lh}(\tau_{2s}))$ and $\tau(\mathrm{lh}(\tau_{2s}) + 1)$ are respectively the sth elements of Z and Y,
 (b) either $\tau \in W_s(Y)$ or there is no Y-regular extension of τ in $W_s(Y)$;
(2) τ_{2s+2} is the least Y-regular τ extending τ_{2s+1}, such that either $\tau(\mathrm{lh}(\tau_{2s+1})) \notin C$ and $\tau \Vdash \mathrm{lh}(\tau_{2s+1}) \in W_s(X)$, or $\tau(\mathrm{lh}(\tau_{2s+1})) \in C$ and $\tau \Vdash \mathrm{lh}(\tau_{2s+1}) \notin W_s(X)$.

For arbitrary Y-regular finite part ρ the set $\{a \mid \exists\tau(\tau \text{ is } Y\text{-regular}, \tau(\mathrm{lh}(\rho) = a, \tau \Vdash \mathrm{lh}(\rho) \in W_s(X))\}$ is e-reducible to Y and hence is not equal to C. Since $J_e(Y) \leq_e Z$ and $C \oplus \overline{C} \leq_e Z$ the sequence $\{\tau_s\}_{s<\omega}$ is computable in Z. As in our reasoning for (6), setting X to be the graph of the function $f = \bigcup_s \tau_s$, we have $Y \leq_e X$ and $J_e(Y) \oplus X \equiv_e J_e(X) \equiv_e Z$. Now assume that $C \leq_e X$. Then $f^{-1}(C) \leq_e X$ and hence $f^{-1}(C) = W_s(X)$ for some natural s. But according to the construction, we have that

$$\mathrm{lh}(\tau_{2s+1}) \in f^{-1}(C) \iff \tau_{2s+2}(\mathrm{lh}(\tau_{2s+1})) \in C \iff \mathrm{lh}(\tau_{2s+1}) \notin W_s(X).$$

This is clearly impossible, and hence $C \not\leq_e X$.

The results so far can be extended to arbitrary finite iterations of the jump operation. We can even require that the jump invert $\mathbf{f}$ have some additional properties. For instance, in addition to $\mathbf{f}^{(n)} = \mathbf{q}$ and $\mathbf{f} \geq \mathbf{a}$, we may require that $\mathbf{f}^{(i)} \geq \mathbf{a}_i$, for a given sequence of degrees $\mathbf{a}_1, \mathbf{a}_2, \ldots, \mathbf{a}_{n-1}$. It is clear that in order for such an $\mathbf{f}$ to exist, the degrees $\mathbf{a}_1, \mathbf{a}_2, \ldots, \mathbf{a}_{n-1}$ must satisfy certain properties. An obvious necessary condition is $\mathbf{a}_i^{(n-i)} \leq \mathbf{q}$ for $1 \leq i \leq n-1$. Before arguing that this condition is also sufficient, we introduce a notation (due to Soskov [56]) that is closely connected to this condition. Let $\mathcal{A} = (A_0, A_1, \ldots, A_{n-1})$ be a sequence of sets of natural numbers. Note that for arbitrary set Q, the condition

$$J_e^{(n-i)}(A_i) \leq_e Q \text{ for } i \leq n$$

is equivalent to the inequality

$$J_e(J_e(\ldots J_e(J_e(J_e(A_0) \oplus A_1) \oplus A_2) \ldots) \oplus A_{n-1}) \leq_e Q. \tag{8}$$

In order to obtain the term on the left-hand side, we define the sequence $P_0(\mathcal{A})$, $P_1(\mathcal{A})$,..., $P_n(\mathcal{A})$ in the following inductive way:

$$P_0(\mathcal{A}) = A_0;$$
$$P_{i+1}(\mathcal{A}) = J_e(P_i(\mathcal{A})) \oplus A_{i+1}.$$

Now inequality (8) takes the form $J_e(P_{n-1}(\mathcal{A})) \leq_e Q$.

Let $\mathcal{A} = (A_0, A_1, \ldots A_{n-1})$ and Q be given and let us furthermore suppose that $J_e(P_{n-1}(\mathcal{A})) \leq_e Q$. Then we can build a total set F such that

$$J_e^{(n)}(F) \equiv_e Q;$$
$$P_i(\mathcal{A}) \leq_e J_e^{(i)}(F) \text{ for } i < n;$$
$$J_e^{(i+1)}(F) \equiv_e J_e(P_{i-1}(\mathcal{A})) \oplus F$$

in the following way.

Use (6) to build a total F_{n-1} such that

$$J_e(F_{n-1}) \equiv_e J_e(P_{n-1}(\mathcal{A})) \oplus F_{n-1} \equiv_e Q,$$
$$P_{n-1}(\mathcal{A}) \leq_e F_{n-1}.$$

Then apply (6) again to obtain F_{n-2}, such that

$$J_e(F_{n-2}) \equiv_e J_e(P_{n-2}(\mathcal{A})) \oplus F_{n-2} \equiv_e F_{n-1},$$
$$P_{n-2}(\mathcal{A}) \leq_e F_{n-2},$$

and note that $J_e^{(2)}(F_{n-2}) \equiv_e J_e(P_{n-1}(\mathcal{A})) \oplus F \equiv_e Q$. Proceed in this way until a total set F_0 with

$$P_i(\mathcal{A}) \leq_e F_0^{(i)} \text{ for all } i \leq n-1,$$

$$J_e^{(i+1)}(F_0) \equiv_e J_e(P_i(\mathcal{A})) \oplus F_0,$$

$$J_e^{(n)}(F_0) \equiv_e Q$$

is obtained.

If we further assume that a set C, such that $C \not\leq_e P_i(\mathcal{A})$ and $C \oplus \overline{C} \leq_e J_e(P_i(\mathcal{A}))$, is given, then at the appropriate step we can use (7) instead of (6) in order to obtain F_0, so that $C \not\leq_e J_e^{(i)}(F_0)$.

In [56] Soskov proves the following theorem, which is a bit more general than the result we have just presented.

Theorem 3.1 *Let $\mathcal{A} = (A_0, A_1, \ldots, A_{n-1})$ $n \geq 1$, and let Q be a total set, such that $J_e(P_{n-1}(\mathcal{A})) \leq_e Q$. Then the enumeration degree of Q contains the nth jump of a total set F such that $P_i(\mathcal{A}) \leq_e J_e^{(i)}(F)$ and $F^{(i+1)} \equiv_e J_e(P_i(\mathcal{A})) \oplus F$. Moreover, for any $i \leq n$ and any set C such that $C \not\leq_e P_i(\mathcal{A})$ and $C \oplus \overline{C} \leq_e Q$, F can be chosen so that $C \not\leq_e F^{(i)}$.*

Instead of using our backwards argument, Soskov constructs directly the set F using a forcing argument in order to control the behaviour of its ith jump. This approach has two advantages. Firstly, this allows him to relax the condition $C \oplus \overline{C} \leq J_e(P_i(\mathcal{A}))$ to the more general $C \oplus \overline{C} \leq_e Q$. Secondly, and more importantly, the construction presented in [56] can be extended to arbitrary transfinite computable iterations of the jump operator. Indeed, in [61] Soskov and Baleva give the following generalisation of Theorem 3.1:

Theorem 3.2 *Let α be a computable ordinal and suppose that we have fixed an ordinal notation in $\mathcal{O}$ for α. Let $\mathcal{A}$ be a sequence of sets of natural numbers with length α. Let Q be a total set such that $P_\alpha(\mathcal{A}) \leq_e Q$. Then there is a total set F such that*

(1) $A_\beta \leq_e J_e^{(\beta)}(F)$ uniformly in $\beta < \alpha$;
(2) $J_e^{(\beta+1)}(F) \equiv_e J_e(P_\beta(\mathcal{A})) \oplus F$ uniformly in $\beta < \alpha$;
(3) $J_e^{(\beta)}(F) \equiv_e P_{<\beta}(\mathcal{A}) \oplus F$ for every limit $\beta \leq \alpha$;
(4) $J_e^{(\alpha)}(F) \equiv_e Q$.

Moreover, if C is a set such that $C \not\leq_e P_\beta(\mathcal{A})$ for some $\beta < \alpha$ and $C \oplus \overline{C} \leq_e Q$, then F can be chosen so that $C \not\leq_e J_e^{(\beta)}(F)$.

We need to give some explanation of the notions and notations used in the formulation of the theorem. We start with the transfinite iteration of the jump operator. For simplicity, let us consider the least limit ordinal ω. The idea is to define the ω-iteration $J_e^{(\omega)}$ of J_e so that $J_e^{(\omega)}(X)$ is the least set with respect to $\leq_e$

such that $J_e^{(n)}(X) \leq_e J_e^{(\omega)}(X)$ uniformly in $n < \omega$ and X. This yields that we have to set $J_e^{(\omega)}(X) = \{\langle x, n\rangle \mid x \in J_e^{(n)}(X)\}$. Having defined $J_e^{(\omega)}$, we define $J_e^{(\omega+m)}$ by simply setting $J_e^{(\omega+m)}(X) = J_e^{(m)}(J_e^{(\omega)}(X))$. Thus we have defined $J_e^{(\beta)}$ for every ordinal $\beta < \omega \cdot 2$. Now the most natural definition of $J_e^{(\omega \cdot 2)}$ would be $J_e^{(\omega \cdot 2)}(X) = \{\langle x, \beta\rangle \mid \beta < \omega \cdot 2 \ \& \ x \in J_e^{(\beta)}(X)\}$. However, the pairing function $\langle \cdot, \cdot \rangle$ is a function from $\omega \times \omega$ onto ω and hence the term $\langle x, \beta\rangle$ is undefined for $\beta \geq \omega$. In order to overcome this problem, we use ordinal notations instead of the ordinals themselves. The system of ordinal notations $\mathcal{O}$ is a partially ordered subset of ω such that for every computable ordinal α there is an element n_α of $\mathcal{O}$, called a notation for α, such that the set $\{m \in \omega \mid m <_{\mathcal{O}} n_\alpha\}$ partially ordered by $<_{\mathcal{O}}$ is isomorphic to α. Further, there are computable functions $p_{\mathcal{O}}$ and $q_{\mathcal{O}}$, such that for any ordinal α and every notation $n_\alpha \in \mathcal{O}$ of α, if $\alpha = \beta + 1$, then $p_{\mathcal{O}}(n_\alpha) <_{\mathcal{O}} n_\alpha$ is an $\mathcal{O}$-notation for β and if α is a limit ordinal then $q_{\mathcal{O}}(n_\alpha)$ is the index of a computable function f such that $f(0) <_{\mathcal{O}} f(1) <_{\mathcal{O}} \cdots <_{\mathcal{O}} f(s) <_{\mathcal{O}} \cdots <_{\mathcal{O}} n_\alpha$ is a sequence of $\mathcal{O}$-notations of ordinals $\alpha_0 < \alpha_1 < \cdots < \alpha_s < \cdots < \alpha$ such that $\alpha = \lim \alpha_s$. Thus having fixed an $\mathcal{O}$-notation n_α of an ordinal α, we can identify all ordinals $\beta < \alpha$ with their $\mathcal{O}$-notations less than n_α with respect to $<_{\mathcal{O}}$. Moreover if $\beta \leq \alpha$ is a limit ordinal, we shall refer to the sequence yielded by the function $q_{\mathcal{O}}$ as $\{\beta(n)\}_{n<\omega}$.

We are ready to give the formal definitions needed for the formulation of Theorem 3.2. Let α be a computable ordinal and fix an $\mathcal{O}$-notation for α. We define $J_e^{(\beta)}(X)$ for an arbitrary set X and ordinal $\beta \leq \alpha$ by the following transfinite induction:

$$J_e^{(0)}(X) = X;$$
$$J_e^{(\beta+1)}(X) = J_e(J_e^{(\beta)}(X));$$
$$J_e^{(\beta)}(X) = \{\langle x, n\rangle \mid x \in J_e^{(\beta(n))}(X)\}, \text{ if } \beta \text{ is a limit ordinal.}$$

Further, let a $\mathcal{A} = \{A_\beta\}_{\beta<\alpha}$ be a sequence of sets of natural numbers. We define the sequence $\{P_\beta(\mathcal{A})\}_{\beta<\alpha}$ with the following transfinite induction on $\beta \leq \alpha$:

$$P_0(\mathcal{A}) = A_0;$$
$$P_{\beta+1}(\mathcal{A}) = J_e(P_\beta(\mathcal{A})) \oplus A_{\beta+1};$$
$$P_\beta(\mathcal{A}) = P_{<\beta}(\mathcal{A}) \oplus A_\beta, \text{ if } \beta \text{ is a limit ordinal,}$$
$$\text{where } P_{<\beta} = \{\langle x, n\rangle \mid x \in P_{\beta(n)}(\mathcal{A})\}.$$

The technique developed by Soskov and Baleva for the construction of the set F in Theorem 3.2 is further developed by Soskov and Kovachev [62] in order to prove the following theorem.

Theorem 3.3 *Let α be a computable ordinal and suppose that we have fixed an ordinal notation in $\mathcal{O}$ of α. Let $\mathcal{A} = \{A_\beta\}_{\beta<\alpha}$ and $\mathcal{C} = \{C_\beta\}_{\beta<\alpha}$ be two sequences of length α such that C_β is not uniformly e-reducible to $P_\beta(\mathcal{A})$ i.e. for every computable function f, there is a $\beta < \alpha$, such that $C_\beta \neq W_{f(\beta)}(P_\beta(\mathcal{A}))$. Then there is a total set F such that A_β is uniformly e-reducible to $J_e^{(\beta)}(F)$, but C_β is not uniformly e-reducible to $J_e^{(\beta)}(F)$.*

Strictly speaking, this is not a real jump inversion theorem, since we do not require the α-jump of F to be a specific set Q. Nevertheless, this theorem is much stronger than Theorem 3.2 in its omitting part. In fact, using Theorem 3.2 we could obtain that C_β is not uniformly e-reducible to $J_e^{(\beta)}(F)$ only in the case when there is a particular $\beta < \alpha$ such that $C_\beta \not\leq_e P_\beta(\mathcal{A})$. On the other hand, Theorem 3.3 allows us to omit even sequences for which $C_\beta \leq_e P_\beta(\mathcal{A})$ for every $\beta < \alpha$, but not uniformly in β. In order to see that this strengthening is not trivial, let us consider the ordinal ω and an arbitrary sequence $\mathcal{A} = \{A_n\}_{n<\omega}$ with length ω. Let us consider the class $\mathcal{Z}$ of all sequences of length ω consisting entirely of c.e. sets (or even finite sets). Clearly, $\mathcal{Z}$ contains continuum many sequences. On the other hand, since there are only countably many computable functions, there can be only countably many sequences $\mathcal{B} = \{B_n\}_{n<\omega}$ for which $B_n \leq_e P_n(\mathcal{A})$ uniformly in n. Thus there must be a sequence $\mathcal{C} \in \mathcal{Z}$ such that C_n is not uniformly e-reducible to $P_n(\mathcal{A})$ (in fact there are continuum many such sequences). Now, in contrast to Theorem 3.2, Theorem 3.3 gives us a total set F for which A_n is uniformly e-reducible to $J_e^{(n)}(F)$, but C_n is not uniformly e-reducible to $J_e^{(n)}(F)$.

Theorem 3.3 is the base for the ω-enumeration reducibility, to which the next section, is dedicated.

4 ω-Enumeration Degrees

Selman's theorem [30] allows us to consider enumeration reducibility as a reducibility generated by the weak or Muchnik reducibility between mass problems. Recall that a mass problem is an arbitrary set (possibly empty) of total functions from ω to ω. Given two mass problems $\mathcal{M}_1$ and $\mathcal{M}_2$, we say that $\mathcal{M}_1$ is *weakly* (or *Muchnik*) reducible to $\mathcal{M}_2$, denoted by $\mathcal{M}_1 \leq_w \mathcal{M}_2$, if for every $f \in \mathcal{M}_2$ there is a $g \in \mathcal{M}_1$, such that $g \leq_T f$. Note that if $\mathcal{M}_1$ and $\mathcal{M}_2$ are upwards closed with respect to Turing reducibility, then $\mathcal{M}_1 \leq_w \mathcal{M}_2$ *iff* $\mathcal{M}_2 \subseteq \mathcal{M}_1$. Now, given a set of natural numbers A, let $\mathcal{M}_A$ denote the mass problem of all functions in which A is computably enumerable, i.e. $\mathcal{M}_A = \{f \in \omega^\omega \mid A \text{ is c.e. in } f\}$. The mass problem $\mathcal{M}_A$ is upwards closed with respect to Turing reducibility. According to Selman's theorem for arbitrary sets A and B,

$$A \leq_e B \iff \forall f \in \omega^\omega (B \text{ is c.e. in } f \Rightarrow A \text{ is c.e. in } f),$$

and hence

$$A \leq_e B \iff \mathcal{M}_A \leq_w \mathcal{M}_B.$$

Following these lines, Soskov [58] considers the following class of mass problems. Let us denote by $\mathcal{S}_\omega$ the set of all denumerable sequences of sets of natural numbers. Given a sequence $\mathcal{A} = \{A_n\}_{n<\omega}$, let $\mathcal{M}_\mathcal{A}$ be the mass problem

$$\mathcal{M}_\mathcal{A} = \{f \in \omega^\omega \mid A_n \text{ is c.e. in } f^{(n)} \text{ uniformly in } n\}. \tag{9}$$

Informally, $\mathcal{M}_\mathcal{A}$ is the mass problem of all functions that encode uniformly the nth element of $\mathcal{A}$ in their nth Turing jump. Now Soskov defines the ω-enumeration reducibility between two elements of $\mathcal{S}_\omega$ by setting

$$\mathcal{A} \leq_\omega \mathcal{B} \iff \mathcal{M}_\mathcal{A} \leq_w \mathcal{M}_\mathcal{B}. \tag{10}$$

Clearly, the relation $\leq_\omega$ is a preorder on $\mathcal{S}_\omega$ and hence it gives rise to an equivalence relation. The equivalence classes under this relation are called ω-enumeration degrees. The partially ordered structure of the ω-enumeration degrees is denoted by $\mathcal{D}_\omega$. It has a least element 0_ω consisting of all sequences that are ω-enumeration reducible to the sequence $\varnothing_\omega = \{\varnothing\}_{n<\omega}$. Further, for any two sequences $\mathcal{A} = \{A_n\}_{n<\omega}$ and $\mathcal{B} = \{B_n\}_{n<\omega}$, the sequence $\mathcal{A} \oplus \mathcal{B} = \{A_n \oplus B_n\}_{n<\omega}$ is such that $\mathcal{M}_{\mathcal{A}\oplus\mathcal{B}} = \mathcal{M}_\mathcal{A} \cap \mathcal{M}_\mathcal{B}$, so the ω-enumeration degree $\mathbf{d}_\omega(\mathcal{A} \oplus \mathcal{B})$ is the least upper bound of $\mathbf{d}_\omega(\mathcal{A})$ and $\mathbf{d}_\omega(\mathcal{B})$. Thus, like $\mathcal{D}_T$ and $\mathcal{D}_e$, $\mathcal{D}_\omega$ is an upper semi-lattice with least element.

The connection between $\mathcal{D}_e$ and $\mathcal{D}_\omega$ goes much further. For any set A let us denote by $A \uparrow \omega$ the sequence $(A, \varnothing, \varnothing, \ldots)$. Note that $\mathcal{M}_{A\uparrow\omega} = \mathcal{M}_A$, so that for any two sets A and B,

$$A \leq_e B \iff A \uparrow \omega \leq_\omega B \uparrow \omega.$$

This means that the mapping $\kappa : \mathcal{D}_e \to \mathcal{D}_\omega$ acting by the rule $\kappa(\mathbf{d}_e(A)) = \mathbf{d}_\omega(A \uparrow \omega)$ is a well-defined embedding of $\mathcal{D}_e$ into $\mathcal{D}_\omega$. Thus $\mathcal{D}_\omega$ may be considered as an extension of the Turing and the enumeration degrees.

Although we refer to $\leq_\omega$ as a reducibility relation, its definition (10) does not give us an actual procedure for calculating the smaller from the bigger term. This, however, is provided by Theorem 3.3. According to it, for every two sequences $\mathcal{A}, \mathcal{B} \in \mathcal{S}_\omega$, if A_n is not uniformly e-reducible to $P_n(\mathcal{B})$, then there is an f witnessing that $\mathcal{M}_\mathcal{B} \not\subseteq \mathcal{M}_\mathcal{A}$. On the other hand, if A_n is uniformly e-reducible to $P_n(\mathcal{B})$, then clearly $\mathcal{M}_\mathcal{B} \subseteq \mathcal{M}_\mathcal{A}$. Since the mass problems of the form $\mathcal{M}_\mathcal{A}$ are upwards closed with respect to $\leq_T$, we have

$$\mathcal{A} \leq_\omega \mathcal{B} \iff A_n \leq_e P_n(\mathcal{B}) \text{ uniformly in } n.$$

Besides the definition of the ω-enumeration reducibility, the ω-enumeration degrees and their basic properties, there are two more important items included in the first paper on the subject. The first one is that, like the structure of the enumeration degrees, $\mathcal{D}_\omega$ is downwards dense. This is proved in two steps: the first one is to show that the ω-enumeration degrees lying beneath $\kappa(0_e{}')$ are dense. The second one is to build a computable function f such that for every sequence $\mathcal{A} = \{A_n\}_{n<\omega}$ either the sequence $\{W_{f(n)}(P_n(\mathcal{A}))\}$ is not ω-enumeration-equivalent to $\mathcal{A}$ or its ω-enumeration degree is less than or equal to $\kappa(0_e{}')$.

The second important thing is the definition of the jump operator on $\mathcal{S}_\omega$: *The jump of a sequence*

$$\mathcal{A} = (A_0, A_1, \ldots, A_n, \ldots) \in \mathcal{S}_\omega$$

is the sequence

$$\mathcal{A}' = (P_1(\mathcal{A}), A_2, A_3, \ldots, A_n, \ldots).$$

A simple application of Theorem 3.1 shows that

$$\mathcal{M}_{\mathcal{A}'} \equiv_w \{f' \mid f \in \mathcal{M}_{\mathcal{A}}\}. \tag{11}$$

From here it follows that the jump operator is monotone and hence it induces a jump operation on the ω-enumeration degrees. The equivalence (11) further implies that the jump operation on $\mathcal{D}_\omega$ agrees with the jump operation on $\mathcal{D}_e$ and the embedding κ.

It turns out that the ω-enumeration degrees behave in an unusual way with respect to the considered jump operation. In [8] Soskov and Ganchev prove the following strong jump inversion theorem:

Theorem 4.1 *For every* $\mathbf{a}, \mathbf{b} \in \mathcal{D}_\omega$ *and every natural n with* $\mathbf{a}^{(n)} \leq \mathbf{b}$*, the system*

$$\left| \begin{array}{ll} \mathbf{x} & \geq \mathbf{a} \\ \mathbf{x}^{(n)} & = \mathbf{b} \end{array} \right. \tag{12}$$

has a least solution.

Note that according to Theorem 3.1 the system (12) can have a least solution in $\mathcal{D}_e$ (or $\mathcal{D}_T$) *iff* $\mathbf{b} = \mathbf{a}^{(n)}$, so $\mathcal{D}_\omega$ is quite different with respect to its jump operation from the classical degree structures.

Theorem 4.1 allows us to consider for every natural n and every $\mathbf{a} \in \mathcal{D}_\omega$ an operation $I^n_{\mathbf{a}}$ defined on the upper cone with least element $\mathbf{a}^{(n)}$ such that $I^n_{\mathbf{a}}(\mathbf{b})$ is the least solution of (12). Using these operations Soskov and Ganchev [8] manage to prove that the range of the embedding κ is first-order definable in the poset of the ω-enumeration degrees augmented by the jump operation, $\mathcal{D}_\omega{}'$, and that the automorphism groups $\mathrm{Aut}(\mathcal{D}_\omega{}')$ and $\mathrm{Aut}(\mathcal{D}_e)$ are isomorphic.

The least jump invert operations play also an important role in the local theory of the ω-enumeration degrees, i.e. the poset $\mathcal{G}_\omega = \{\mathbf{a} \in \mathcal{D}_\omega \mid a \leq 0_\omega{}'\}$. For every natural n, let us set $\mathbf{o}_n = I^n_{0_\omega}(0_\omega{}^{(n+1)})$, i.e. $\mathbf{o}_n$ denotes least ω-enumeration degree, such that $\mathbf{o}_n{}^{(n)} = 0_\omega{}^{(n+1)}$. Clearly, $\mathbf{o}_n{}^{(n+1)} = \mathbf{o}_{n+1}{}^{(n+1)}$, so that

$$0_\omega{}' = \mathbf{o}_0 \geq \mathbf{o}_1 \geq \cdots \geq \mathbf{o}_n \geq \ldots \tag{13}$$

is a sequence of degrees in $\mathcal{G}_\omega$. Furthermore, for each natural n, $\mathbf{o}_n$ is the least element of H_n—the collection of all high-n ω-enumeration degrees. Since for each n there is a Turing, and hence an enumeration and hence an ω-enumeration degree which is high-$n+1$ but not high-n, the sequence (13) is strictly decreasing. Nevertheless, it does not converge to the least degree 0_ω. In fact, the set $\mathcal{AZ} = \{\mathbf{x} \mid \forall n(\mathbf{x} \leq \mathbf{o}_n)\}$ is a countable ideal. The elements of $\mathcal{AZ}$ are referred to as *almost zero* (or simply *a.z.*) degrees. The reason for this is that a degree $\mathbf{a}$ is *a.z.* *iff* there is a sequence $\{A_n\}_{n<\omega} \in \mathbf{a}$ with $A_n \leq_e J^{(n)}_e(\emptyset)$ for every natural n. Note that $\{A_n\}_{n<\omega} \in 0_\omega$ *iff* $A_n \leq_e J^{(n)}_e(\emptyset)$ uniformly in n, so the only difference between a sequence generating the zero degree and a sequence generating a non-zero *a.z.* degree is the uniformity condition.

Surprisingly, the *a.z.* degrees in $\mathcal{G}_\omega$ are intermediate, i.e. are neither low-n nor high-n for any n. Moreover, Soskov and Ganchev proved that a degree in $\mathcal{G}_\omega$ is intermediate *iff* it does majorise a non-zero *a.z.* degree, but does not majorise every *a.z.* degree. In particular, a degree in $\mathcal{G}_\omega$ is high-n for some n *iff* it majorises every *a.z.* degree, whereas a degree in $\mathcal{G}_\omega$ is low-n for some n *iff* it majorises no non-zero *a.z.* degree.

Besides defining the high-n and the *a.z.* degrees, the degrees $\mathbf{o}_n$ are also connected with the low-n degrees. In fact, a simple argument shows that a degree in $\mathcal{G}_\omega$ is low-n *iff* it forms a minimal pair with $\mathbf{o}_n$. Further, the degrees $\mathbf{o}_n$ supply intervals isomorphic to jump intervals of enumeration and ω-enumeration degrees. More precisely, for every natural n,

$$\mathcal{D}_\omega[\mathbf{o}_{n+1}, \mathbf{o}_n] \cong \mathcal{D}_e[0_e{}^{(n)}, 0_e{}^{(n+1)}], \tag{14}$$

$$\mathcal{D}_\omega[0_\omega, \mathbf{o}_n] \cong \mathcal{D}_\omega[0_\omega{}^{(n)}, 0_\omega{}^{(n+1)}]. \tag{15}$$

It turns out that each of the degrees $\mathbf{o}_n$ is first-order definable in the poset $\mathcal{G}_\omega$ (Ganchev and M. Soskova [9]), so that each of the jump classes L_n and H_n is first order definable in $\mathcal{G}_\omega$ for every natural n. The key tool for this definability result is the Kalimullin pairs in $\mathcal{G}_\omega$. According to the definition, the pair $\{\mathbf{a}, \mathbf{b}\} \subseteq \mathcal{G}_\omega$ is a Kalimullin pair (or simply a $\mathcal{K}$-pair) in $\mathcal{G}_\omega$ if

$$\forall \mathbf{x} \in \mathcal{G}_\omega(\mathbf{x} = (\mathbf{x} \vee \mathbf{a}) \wedge (\mathbf{x} \vee \mathbf{b})).$$

Kalimullin pairs were first introduced by Kalimullin [16] for the structure of the enumeration degrees. He considered pairs $\{\mathbf{a}, \mathbf{b}\}$ of enumeration degrees such that

$$\forall \mathbf{x} \in \mathcal{D}_e(\mathbf{x} = (\mathbf{x} \vee \mathbf{a}) \wedge (\mathbf{x} \vee \mathbf{b})). \tag{16}$$

The property making such pairs very useful is that a pair $\{\mathbf{a},\mathbf{b}\} \subseteq \mathcal{D}_e$ has the property (16) if and only if there are sets $A \in \mathbf{a}$, $B \in \mathbf{b}$ and a c.e. set W, such that

$$A \times B \subseteq W \;\&\; \overline{A} \times \overline{B} \subseteq \overline{W}. \tag{17}$$

In order to illustrate how this property is used note that if B is not c.e., then $A = \{x \mid \exists y \in \overline{B}(\langle x,y\rangle \in W)\}$, and, symmetrically, $\overline{A} = \{x \mid \exists y \in B(\langle x,y\rangle \in \overline{W})\}$. Thus if B is not c.e., then $A \leq_e \overline{B} \oplus W$ and $\overline{A} \leq_e B \oplus \overline{W}$. From here it follows that if $\mathbf{a}$, $\mathbf{b}$ and $\mathbf{c}$ are non-zero enumeration degrees, such that each of the pairs $\{\mathbf{a},\mathbf{b}\}$, $\{\mathbf{b},\mathbf{c}\}$ and $\{\mathbf{c},\mathbf{a}\}$ is a $\mathcal{K}$-pair, then $\mathbf{a} \vee \mathbf{b} \vee \mathbf{c} \leq 0_e{}'$. On the other, hand there is a triple $\mathbf{a}$, $\mathbf{b}$ and $\mathbf{c}$ like above such that $\mathbf{a} \vee \mathbf{b} \vee \mathbf{c} = 0_e{}'$ (Kalimullin [16]), so that $0_e{}'$ is the largest degree, which is the join of a triple of $\mathcal{K}$-pairs.

A further investigation of the $\mathcal{K}$-pairs in $\mathcal{D}_e$ revealed their usefulness in a series of definability results in the local structure of the enumeration degrees $\mathcal{G}_e$ (Ganchev and M. Soskova [10–13]). The two most important ones are the first-order definabilities in $\mathcal{G}_e$ of a standard model of arithmetic and of the total degrees.

As far as the poset $\mathcal{G}_\omega$ is concerned, there is still no full characterization of the $\mathcal{K}$-pairs similar to (17). The partial results so far show that we can distinguish between two kinds of Kallimulin pairs in $\mathcal{G}_\omega$. The first kind is the one inherited from the Kalimullin pairs in $\mathcal{D}_e$. They have the form $\{I^n_{0_\omega}(\kappa(\mathbf{a})), I^n_{0_\omega}(\kappa(\mathbf{b}))\}$, where $\{\mathbf{a},\mathbf{b}\} \subseteq \mathcal{D}_e$ is a $\mathcal{K}$-pair relative to the cone of the enumeration degrees above $0_e{}^{(n)}$. These are the $\mathcal{K}$-pairs in $\mathcal{G}_\omega$ that play the key role in defining the degrees $\mathbf{o}_n$.

The second kind consists of the Kallimulin pairs consisting of two *a.z.* ω-enumeration degrees. In order to obtain a characterization of the sequences generating such pairs, Soskov and M. Soskova [63] take the following line of thought.

Let us fix a non-zero *a.z.* $\mathbf{a}$ and a sequence $\{A_n\}_{n<\omega} \in \mathbf{a}$. Recall that this implies that $A_n \leq_e J_e^{(n)}(\varnothing)$ for every n, but not uniformly in n. Consider a sequence $i_0, i_1, \ldots, i_n, \ldots$ such that, for every n, $A_n = W_{i_n}(J_e^{(n)}(\varnothing))$. It is clear that i_n cannot be computed in $J_e^{(n)}(\varnothing)$ uniformly in n. However, it could be the case that i_n can be computed in $J_e^{(n+1)}(\varnothing)$ uniformly in n. The latter is equivalent to the sequence $\mathcal{I} = \{\{i_n\}\}_{n<\omega}$ being ω-enumeration reducible to $\varnothing_\omega{}'$, i.e. $\mathbf{d}_\omega(\mathcal{I})$ being a non-zero *a.z.* degree in $\mathcal{G}_\omega$. If this is the case Soskov and M. Soskova call $\mathbf{a}$ a *super a.z.* degree. In other words, a degree $\mathbf{a}$ is *super a.z.* iff there is a sequence $\{A_n\}_{n<\omega} \in \mathbf{a}$ such that we can compute uniformly in $J_e^{(n+1)}(\varnothing)$ an index i_n such that $A_n = W_{i_n}(J_e^{(n)}(\varnothing))$.

It turns out [63] that *super a.z.* degrees exist, but not all *a.z.* degrees are *super a.z.* Moreover, every degree majorised by a *super a.z.* degree is a *super a.z.* degree, and the l.u.b of two *super a.z.* degrees is again a *super a.z.* degree. Thus the *super a.z.* degrees form a proper subideal of the *a.z.* degrees.

The $\mathcal{K}$-pairs of *super a.z.* degrees admit a characterization very close to the one for the $\mathcal{K}$-pairs of enumeration degrees. In fact, a pair $\{\mathbf{a},\mathbf{b}\}$ of *super a.z.* degrees is a $\mathcal{K}$-pair if and only if there are sequences $\{A_n\}_{n<\omega} \in \mathbf{a}$, $\{B_n\}_{n<\omega} \in \mathbf{b}$ and $\{R_n\}_{n<\omega} \in 0_\omega$ such that

$$\forall n \in \omega(A_n \times B_n \subseteq R_n \;\&\; \overline{A_n} \times \overline{B_n} \subseteq \overline{R_n}).$$

This characterization gives the tool needed for the construction of non-zero $\mathcal{K}$-pairs of *super a.z.* degrees. In fact, using a priority argument Soskov and M. Soskova prove that each interval of *super a.z.* degrees contains an independent $\mathcal{K}$-system, i.e. a sequence of degrees generated by a sequence of sequences $\mathcal{A}_0, \mathcal{A}_1, \ldots, \mathcal{A}_n, \ldots$ such that $\{\mathbf{d}_\omega(\mathcal{A}_i), \mathbf{d}_\omega(\bigoplus_{n\neq i} \mathcal{A}_n)\}$ is a $\mathcal{K}$-pair for each $i \in \omega$. As shown by Ganchev and Soskova [10], the existence of an independent $\mathcal{K}$-system in a given interval suffices for every countable distributive lattice to be embedded in that interval. Thus every countable distributive lattice is embeddable in each interval of *super a.z.* degrees, showing the high structural complexity of the ideal of the *super a.z.* degrees.

A similar but weaker result has been proven by Soskov and M. Soskova [68] for the whole structure $\mathcal{G}_\omega$. Using once again a priority argument on sequences of sets of natural numbers, they prove that each interval in $\mathcal{G}_\omega$ contains an independent system, i.e. a sequence of degrees generated by a sequence of sequences $\mathcal{A}_0, \mathcal{A}_1, \ldots, \mathcal{A}_n, \ldots$ such that $\mathcal{A}_i \not\leq_\omega \bigoplus_{n\neq i} \mathcal{A}_n$ for each $i \in \omega$. The existence of an independent system guarantees the embeddability of every countable partial order (Sacks [27]). Thus every partial order is embeddable in every interval in $\mathcal{G}_\omega$. In particular, since $\mathcal{D}_\omega[\mathbf{o}_1, \mathbf{o}_0] \cong \mathcal{D}_e[0_e, 0_e'] = \mathcal{G}_e$ (see (14)), every countable partial order is embeddable in every interval in the local structure of the enumeration degrees.[12]

The results so far reveal that the poset of the ω-enumeration degrees is a very rich and interesting extension of the poset of the enumeration degrees. The two structures are closely connected, but are very different with respect to the behavior of the respective jump operations. In fact, the least jump invert theorem makes $\mathcal{D}_\omega$ far more well-arranged than its enumeration counterpart. Further, the local substructure $\mathcal{G}_\omega$ of the ω-enumeration degrees exhibits richness unmatched by the local structures of the enumeration and of the Turing degrees. However, the definition of the ω-enumeration reducibility is not motivated by a natural computation procedure. In fact, its main motivations are Selman's theorem for enumeration reducibility and the Soskov–Kovachev jump inversion theorem (Theorem 3.3). This poses two natural questions: Firstly, why use the relation *c.e. in* in the definition of the mass problem $\mathcal{M}_\mathcal{A}$ associated to the sequence $\mathcal{A}$ (see (9))? The answer is that there is no particular reason for using this specific relation. We could use any known reducibility relation, provided that we can prove for it a theorem analogous to Theorem 3.3. As a result, we would obtain a degree structure, which would be situated somewhere between the degree structure defined by the original relation and the poset of the Muchnik degrees. This intermediate structure will inherit features from both the small and the big structures. Most probably the least jump inversion theorem will be valid in the new structure and it will have a very rich local structure. Indeed, in a recent paper, Ganchev and Sariev [7] have substituted the relation *c.e. in* with the relation *Turing reducible to*. The resulting reducibility relation between sequences, called ω-Turing reducibility, induces a poset of degrees $\mathcal{D}_{\omega,T}$, which is an extension of the poset $\mathcal{D}_T$ of the Turing degrees. The behavior of $\mathcal{D}_{\omega,T}$ with respect to $\mathcal{D}_T$ is very similar to the behavior of $\mathcal{D}_\omega$ with respect to $\mathcal{D}_e$. Indeed, $\mathcal{D}_{\omega,T}$ contains a definable copy of

[12]This result was first proven by Bianchini [1].

$\mathcal{D}_T$ and has a local structure very similar to the structure of $\mathcal{G}_\omega$ described above. All these are in fact properties inherited from the poset of the Muchnik degrees. As for the properties inherited from $\mathcal{D}_T$, $\mathcal{D}_{\omega,T}$ features minimal degrees, which do not appear in the structure of the ω-enumeration degrees.

The second natural question arising from the definition of the ω-enumeration reducibility is about why we consider sequences with length ω. Here the answer is a bit more complicated. The main reason for considering sequences of length ω, instead of sequences with arbitrary computable length, is the ordinal notations described in the previous section. Recall that in order to work with infinite ordinals less than or equal to a computable ordinal $\alpha > \omega$, we must fix an ordinal notation in $\mathcal{O}$ for α. All ordinals greater than or equal to ω have denumerably many notations in $\mathcal{O}$, so that we have infinitely many choices for the ordinal notation of α. Thus in order to define a reducibility relation on the set $\mathcal{S}_\alpha = \{\{A_\beta\}_{\beta<\alpha} \mid \forall\beta < \alpha(A_\beta \subseteq \omega)\}$, we must first fix an ordinal notation $n_\alpha \in \mathcal{O}$ for α, thus fixing ordinal notations n_β for every ordinal $\beta < \alpha$. Then to every sequence $\mathcal{A} = \{A_\beta\}_{\beta<\alpha}$ in $\mathcal{S}_\alpha$, we associate the mass problem $\mathcal{M}^{n_\alpha}_{\mathcal{A}} = \{f \in \omega^\omega \mid A_\beta$ is c.e. in $f^{(n_\beta)}$ uniformly in $\beta < \alpha\}$, and for every $\mathcal{A}, \mathcal{B} \in \mathcal{S}_\alpha$, we set

$$\mathcal{A} \leq^{n_\alpha}_\alpha \mathcal{B} \iff \mathcal{M}^{n_\alpha}_{\mathcal{B}} \subseteq \mathcal{M}^{n_\alpha}_{\mathcal{A}}.$$

Thus we define a preorder relation on $\mathcal{S}_\alpha$ that, of course, gives rise to a degree structure $\mathcal{D}^{n_\alpha}_\alpha$. However, we would like to define a degree structure $\mathcal{D}_\alpha$ not dependent on the choice of the ordinal notation of α. Clearly, this is trivially possible if for every notation $k_\alpha \in \mathcal{O}$ of α, we have

$$\mathcal{A} \leq^{n_\alpha}_\alpha \mathcal{B} \iff \mathcal{A} \leq^{k_\alpha}_\alpha \mathcal{B}. \tag{18}$$

However, most probably the above equivalence is not true. Indeed, note that for the sequence $\mathcal{N} = \{\{n_\beta\}\}_{\beta<\alpha}$ of the notations of the ordinals $\beta < \alpha$ yielded by n_α, the mass problem $\mathcal{M}^{n_\alpha}_{\mathcal{N}}$ coincides with ω^ω. Thus a necessary condition for (18) is the equality

$$\mathcal{M}^{n_\alpha}_{\mathcal{K}} = \omega^\omega \tag{19}$$

for every sequence $\mathcal{K} = \{\{k_\beta\}\}_{\beta<\alpha}$ consisting of the notations of the ordinals $\beta < \alpha$ yielded by a notation k_α for α. A simple argument shows that this condition is also sufficient.

Equality (19) means that k_β can be computed in $\varnothing^{(n_\beta)}$ uniformly in $\beta < \alpha$. However, the best result so far (Spector) shows that in order to compute k_β, we must use oracle $\varnothing^{(n_\beta+2)}$, which is two Turing jumps above the one that we need.

Question 4.2 (Soskov) Prove either equivalence (18) or $\mathcal{D}^{n_\alpha}_\alpha \cong \mathcal{D}^{k_\alpha}_\alpha$.

5 Degree Spectra

Another reducibility, inherited from the Muchnik reducibility, Soskov was particularly interested in was the reducibility between abstract countable structures induced by their degree spectra. The degree spectrum of an abstract countable structure has been introduced by Richter [26] in order to measure its computability power. Suppose that $\mathfrak{A}$ is a countable structure over a computable, possibly infinite, language $\mathcal{L}$. A presentation of $\mathfrak{A}$ is just an isomorphic copy of $\mathfrak{A}$ with domain ω (or an initial segment of ω). The degree of a given presentation of $\mathfrak{A}$ is the Turing degree of the join of its relations and the graphs of its operations (if $\mathcal{L}$ is infinite we take the uniform join of the relations and the graphs of the operations). Finally, the spectrum $Sp(\mathfrak{A})$ of $\mathfrak{A}$ is the collection of the degrees of its presentations. More formally, if $\mathfrak{A} = (A; R_0, R_1, \ldots)$ (here, for simplicity, we take a purely relational denumerable structure), then

$$Sp(\mathfrak{A}) = \{\mathbf{d}_T(f^{-1}(\mathfrak{A})) \mid f \text{ is a bijection from } \omega \text{ onto } A\}, \tag{20}$$

where $f^{-1}(\mathfrak{A})$ stands for the set $f^{-1}(R_0) \oplus f^{-1}(R_1) \oplus \cdots$. A structure is said to have a degree if its spectrum has a least element. A result by Richter [26] is that the spectrum of every linear order contains a minimal pair, so if a linear order has a degree, it must be 0_T. On the other hand, there are continuum many non-isomorphic countable linear orders so that there must be a linear order that does not have a degree. Thus, in general, we need the whole spectrum of a structure in order to characterize its effective complexity.

Clearly, considering the spectrum of $\mathfrak{A}$ is equivalent to considering the mass problem

$$\mathcal{M}_{\mathfrak{A}} = \{g \in \omega^\omega \mid \mathbf{d}_T(g) \in Sp(\mathfrak{A})\}$$

and hence the spectra of structures induce a reducibility relation between abstract structures through Muchnik reducibility. More precisely, given abstract structures $\mathfrak{A}$ and $\mathfrak{B}$, we say that $\mathfrak{A} \leq_w \mathfrak{B}$ *iff* $\mathcal{M}_{\mathfrak{A}} \leq_w \mathcal{M}_{\mathfrak{B}}$. In general, the mass problem $\mathcal{M}_{\mathfrak{A}}$ is upwards closed with respect to Turing reducibility, as according to a result by Knight [17], if $\mathfrak{A}$ is not *trivial*,[13] then $\mathcal{M}_{\mathfrak{A}}$ consists of all functions g that can compute a presentation of $\mathfrak{A}$. Thus for every two nontrivial structures $\mathfrak{A}$ and $\mathfrak{B}$, we have that

$$\mathfrak{A} \leq_w \mathfrak{B} \iff Sp(\mathfrak{B}) \subseteq Sp(\mathfrak{A}).$$

[13] A structure is called trivial if there are finitely many elements, such that every permutation of the domain leaving these elements fixed is an automorphism. For example any complete graph is a trivial structure.

5.1 Enumeration Spectra

In his first paper on the subject of degree spectra [57], Soskov proposes two slight modifications in the definition of the spectrum (20). The first one is to consider surjective mappings from the set of the natural numbers onto the domain of the structure, instead of bijective ones. In other words, we substitute the bijections with enumerations of the structure. Having taken this first step, it is more natural to work with the enumeration reducibility instead of the Turing reducibility. Thus we come to the definition of the *degree spectrum* of an abstract structure proposed by Soskov in [57]: Given an abstract structure $\mathfrak{A}$, its degree spectrum (which we shall refer to as *enumeration spectrum*) is the set

$$Sp_e(\mathfrak{A}) = \{\mathbf{d}_e(f^{-1}(\mathfrak{A})) \vee \mathbf{d}_e(f^{-1}(=)) \vee \mathbf{d}_e(f^{-1}(\neq)) \mid f \text{ is an enumeration of } \mathfrak{A}\}.$$

Clearly, the enumeration spectrum differs from the spectrum of an abstract structure. More precisely, in general the enumeration spectrum is not the image of the spectrum under the natural embedding ι. This is due to the fact, that when using Turing reducibility, we have information of both the positive and the negative information contained in the relations of the structure, whereas enumeration reducibility takes care only of the positive information. Thus, it may be the case that the enumeration spectrum contains non-total degrees, i.e. degrees that are not images of Turing degrees under the embedding ι. This, in fact, is the case with any structure of the form $\mathfrak{A}_P = (\omega; R_P)$ where R_P is a unary predicate corresponding to a set P with non-total enumeration degree. Indeed, if we consider the identity mapping from ω onto ω, we obtain that $\mathbf{d}_e(P) \in Sp_e(\mathfrak{A})$, so that $Sp_e(\mathfrak{A}) \neq \iota(Sp(\mathfrak{A}))$.

On the other hand, the enumeration spectrum is not obliged to contain a non-total degree. In fact, the relations in the structure could be such that we could determine the negative information by asking questions only about the positive information. This, for example, is the case for any purely operational structure (a structure without relations). To be more precise, let us consider a group $\mathfrak{G} = (G; +_G)$. Let f be an enumeration of $\mathfrak{G}$. We have that

$$\langle x,y,z\rangle \notin f^{-1}(+_G) \iff f(x) +_G f(y) \neq f(z) \iff$$

$$\exists z' \left(z \neq z' \;\&\; f(x) +_G f(y) = f(z')\right) \iff \exists z' \left(z \neq z' \;\&\; \langle x,y,z'\rangle \in f^{-1}(+_G)\right),$$

so that $f^{-1}(+_G)$, which is in fact $f^{-1}(\mathfrak{G})$, is a total set and hence $\mathbf{d}_e(f^{-1}(\mathfrak{G}))$ is a total degree. Thus the enumeration spectrum of $\mathfrak{G}$ consists entirely of total degrees. Soskov refers to structures with this property as total structures.

The first observation Soskov made about the enumeration spectrum is that it is always upwards closed with respect to total enumeration degrees, i.e. if $\mathbf{a} \in Sp_e(\mathfrak{A})$ and $\mathbf{b}$ is a total enumeration degree with $\mathbf{a} \leq \mathbf{b}$, then $\mathbf{b} \in Sp_e(\mathfrak{A})$. Recall that the total degrees are the enumeration degrees corresponding to the Turing degrees under the natural embedding ι, so that with respect to the enumeration spectrum there is no difference between trivial and non-trivial structures. Secondly, the enumeration spectrum does not affect the notion of degree of a structure, since for every abstract structure $\mathfrak{A}$ and any enumeration f of $\mathfrak{A}$ there is a bijective enumeration h of $\mathfrak{A}$, such that $h^{-1}(\mathfrak{A}) \leq_e f^{-1}(\mathfrak{A})$. These two results imply that, in general, the total degrees corresponding to the Turing degrees in the spectrum of a structure are contained in its enumeration spectrum, and, moreover in the particular case of non-trivial structures, we have

$$\iota(Sp(\mathfrak{A})) = \{\mathbf{x} \in Sp_e(\mathfrak{A}) \mid \mathbf{x} \text{ is total}\}.$$

Next, Soskov introduced the notion of *co-spectrum* of a structure by setting it to be the set of all enumeration degrees that are lower bounds for the enumeration spectrum of the structure, i.e.

$$CoSp_e(\mathfrak{A}) = \{\mathbf{x} \in \mathcal{D}_e \mid \forall \mathbf{a} \in Sp_e(\mathfrak{A})(\mathbf{x} \leq \mathbf{a})\}.$$

Clearly, the co-spectrum of any structure is a countable ideal of enumeration degrees. Moreover, it is exactly the ideal of the lower bounds of the total degrees corresponding to the Turing degrees in the spectrum of the structure.

More interesting questions are what properties the co-spectrum has with respect to the (enumeration) spectrum and what countable ideals of enumeration degrees are co-spectra of abstract structures. It turns out that the co-spectra behave similarly to the ordinary ideals of enumeration degrees. In fact, Soskov [57] proved that, for every structure $\mathfrak{A}$, there are total degrees **a** and **b** in the enumeration spectrum of $\mathfrak{A}$, such that for every enumeration degree **x**,

$$\mathbf{x} \in CoSp_e(\mathfrak{A}) \iff \mathbf{x} \leq \mathbf{a} \;\&\; \mathbf{x} \leq \mathbf{b}.$$

In other words, the enumeration spectrum of $\mathfrak{A}$ contains a minimal pair of total degrees for the co-spectrum of $\mathfrak{A}$. Further, for every abstract structure, there is an enumeration degree separating the total degrees contained in the enumeration spectrum of the structure from the ones contained in the co-spectrum. More precisely, for any abstract structure $\mathfrak{A}$, there is an enumeration degree $\mathbf{q} \notin CoSp_e(\mathfrak{A})$, such that for any total degree **x**, if **x** is comparable with **q**, then either $\mathbf{x} \in Sp_e(\mathfrak{A})$ or $\mathbf{x} \in CoSp_e(\mathfrak{A})$ (compare this to the result of Slaman and Sorbi [34] about the existence of quasi-minimal degrees over every ideal of enumeration degrees).

The existence of a degree separating the total degrees in the enumeration spectrum from the ones in the co-spectrum, as noticed by Soskov, implies that if an abstract structure does not have a degree, then its enumeration spectrum cannot be the union of countably many upper cones with total degrees as least elements. Indeed, suppose that $\mathbf{b}_0, \mathbf{b}_1, \ldots, \mathbf{b}_n, \ldots$ are total degrees and $\mathfrak{A}$ is an abstract structure with $Sp_e(\mathfrak{A}) = \{\mathbf{x} \mid \mathbf{b}_n \leq \mathbf{x} \text{ for some natural } n\}$. Let **q** be a degree separating the spectrum of $\mathfrak{A}$ from its co-spectrum. Now assume that $\mathbf{b}_n \leq \mathbf{q}$ for no natural n. Then there is a total degree, **x**, such that $\mathbf{q} \leq \mathbf{x}$ and $\mathbf{b}_n \not\leq \mathbf{x}$ for any natural n. But the first one implies that $\mathbf{x} \in Sp_e(\mathfrak{A})$, whereas the second one implies that $\mathbf{x} \notin Sp_e(\mathfrak{A})$, which is a contradiction. Thus $\mathbf{b}_n \leq \mathbf{q}$ for some n, and hence $\mathbf{b}_n$ lies in the co-spectrum of $\mathfrak{A}$. Thus $\mathbf{b}_n$ is the least element of the spectrum and hence the degree of $\mathfrak{A}$.

It turns out that the similarities between the co-spectra and the ordinary countable ideals of enumeration degrees are not accidental. In fact, Soskov [57] showed that every countable ideal of enumeration degrees is the co-spectrum of an abstract structure. With this we may consider the problem of the characterization of the co-spectra of abstract structures closed.

5.2 *Joint Spectra*

In [66] Soskov and A. Soskova propose a generalisation of the notions of enumeration degree spectra and co-spectra of abstract structures. This generalisation mimics the concepts invented by Soskov in his jump inversion theorems (see Sect. 3). Suppose that instead of one abstract structure we are a given a sequence $\overrightarrow{\mathfrak{A}} = (\mathfrak{A}_0, \mathfrak{A}_1, \dots, \mathfrak{A}_n)$ for some natural number n. We define the *joint* (enumeration) spectrum of $\overrightarrow{\mathfrak{A}}$ to be

$$Sp(\overrightarrow{\mathfrak{A}}) = \{\mathbf{x} \in \mathcal{D}_e \mid \forall k \leq n\ (\mathbf{x}^k \in Sp_e(\mathfrak{A}_k))\} \tag{21}$$

and, respectively, the joint co-spectrum to be

$$CoSp(\overrightarrow{\mathfrak{A}}) = \{\mathbf{x} \in \mathcal{D}_e \mid \forall \mathbf{a} \in Sp(\overrightarrow{\mathfrak{A}})(\mathbf{x} \leq \mathbf{a})\}. \tag{22}$$

It turns out that similarly to the enumeration spectrum of an abstract structure, the joint spectrum of a sequence of structures is upwards closed with respect to total enumeration degrees. Further, as shown by A. Soskova [64], the joint spectrum of any sequence of structures contains a minimal pair for the joint co-spectrum and, moreover, there is a degree separating the total degrees in the joint spectrum from the ones in the joint co-spectrum of the sequence.

Clearly, definitions (21) and (22) can be extended to sequences with arbitrary computable length. However, in this way we cannot capture the notion of uniformity that turned out to be important in the case of the ω-enumeration degrees. In order to overcome this obstacle, Soskov [59] proposes another definition for the joint spectrum for sequences of length ω. Let $\overrightarrow{\mathfrak{A}}$ be a sequence of length ω of denumerable abstract structures. Define the joint spectrum $JSp(\overrightarrow{\mathfrak{A}})$ of $\overrightarrow{\mathfrak{A}}$ to be the collection of all Turing degrees containing a set, say B, for which there are enumerations $f_0, f_1, \dots, f_n, \dots$ of $\mathfrak{A}_0, \mathfrak{A}_1, \dots, \mathfrak{A}_n, \dots$ respectively, such that $f_n^{-1}(\mathfrak{A}_n)$ is c.e. in $B^{(n)}$ uniformly in n.

Note that the definition of the joint spectrum of sequences of length ω is an extension of the definition of the joint spectrum for finite sequences. Indeed, let $\overrightarrow{\mathfrak{A}} = (\mathfrak{A}_0, \mathfrak{A}_1, \dots, \mathfrak{A}_n)$ and let $\overrightarrow{\mathfrak{A}} \uparrow \omega = \{\mathfrak{A}_k\}_{k<\omega}$, where $\mathfrak{A} = (\omega, =)$ for $k > n$.

Then

$$JSp(\overrightarrow{\mathfrak{A}} \uparrow \omega) = \{\iota^{-1}(\mathbf{x}) \mid \mathbf{x} \in Sp(\overrightarrow{\mathfrak{A}})\ \&\ \mathbf{x} \text{ is total}\}.$$

The definition of $JSp(\overrightarrow{\mathfrak{A}})$ is very similar to the definition of the mass problems associated to the sequences of sets of natural numbers in the definition of the ω-enumeration reducibility. In fact, as proven by Soskov [59], for every sequence $\mathcal{A}$ of sets of natural numbers, there is a sequence of structures $\overrightarrow{\mathfrak{A}}$, such that

$$JSp(\overrightarrow{\mathfrak{A}}) = \{\mathbf{d}_T(B) \mid B \in \mathcal{M}_{\mathcal{A}}\}.$$

As we shall see later, this will imply that the structure of the ω-enumeration degrees is embeddable in the structure of spectra of abstract structures.

5.3 *Jump Spectra*

Recall that a structure is said to have a degree if its spectrum has a least element. There are a lot of examples of structures not having a degree. For instance, Coles et al. [2] show that there are subgroups of the additive group of the rational numbers such that their spectrum does not have a least element. Indeed, in order to characterize the spectrum of a subgroup $\mathfrak{G}$ of the additive group of the rational numbers, it is enough to fix a non-zero element a of $\mathfrak{G}$ and consider the set

$$S_a(\mathfrak{G}) = \{\langle p_i, k\rangle \mid \exists x \in \mathfrak{G}(p_i^k.x = a)\},$$

where p_0, $p_1, \ldots$, $p_n, \ldots$ is the list of all prime numbers. It turns out that the spectrum of $\mathfrak{G}$ is the collection of all Turing degrees in which $S_a(\mathfrak{G})$ is computably enumerable. Thus G has a degree if and only if the set $S_a(\mathfrak{G})$ is total. On the other hand, for every set, there is a subgroup $\mathfrak{G}$ of the additive group of the rational numbers such that the set $S_a(\mathfrak{G})$ is enumeration-equivalent to the given set. Hence if we choose the set so that its enumeration degree is not total, then the respective group $\mathfrak{G}$ would not have a degree. On the other hand, the characterization of the spectrum of $\mathfrak{G}$ and Theorem 3.1 imply that the Turing degree corresponding to the jump of the enumeration degree of $S_a(\mathfrak{G})$ is the least among all the jumps of the degrees in $Sp(\mathfrak{G})$. This makes it natural to consider the so-called jump spectrum of a structure, which is the set

$$Sp(\mathfrak{A})' = \{\mathbf{x}' \mid \mathbf{x} \in Sp(\mathfrak{A})\}.$$

A structure is said to have a jump degree if its jump spectrum has a least element. Thus every subgroup of the additive group of the rational numbers has a jump degree although it might not have a degree.

A natural question arising immediately is: *Is every jump spectrum a spectrum of a structure?* An affirmative answer has been given independently by Soskov and Soskova [67] and Montalban [21]. The idea realized by Soskov and Soskova is when given an abstract structure $\mathfrak{A}$, to first consider its Moschovakis extension $\mathfrak{A}^*$ (see Sect. 2.1). The structure $\mathfrak{A}^*$ is not trivial, so its spectrum may be not equal to the spectrum of $\mathfrak{A}$. However, $\mathfrak{A} \equiv_w \mathfrak{A}^*$, so that the Moschovakis extension does not affect the reducibility we care about. Moreover, if the structure $\mathfrak{A}$ is not trivial then $Sp(\mathfrak{A}) = Sp(\mathfrak{A}^*)$. Now the jump $\mathfrak{A}'$ of the structure $\mathfrak{A}$ is defined to be the structure $\mathfrak{A}^*$ augmented by an universal semi-search computable predicate. This turns out to be sufficient for $Sp(\mathfrak{A}^*)' = Sp(\mathfrak{A}')$. Thus the jump spectrum of a nontrivial structure is the spectrum of a structure. In general, we have that jump spectrum of every structure is Muchnik-equivalent to the spectrum of some structure.

Following the lines of the usual investigation concerning a jump operation, the next natural question we need to consider is what can we say about inverting the jump. More precisely, we would like to see whether, given a structure $\mathfrak{B}$ with $Sp(\mathfrak{B}) \subseteq \{\mathbf{x} \in \mathcal{D}_T \mid x \geq 0_T'\}$, there is a structure $\mathfrak{A}$ with $\mathfrak{A}' \equiv_w \mathfrak{B}$. Using the technique of Marker extension, Soskov and Soskova [67] manage to prove a far more general result. Namely, for every two abstract structures $\mathfrak{B}$ and $\mathfrak{C}$ with $\mathfrak{C}' \leq_w \mathfrak{B}$, there is a structure $\mathfrak{A}$ such that $\mathfrak{C} \leq_w \mathfrak{A}$ and $\mathfrak{A}' \equiv_w \mathfrak{B}$.

Clearly, the jump inversion can be iterated finitely many times provided that the spectrum of the structure is contained in the degrees above an appropriate jump of the zero degree. As an application, we can easily provide examples of structures having $(n+1)$-jump degree but not having a k-jump degree for any $k \leq n$. Indeed, recall that for any subgroup $\mathfrak{G}$ of the additive group of the rational numbers, $\mathfrak{G}$ always has a jump degree, whereas it has a degree if and only if $\mathbf{d}_e(S_a(\mathfrak{G}))$ is total and for any set A there is a subgroup $\mathfrak{G}$ of the additive group of the rational numbers such that $S_a(\mathfrak{G}) \equiv_e A$. Now, let us fix a natural number n and a set A, such that $\mathbf{d}_e(A)$ is not total and $\mathbf{d}_e(A) \geq 0_e^{(n)}$. Let us consider a group $\mathfrak{G}$ with $S_a(\mathfrak{G}) \equiv_e A$. Then $\mathfrak{G}$ does not have a degree and $Sp(\mathfrak{G}) \subsetneq \{\mathbf{x} \in \mathcal{D}_T \mid \mathbf{x} \geq 0_T^{(n)}\}$. Now the jump inversion theorem provides us a sequence of structures $\mathfrak{A}_0 = \mathfrak{G}, \mathfrak{A}_1, \ldots, \mathfrak{A}_n$ with $\mathfrak{A}_k = {\mathfrak{A}_{k+1}}'$. Since $\mathfrak{A}_0$ does not have a degree, clearly neither of the structures $\mathfrak{A}_i$ has a degree. Thus $\mathfrak{A}_n$ is the desired structure having $(n+1)$-jump degree, but not having a k-degree for any $k \leq n$.

The next natural question one might consider is, can we invert arbitrary computable jumps as in the case of Turing and enumeration degrees? More precisely, given a computable ordinal, say α, and a structure, say $\mathfrak{B}$, having a spectrum that is contained in the degrees that are above the α-jump of the zero degree, is there a structure $\mathfrak{A}$ such that $Sp(\mathfrak{B}) = \{\mathbf{x}^{(\alpha)} \mid \mathbf{x} \in Sp(\mathfrak{A})\}$? This time the answer turns out to be negative. In fact, jump inversion fails for the first limit ordinal, ω. In order to prove this, Soskov considers what he calls the ω-co-spectrum of a structure. The ω-co-spectrum of a structure is simply the set of enumeration degrees

$$CoSp_\omega(\mathfrak{A}) = \{\mathbf{y} \in \mathcal{D}_e \mid \forall \mathbf{x} \in Sp(\mathfrak{A})(\mathbf{y} \leq \iota(\mathbf{x}^{(\omega)}))\}.$$

Now given a structure $\mathfrak{B}$, if there is a structure $\mathfrak{A}$ with $Sp(\mathfrak{B}) = \{\mathbf{x}^{(\omega)} \mid \mathbf{x} \in Sp(\mathfrak{A})\}$, then it would be the case $CoSp(\mathfrak{B}) = CoSp_\omega(\mathfrak{A})$. Recall that any countable ideal can be a co-spectrum of a structure, so if we have ω-jump inversion, then any countable ideal containing $0^{(\omega)}$ would be the ω-co-spectrum of a structure. But this turns out to be impossible since Soskov proved [60] that for every structure $\mathfrak{A}$ the total degrees are upwards dense in its ω-co-spectrum, i.e. for every $\mathbf{a} \in CoSp_\omega(\mathfrak{A})$, there is a total $\mathbf{b} \in CoSp_\omega(\mathfrak{A})$ such that $\mathbf{a} \leq \mathbf{b}$. Hence a principal ideal having a non-total degree as top element cannot be the ω-co-spectrum of a structure, and hence ω-jump inversion is impossible in general.

In his last paper [59] Soskov makes a thorough investigation of the Marker extension used for the proof of the jump inversion theorem. These investigations revealed that the Marker extension can be used to prove the following very general result: For every sequence $\overrightarrow{\mathfrak{B}} = \{\mathfrak{B}_n\}_{n<\omega}$, there is a structure $\mathfrak{A}$ such that

$$Sp(\mathfrak{A}) = JSp(\overrightarrow{\mathfrak{B}}). \tag{23}$$

First of all recall, that for every sequence $\mathcal{A}$ of sets of natural numbers, there is a sequence $\overrightarrow{\mathfrak{B}}$, such that

$$\mathcal{M}_\mathcal{A} = \{X \mid \mathbf{d}_T(X) \in JSp(\overrightarrow{\mathfrak{B}})\}.$$

This, together with (23), implies that we can associate a structure $\mathfrak{A}_\mathcal{A}$ with every sequence $\mathcal{A}$, so that

$$\mathcal{M}_\mathcal{A} = \{X \mid \mathbf{d}_T(X) \in Sp(\mathfrak{A}_\mathcal{A})\}.$$

Thus we have that

$$\mathcal{A} \leq_\omega \mathcal{B} \iff \mathfrak{A}_\mathcal{A} \leq_w \mathfrak{A}_\mathcal{B},$$

and hence the structure of the ω-enumeration degrees is embeddable in the structure of spectra of structures.

Secondly, (23) provides a least jump inversion theorem for the structure of spectra of structures. Indeed, suppose that $\mathfrak{B}$ is such that $Sp(\mathfrak{B}) \subseteq \{\mathbf{x} \in \mathcal{D}_T \mid \mathbf{x} \geq 0'_T\}$. Consider the sequence $\overrightarrow{\mathfrak{B}} = \{\mathfrak{B}_n\}_{n<\omega}$, where $\mathfrak{B}_1 = \mathfrak{B}$ and $\mathfrak{B}_n = (\omega; =)$ for $n \neq 1$. Then

$$JSp(\overrightarrow{\mathfrak{B}}) = \{\mathbf{x} \in \mathcal{D}_T \mid \mathbf{x}' \in Sp(\mathfrak{B})\}.$$

Now let $\mathfrak{A}$ be such that $Sp(\mathfrak{A}) = JSp(\overrightarrow{\mathfrak{B}})$. Then clearly $\mathfrak{A}' \equiv_w \mathfrak{B}$ and, for every $\mathfrak{C}$ with $\mathfrak{C}' \equiv_w \mathfrak{B}$, we have $\mathfrak{A} \leq_w \mathfrak{C}$.

Finally, analyzing the Wehner construction of a structure with a spectrum consisting of all non-low Turing degrees, Soskov constructs a sequence $\overrightarrow{\mathfrak{B}}$ such that

$$JSp(\overrightarrow{\mathfrak{B}}) = \{\mathbf{x} \in \mathcal{D}_T \mid \forall n(\mathbf{x}^{(n)} \neq 0_T^{(n)}\}.$$

Using once again (23), we obtain a structure $\mathfrak{A}$ whose spectrum is the collection of all Turing degrees, which are low-n for no natural n.

References

1. C. Bianchini, Bounding Enumeration Degrees, Ph.D. thesis, University of Siena, 2000
2. R. Coles, R. Downey, T. Slaman, Every set has a least jump enumeration. J. Lond. Math. Soc. **62**, 641–649 (1998)
3. S.B. Cooper, Partial degrees and the density problem. Part 2: The enumeration degrees of the Σ_2 sets are dense. J. Symb. Log. **49**, 503–513 (1984)
4. A.P. Ershov, Abstract computability on algebraic structures, in *Algorithms in Modern Mathematics and Computer Science*. Lecture Notes in Computer Science, vol. 122 (Springer, Berlin, 1981), pp. 397–420
5. R.M. Friedberg, A criterion for completeness of degrees of unsolvability. J. Symb. Log. **22**, 159–160 (1957)
6. H. Friedman, Algorithmic procedures, generalized Turing algorithms and elementary recursion theory, in *Logic Colloquium'69*, ed. by R.O. Gandy, C.E.M. Yates (North-Holland, Amsterdam, 1971), pp. 361–389
7. H. Ganchev, A.C. Sariev, The ω-Turing degrees. Ann. Pure Appl. Log. **65**(9), 1512–1532 (2014)
8. H. Ganchev, I.N. Soskov, The jump operator on the ω-enumeration degrees. Ann. Pure Appl. Log. **160**(3), 289–301 (2009)
9. H. Ganchev, M.I. Soskova, The high/low hierarchy in the local structure of the ω-enumeration degrees. Ann. Pure Appl. Log. **163**, 547–566 (2012)
10. H. Ganchev, M.I. Soskova, Embedding distributive lattices in the Σ_2^0 enumeration degrees. J. Log. Comput. **22**, 779–792 (2012)
11. H. Ganchev, M.I. Soskova, Cupping and definability in the local structure of the enumeration degrees. J. Symb. Log. **77**, 133–158 (2012)
12. H. Ganchev, M.I. Soskova, Interpreting true arithmetic in the local structure of the enumeration degrees. J. Symb. Log. **77**, 1184–1194 (2012)
13. H. Ganchev, M.I. Soskova, Definability via Kalimullin pairs in the structure of the enumeration degrees. Trans. Am. Math. Soc. **367**, 4873–4893 (2015)
14. S.S. Goncharov, V.S. Harizanov, J.F. Knight, R.A. Shore, Π_1^1 relations and paths through $\mathcal{O}$. J. Symb. Log. **69**, 585–611 (2004)
15. C.E. Gordon, Prime and search computability characterized as definability in certain sublanguages of constructible $L_{\omega_1,\omega}$. Trans. Am. Math. Soc. **197**, 391–407 (1974)
16. I.S. Kalimullin, Definability of the jump operator in the enumeration degrees. J. Math. Log. **3**, 257–267 (2003)
17. J.F. Knight, Degrees coded in jumps of orderings. J. Symb. Log. **51**, 1034–1042 (1986)
18. D. Lacombe, Deux généralisations de la notion de récursivité relative. C. R. Acad. Sci. Paris **258**, 3410–3413 (1964)
19. K. McEvoy, Jumps of quasi-minimal enumeration degrees. J. Symb. Log. **50**, 839–848 (1985)
20. K. McEvoy, S.B. Cooper, On minimal pairs of enumeration degrees. J. Symb. Log. **50**, 983–1001 (1985)

21. A. Montalban, Notes on the jump of a structure, in *Mathematical Theory and Computational Practice* (Springer, Berlin, 2009), pp. 372–378
22. Y.N. Moschovakis, Abstract first order computability. I. Trans. Am. Math. Soc. **138**, 427–464 (1969)
23. Y.N. Moschovakis, Abstract computability and invariant definability. J. Symb. Log. **34**, 605–633 (1969)
24. Y.N. Moschovakis, *Elementary Induction on Abstract Structures* (North-Holland, Amsterdam, 1974)
25. V.A. Nepomniaschy, Criteria for the algorithmic completeness of systems of operations, in *Programming Theory. Proceedings of a Symposium* (Novosibirsk, 7–11 Aug 1972), Part 1, ed. by V.A. Nepomnyaschy (Computer Centre of the Siberian Branch of the Soviet Academy of Sciences, Novosibirsk, 1972), pp. 267–279 [in Russian]
26. L.J. Richter, Degrees of Structures. J. Symb. Log. **46**, 723–731 (1981)
27. G.E. Sacks, *Degrees of Unsolvability* (Princeton University Press, Princeton, 1963)
28. L.E. Sanchis, Hyperenumeration reducibility. Notre Dame J. Formal Logic **19**, 405–415 (1978)
29. L.E. Sanchis, Reducibilities in two models of combinatory logic. J. Symb. Log. **44**, 221–234 (1979)
30. A.L. Selman, Arithmetical reducibilities I. Z. Math. Logik Grundlag. Math. **17**, 335–350 (1971)
31. J.C. Shepherdson, Computation over abstract structures, in *Logic Colloquium'73*, ed. by H.E. Rose, J.C. Shepherdson (North-Holland, Amsterdam, 1975), pp. 445–513
32. D. Skordev, Combinatory spaces and recursiveness in them (Publisher House of the Bulgarian Academy of Science, Sofia 1980) [in Russian]
33. D. Skordev, *Computability in Combinatory Spaces. An Algebraic Generalization of Abstract First Order Computability* (Kluwer Academic Publishers, Dordrecht, 1992)
34. T.A. Slaman, A. Sorbi, Quasi-minimal enumeration degrees and minimal Turing degrees. Mat. Pura Appl. **174**, 97–120 (1998)
35. I.N. Soskov, Prime computable functions of finitely many arguments with argument and function values in the basic set. M.Sc. thesis, Sofia University, 1979 [in Bulgarian]
36. I.N. Soskov, Prime computable functions in the basic set, in *Mathematical Logic. Proceedings of the Conference on Mathematical Logic. Dedicated to the memory of A.A. Markov (1903–1979)*, (Sofia, 22–23 Sept 1980) ed. by D. Skordev (Publishers House of the Bulgarian Academy of Sciences, Sofia, 1984), pp. 112–138 [in Russian]
37. I.N. Soskov, An example of a basis which is Moschovakis complete without being program complete, in *Mathematical Theory and Practice of Program Systems*, ed. by A.P. Ershov (Computer Centre of the Siberian Branch of the Soviet Academy of Sciences, Novosibirsk, 1982), pp. 26–33 [in Russian]
38. I.N. Soskov, Computability in algebraic systems. C.R. Acad. Bulg. Sci. **36**, 301–304 (1983) [in Russian]
39. I.N. Soskov, Algorithmically complete algebraic systems, C. R. Acad. Bulg. Sci. **36**, 729–731 (1983) [in Russian]
40. I.N. Soskov, Computability over partial algebraic systems. Ph.D. thesis, Sofia University, 1983 [in Bulgarian]
41. I.N. Soskov, The connection between prime computability and recursiveness in functional combinatory spaces, in *Mathematical Theory of Programming*, ed. by A.P. Ershov, D. Skordev. Computer Centre of the Siberian Branch of the Soviet Academy of Sciences, Novosibirsk, 1985, pp. 4–11 [in Russian]
42. I.N. Soskov, Prime computability on partial structures, in *Mathematical Logic and its Applications. Proceedings of an Advanced International Summer School and Conference on Mathematical Logic and Its Applications in honor of the 80th anniversary of Kurt Gödel's birth* (Druzhba, Bulgaria, 24 Sept–4 Oct 1986) ed. by D. Skordev (Plenum Press, New York, 1987), pp. 341–350
43. I.N. Soskov, Definability via enumerations. J. Symb. Log. **54**, 428–440 (1989)

44. I.N. Soskov, An external characterization of the prime computability. Ann. Univ. Sofia **83**, livre 1 – Math. 89–111 (1989)
45. I.N. Soskov, Maximal concepts of computability and maximal programming languages, Technical Report, Contract no. 247/1987 with the Ministry of Culture, Science and Education, Laboratory for Applied Logic, Sofia University (1989)
46. I.N. Soskov, Horn clause programs on abstract structures with parameters (extended abstract). Ann. Univ. Sofia **84**, livre 1 – Math. 53–61 (1990)
47. I.N. Soskov, On the computational power of the logic programs, in *Mathematical Logic. Proceedings of the Summer School and Conference on Math. Logic, honourably dedicated to the 90th anniversary of Arend Heyting (1898–1980)* (Chaika (near Varna), Bulgaria, 13–23 Sept 1988) ed. by P. Petkov (Plenum Press, New York, 1990), pp. 117–137
48. I.N. Soskov, Computability by means of effectively definable schemes and definability via enumerations. J. Symb. Log. **55**, 430–431 (1990) [abstract]
49. I.N. Soskov, Computability by means of effectively definable schemes and definability via enumerations. Arch. Math. Logic **29**, 187–200 (1990)
50. I.N. Soskov, Second order definability via enumerations. Zeitschr. f. math Logik und Grundlagen d. Math. **37**, 45–54 (1991)
51. I.N. Soskov, Maximal concepts of computability on abstract structures. J. Symb. Log. **57**, 337–338 (1992) [abstract]
52. I.N. Soskov, Intrinsically Π^1_1-relations. Math. Log. Q. **42**, 109–126 (1996)
53. I.N. Soskov, Intrinsically hyperarithmetical sets. Math. Log. Q. **42**, 469–480 (1996)
54. I.N. Soskov, Constructing minimal pairs of degrees. Ann. Univ. Sofia **88**, 101–112 (1997)
55. I.N. Soskov, Abstract computability and definability: external approach. Dr. Hab. thesis, Sofia University, 2000 [in Bulgarian]
56. I.N. Soskov, A jump inversion theorem for the enumeration jump. Arch. Math. Log. **39**, 417–437 (2000)
57. I.N. Soskov, Degree spectra and co-spectra of structures. Ann. Univ. Sofia **96**, 45–68 (2004)
58. I.N. Soskov, The ω-enumeration degrees. J. Log. Comput. **17**, 1193–1214 (2007)
59. I.N. Soskov, Effective properties of Marker's extensions. J. Log. Comput. **23**(6), 1335–1367 (2013)
60. I.N. Soskov, A note on ω-jump inversion of degree spectra of structures, in *The Nature of Computation. Logic, Algorithms, Applications*. Lecture Notes in Computer Science, vol. 7921 (Springer, Berlin, 2013), pp. 365–370
61. I.N. Soskov, V. Baleva, Regular enumerations. J. Symb. Log. **67**, 1323–1343 (2002)
62. I. Soskov, B. Kovachev, Uniform regular enumerations. Math. Struct. Comput. Sci. **16**(5), 901–924 (2006)
63. I.N. Soskov, M.I. Soskova, Kalimullin pairs of Σ^0_2 omega-enumeration degrees. J. Softw. Inf. **5**, 637–658 (2011)
64. A.A. Soskova, Minimal pairs and quasi-minimal degrees for the joint spectra of structures, in *New Computational Paradigms*. Lecture Notes in Computer Science, vol. 3526 (Springer, Berlin, 2005), pp. 451–460
65. A.A. Soskova, I.N. Soskov, Effective enumerations of abstract structures, in *Mathematical Logic. Proceedings of the Summer School and Conference on Mathematical Logic, honourably dedicated to the 90th anniversary of Arend Heyting (1898–1980)* (Chaika (near Varna), Bulgaria, 13–23 Sept 1988) ed. by P. Petkov (Plenum Press, New York, 1990), pp. 361–372
66. A.A. Soskova, I.N. Soskov, Co-spectra of joint spectra of structures. Ann. Sofia Univ. Fac. Math. Inf. **96**, 35–44 (2004)
67. A.A. Soskova, I.N. Soskov, A jump inversion theorem for the degree spectra. J. Log. Comput. **19**, 199–215 (2009)
68. M.I. Soskova, I.N. Soskov, Embedding countable partial orderings into the enumeration and ω-enumeration degrees. J. Log. Comput. **22**, 927–952 (2012)

Physical Logic

Kate Clements, Fay Dowker, and Petros Wallden

Abstract In R. D. Sorkin's framework for logic in physics a clear separation is made between the collection of unasserted propositions about the physical world and the affirmation or denial of these propositions by the physical world. The unasserted propositions form a Boolean algebra because they correspond to subsets of an underlying set of spacetime histories. *Physical* rules of inference apply not to the propositions in themselves but to the affirmation and denial of these propositions by the actual world. This *physical logic* may or may not respect the propositions' underlying Boolean structure. We prove that this logic is Boolean if and only if the following three axioms hold: (i) The world is affirmed, (ii) Modus Ponens and (iii) If a proposition is denied then its negation, or complement, is affirmed. When a physical system is governed by a dynamical law in the form of a quantum measure with the rule that events of zero measure are denied, the axioms (i)–(iii) prove to be too rigid and need to be modified. One promising scheme for quantum mechanics as quantum measure theory corresponds to replacing axiom (iii) with axiom (iv). Nature is as fine grained as the dynamics allows.

1 Introduction

The view that the mode of reasoning we use for classical physics is not appropriate when discussing a quantum system is widespread, if not mainstream. For example, in the Quantum Mechanics volume of his *Lectures on Physics*, R. P. Feynman refers

K. Clements
Blackett Laboratory, Imperial College, London SW7 2AZ, UK

F. Dowker (✉)
Blackett Laboratory, Imperial College, London SW7 2AZ, UK

Perimeter Institute, 39 Caroline St. N., Waterloo, ON, Canada N2L 2Y5

Institute for Quantum Computing, University of Waterloo, ON, Canada N2l 3G1
e-mail: f.dowker@imperial.ac.uk

P. Wallden
School of Informatics, University of Edinburgh, Edinburgh EH8 9AB, UK

IPaQS, Heriot-Watt University, Edinburgh EH14 1AS, UK

S.B. Cooper, M.I. Soskova (eds.), *The Incomputable*, Theory and Applications of Computability, DOI 10.1007/978-3-319-43669-2_3

to "the logical tightrope on which we must walk if we wish to describe nature successfully" [1]. In order to investigate the nature of this "tightrope" further, in a systematic way, we need a framework for logic that is relevant for *physics* (rather than, say, mathematics or language) and within which the logic used for classical physics can be identified, characterised, assessed and, if necessary, replaced. Recently, a unifying foundation for physical theories with spacetime character—Generalised Measure Theory (GMT)—which provides just such a framework has been set out [2–5]. The key to the clarity that this formalism brings to the study of deductive inference in physics is the distinction it makes between the *assertion* of propositions about the physical world and the propositions themselves, the latter corresponding merely to questions waiting to be answered [6, 7]. Identifying the answers to the questions as the physical content of the theory, as explained below, makes it a small step to consider the possibility of non-standard rules of inference; to do so is to open a new window on the variety of antinomies with which quantum mechanics is so infamously plagued (or blessed) [4, 8–11].

We begin in Sect. 2 by identifying three basic structures that constitute a general framework for reasoning about the physical world. Taking the revolution of relativity seriously, we assume that the physical theory has a *spacetime* character in the sense that it is based on a set of spacetime histories which represent the finest grained descriptions of the system conceivable within the theory. In Sect. 3 we give names to certain rules of inference and situate classical, Boolean rules of inference within this framework. In Sect. 4 we investigate which rules are implied by which others, abstractly, by mathematical manipulation alone, setting aside the question of which might be necessary or desirable for physics. We focus on the rule of inference known as *modus ponendo ponens* (*modus ponens* for short), the basis for deductive proof without which the ability to reason at all might seem to be compromised from the outset.[1] We will show that—on the mildest conceivable assumption that something happens in the world—modus ponens implies Boolean logic *if* it is supplemented by the rule "If a proposition is denied by the physical world, then its negation is affirmed." These results are independent of whether the theory is classical, quantum or transquantum in Sorkin's hierarchy of physical theories [2]. We will show that one currently favoured scheme for interpreting quantum theory, the *multiplicative scheme*, coincides with the adoption of modus ponens, together with a condition of finest grainedness.

2 The Threefold Structure

The details of any logical scheme for physics—in particular, the events about which it is intended to reason—will plainly depend on the system one has in mind. Nonetheless, one can describe a class of schemes rather generally in terms of three

[1] Lewis Carroll gives in [12] a witty account of the implications of a failure to take up modus ponens explicitly as a rule of inference.

components [5]:

(i) the set, $\mathfrak{A}$, of all *questions* that can be asked about the system;
(ii) the set, S, of possible *answers* to those questions; and
(iii) a collection, $\mathfrak{A}^*$, of *answering maps* $\phi : \mathfrak{A} \rightarrow S$, exactly one of which corresponds to the physical world.

While such a framework may not be the most general that could be conceived, it is broad enough to encompass all classical theories, including stochastic theories such as Brownian motion. In such a classical physical theory $\mathfrak{A}$ is a Boolean algebra, $S = \mathbb{Z}_2 = \{0, 1\}$ and ϕ is a homomorphism from $\mathfrak{A}$ into $\mathbb{Z}_2$, as we describe below. To make the framework general enough to include quantum theories one might consider altering any or all of these three classical ingredients. It is remarkable that the only change that appears to be necessary in order to accommodate quantum theories in a spacetime approach based on the Dirac–Feynman path integral is to free ϕ from the constraint that it be a homomorphism [4, 5, 7]. We will assume the following about the three components.

(i) $\mathfrak{A}$ is a Boolean algebra, which we refer to as the *event algebra*. The elements of $\mathfrak{A}$ are equivalently and interchangeably referred to as *propositions* or *events*, where it is understood that these terms refer to *unasserted* propositions. An event A can also be thought of as corresponding to the *question*, "Does A happen?". The elements of $\mathfrak{A}$ are subsets of the set, Ω, of spacetime histories of the physical system. For instance, in Brownian motion Ω is the set of Wiener paths. Use of the term 'event' to refer to a subset of Ω is standard for stochastic processes. In the quantal case, Ω is the set of histories summed over in the path integral, for example the particle trajectories in non-relativistic quantum mechanics. An example of an event in that case is the set of all trajectories in which the particle passes through some specified region of spacetime.

The Boolean operations of meet $\wedge$ and join $\vee$ are identified with the set operations of intersection $\cap$ and union $\cup$, respectively. A note of warning: using in this context the words *and* and *or* to denote the algebra elements that result from these set operations can lead to ambiguity. In this paper we will try to eliminate the ambiguity by the use of single inverted commas, so that 'A or B' denotes the event $A \vee B$; 'A and B', the event $A \wedge B$.

The zero element $\emptyset \in \mathfrak{A}$ is the empty set and the unit element $\mathbf{1} \in \mathfrak{A}$ is Ω itself. The operations of multiplication and addition of algebra elements are, respectively,

$$AB := A \cap B, \quad \forall A, B \in \mathfrak{A};$$

$$A + B := (A \setminus B) \cup (B \setminus A), \quad \forall A, B \in \mathfrak{A}\,.$$

With these operations, $\mathfrak{A}$ *is* an algebra in the sense of being a vector space over the finite field $\mathbb{Z}_2$. A useful expression of the subset property is: $A \subseteq B \Leftrightarrow$

$AB = A$. We have, for all A in $\mathfrak{A}$,

$$AA = A; \tag{1}$$

$$A + A = \emptyset; \text{ and} \tag{2}$$

$$\neg A := \Omega \setminus A = \mathbf{1} + A\,. \tag{3}$$

The event $\mathbf{1} + A$ may be referred to as $\neg A$, as the *complement* of A, or, again with single inverted commas, as 'not A'.[2]

(ii) Together with the algebra of questions comes the space of potential *answers* that the physical system can provide to those questions. Whilst one can envisage any number of generalisations, with intermediate truth values for example, we follow Sorkin and keep as the answer space that of classical logic, namely the Boolean algebra $\mathbb{Z}_2 \equiv \{0, 1\} \equiv \{\text{false}, \text{true}\} \equiv \{\text{no}, \text{yes}\}$. To answer the question $A \in \mathfrak{A}$ with 1 (0) is to assert that the event A does (does not) happen; equivalently, we say that A is affirmed (denied).

(iii) Finally, one has the set $\mathfrak{A}^*$ of allowed *answering maps*, also called *co-events*,[3] $\phi : \mathfrak{A} \to \mathbb{Z}_2$. We assume that a co-event is a non-constant map: $\phi \neq 0$ and $\phi \neq 1$. That is, a co-event must affirm at least one event and deny at least one event. To specify a co-event is to answer every physical question about the system, and thus to give as complete an account of what happens as one's theory permits. The physical world corresponds to exactly one co-event from $\mathfrak{A}^*$. In other words, the physical world provides (or is equivalent to) a definite answer to every question and $\mathfrak{A}^*$ is the set of possible physical worlds.

This threefold structure of event algebra $\mathfrak{A}$, answer space $\mathbb{Z}_2$ and collection of answering maps or co-events $\phi : \mathfrak{A} \to \mathbb{Z}_2$ makes sense out of possibilities that seem otherwise non-sensical [6]. The threefold structure is appropriate to physics, where perfectly sensible, meaningful events are not in themselves true or false (unlike, for example, and in one view, mathematical statements). Each event will either happen or not *in the physical world*, but which it is contingent.

It seems an appropriate point to note that the threefold structure is also apparent in the framework for interpreting quantum physics that has come commonly to be known as "Quantum Logic." There are major differences with Sorkin's framework, however. In place of the Boolean algebra of propositions there is in Quantum Logic an orthocomplemented lattice of subspaces of a Hilbert space. In place of the set of yes/no answers $\mathbb{Z}_2$, there is the set of probabilities, real numbers between 0 and 1. And in place of the co-event, there is the state which maps each subspace of Hilbert space to a probability. Thus, at the very outset, the space of propositions in Quantum Logic has a non-Boolean character due to the focus on Hilbert space as the arena for the physics. Quantum Logic sprang from a canonical approach to

---

[2]See [6] for a discussion of the ambiguity in the phrase "not A".

[3]The notation $\mathfrak{A}^*$ reflects the nature of the co-event space as dual to the event algebra.

quantum theory in which Hilbert space is fundamental. Hilbert space has no place in classical physics, hence the starting point—the set of unasserted propositions—for Quantum Logic is different than in classical physics. In contrast, in a path integral approach to quantum physics as adopted by Sorkin, the axiomatic basis is a set of spacetime histories just as it is in classical physics and therefore the structure of the set of unasserted propositions is the same in both classical and quantum physics: this is a *unifying* framework.

3 Rules of Inference

If we somehow came to know the co-event that corresponds to the whole universe, then there would be no need for rules of inference: we would know everything already. Rules of inference are needed because our knowledge is partial and limited and to extend that knowledge further we need to be able to deduce new facts from established ones. As stressed by Sorkin, in this view dynamical laws in physics are rules of inference [7]: using the laws of gravity, we can infer from the position of the moon tonight its position yesterday and its position tomorrow.

For the purposes of this paper we call any condition restricting the collection of allowed co-events a rule of inference.[4] One could begin by considering the set of *all* non-constant maps from $\mathfrak{A}$ into $\mathbb{Z}_2$; a rule of inference is then any axiom that reduces this set. One axiom that has been suggested [4, 5] is that of *preclusion*, the axiom that an event of zero measure is denied. Explicitly, if μ is the (classical, quantum or transquantum) measure on the event algebra, encoding both the dynamics and the initial conditions for the system, then $\mu(A) = 0 \Rightarrow \phi(A) = 0$. We will return to preclusion in a later Section; for the time being, our attention will focus on rather more structural axioms.

First, let us define some properties of co-events that it might be desirable. In all the following definitions, ϕ is a co-event and we recall that ϕ is assumed not to be the zero map or unit map. We begin with properties that reflect the algebraic structure of $\mathfrak{A}$ itself.

Definition 1 ϕ is *zero-preserving* if

$$\phi(\emptyset) = 0\,. \tag{4}$$

Definition 2 ϕ is *unital* if

$$\phi(\mathbf{1}) = 1\,. \tag{5}$$

One could call this condition "the world is affirmed."

[4]An alternative is to call any condition on the allowed co-events a dynamical law.

Definition 3 ϕ is *multiplicative* if

$$\phi(AB) = \phi(A)\phi(B), \quad \forall A, B \in \mathfrak{A} . \tag{6}$$

Definition 4 ϕ is *additive* or *linear* if

$$\phi(A + B) = \phi(A) + \phi(B), \quad \forall A, B \in \mathfrak{A} . \tag{7}$$

A further set of conditions is motivated directly as the formalisation of the rules of inference that we use in classical reasoning. As mentioned in the introduction, arguably the most desirable among these is modus ponens, commonly stated thus:

MP: *If A implies B and A, then B.*

However, it is now easy to appreciate why care must be taken in distinguishing (mere unasserted) events from statements about the physical world, i.e. affirmed or denied events. The rules of inference we are interested in here are those that tell us how to deduce statements about what happens in the physical world from other such statements. To render modus ponens fully in terms of the threefold framework for physics, we re-express it as

MP: *If 'A implies B' is affirmed and A is affirmed, then B is affirmed,*

where 'A implies B' is an event, an element of $\mathfrak{A}$ which we denote symbolically as $A \to B := \neg(A \wedge (\neg B))$. We have,

$$\begin{aligned} A \to B &= \mathbf{1} + A(\mathbf{1} + B) \\ &= \mathbf{1} + A + AB, \end{aligned} \tag{8}$$

which small manipulation shows, incidentally, how much easier it is to work with the arithmetic form of the operations than with $\wedge$, $\vee$ and $\neg$. The condition of modus ponens is then:

Definition 5 ϕ is MP if

$$\phi(A \to B) = 1 \,, \, \phi(A) = 1 \; \Rightarrow \; \phi(B) = 1, \quad \forall A, B \in \mathfrak{A} . \tag{9}$$

Distinct from MP and from each other are the two strains of "proof by contradiction," which we shall call C1 and C2. In words, we can state them as follows:

C1: *If event A is affirmed, then its complement is denied.*

C2: *If event A is denied, then its complement is affirmed*

We point out that C1 and C2 are referred to as the *law of contradiction* and the *law of the excluded middle*, respectively, in [13].[5] These two conditions are distinct and C2 turns out to play a more important role in our analysis. As we will see in the next few sections, C2 is independent of MP and multiplicativity, while C1 is essentially implied by these conditions.

The formal definitions of conditions C1 and C2 on co-events are:

Definition 6 ϕ is C1 if

$$\phi(A) = 1 \Rightarrow \phi(\mathbf{1} + A) = 0, \quad \forall A \in \mathfrak{A}. \tag{10}$$

Definition 7 ϕ is C2 if

$$\phi(A) = 0 \Rightarrow \phi(\mathbf{1} + A) = 1, \quad \forall A \in \mathfrak{A}. \tag{11}$$

3.1 An Example: Classical Physics, Classical Logic

In classical physics we use classical logic because, in classical physics, One History Happens. Indeed, the rules of inference known collectively as classical, Boolean logic follow from the axiom that the physical world corresponds to exactly one history in Ω [5]. In a later section we will give a list of equivalent forms of this axiom in the case of finite Ω; here, we note only that if the physical world corresponds to history $\gamma \in \Omega$ then all physical questions can be answered. In other words, γ gives rise to a co-event $\gamma^* : \mathfrak{A} \to \mathbb{Z}_2$, as

$$\gamma^*(A) = \begin{cases} 1 & \text{if } \gamma \in A \\ 0 & \text{otherwise,} \end{cases} \tag{12}$$

$$\forall A \in \mathfrak{A}\,.$$

It can be shown that such a γ^* is both multiplicative and additive, i.e. it is a homomorphism from $\mathfrak{A}$ into $\mathbb{Z}_2$. It is easy to see that γ^* is zero-preserving and unital, and one can further use its homomorphicity to prove it is C1, C2 and MP. For example, we have:

Lemma 1 *If co-event ϕ is additive and unital then it is zero-preserving, C1 and C2.*

[5]Note also that at the level of the Boolean algebra of events we always have $\neg\neg A = A$ and, moreover, if "the law of the excluded middle" is taken to mean that every event is either affirmed or denied then our framework respects it by fiat because that is just the statement that ϕ is a map to $\mathbb{Z}_2$ [6]. This illustrates how careful one must be to be clear.

Proof Let ϕ be additive and unital. Then

$$1 = \phi(\mathbf{1}) = \phi(\mathbf{1} + A + A) = \phi(\mathbf{1} + A) + \phi(A), \quad \forall A \in \mathfrak{A},$$

which implies that exactly one of $\phi(\mathbf{1} + A)$ and $\phi(A)$ is equal to 1. So ϕ is C1 and C2, and $\phi(\emptyset) = 0$. □

If One History Happens, as in classical physics, the physical world fully respects the Boolean structure of the event algebra, and the logical connectives *and*, *or*, *not* and so forth may be used carelessly, without the need to specify whether they refer to asserted or to unasserted propositions. One doesn't have to mind one's logical Ps and Qs over (potentially ambiguous) statements such as "A or B happens" when ϕ is a homomorphism:

Lemma 2 *If co-event ϕ is a homomorphism then $\phi(A \vee B) = 1 \iff \phi(A) = 1$ or $\phi(B) = 1$.*

Proof

$$\begin{aligned}
&\phi(A \vee B) = 1 \\
\iff &\phi(AB + A + B) = 1 \\
\iff &\phi(A)\phi(B) + \phi(A) + \phi(B) = 1 \\
\iff &(\phi(A) + 1)(\phi(B) + 1) = 0\,.
\end{aligned}$$

□

So no ambiguity arises because 'A or B' happens if and only if A happens or B happens.

4 Results

Theorem 1 *The following conditions on a co-event ϕ are equivalent:*

(i) *ϕ is MP and unital;*
(ii) *$\phi^{-1}(1) := \{A \in \mathfrak{A} \mid \phi(A) = 1\}$ is a filter*[6]*;*
(iii) *ϕ is multiplicative.*

Proof (i) $\Rightarrow$ (ii)
Let ϕ be MP and unital.

[6]We assume a filter is non-empty and not equal to the whole of $\mathfrak{A}$.

First, we show that the superset of an affirmed event is affirmed. Let $\phi(A) = 1$ and B be such that $AB = A$. Then

$$\begin{aligned}\phi(A \to B) &= \phi(1 + A + A) \\ &= \phi(1) \\ &= 1\end{aligned}$$

by unitality. By MP it follows that $\phi(B) = 1$.

Now we show that the intersection of two affirmed events is also affirmed. Let $\phi(C) = \phi(D) = 1$. We have that $D(C \to CD) = D(1 + C + CD) = D$, and so $\phi(D(C \to CD)) = 1$. By the first part of the proof, this implies that $\phi(C \to CD) = 1$, so that by MP, $\phi(CD) = 1$.

Finally, ϕ is unital, so $\phi^{-1}(1)$ is non-empty and $\phi \neq 1$ so $\phi^{-1}(1)$ is not equal to $\mathfrak{A}$.

(ii) $\Rightarrow$ (iii)

Let $\phi^{-1}(1)$ be a filter and $A, B \in \mathfrak{A}$. Then there are two cases to check.

(a) If $\phi(A) = \phi(B) = 1$ then the filter property implies that $\phi(AB) = 1$.
(b) Assume without loss of generality that $\phi(A) = 0$. Since A is a superset of AB, we must therefore have that $\phi(AB) = 0$; otherwise, the filter property would lead to the conclusion that $\phi(A) = 1$, a contradiction.

So ϕ is multiplicative.

(iii) $\Rightarrow$ (i)

Let ϕ be multiplicative. Since $\phi \neq 0$, $\exists X \in \mathfrak{A}$ s.t. $\phi(X) = 1$. Then,

$$1 = \phi(X) = \phi(\mathbf{1} \cdot X) = \phi(\mathbf{1})\phi(X) = \phi(\mathbf{1}),$$

so ϕ is unital.

Now suppose $\phi(A) = \phi(A \to B) = 1$. We have that $A(A \to B) = A(1 + A + AB) = AB$, and thus $\phi(AB) = \phi(A)\phi(A \to B) = 1$. It follows that $\phi(A)\phi(B) = 1$, so that $\phi(B) = 1$. So ϕ is MP. □

Note, however, the following.

Remark MP alone is not enough to guarantee multiplicativity, as shown by the following example. Consider the event algebra $\mathfrak{A} = \{\emptyset, \mathbf{1}\}$, and the co-event

$$\begin{aligned}\phi(\emptyset) &= 1; \\ \phi(\mathbf{1}) &= 0.\end{aligned}$$

MP is trivially satisfied: $\phi(\emptyset \to \mathbf{1}) = \phi(\mathbf{1} + \emptyset + \emptyset) = \phi(\mathbf{1}) = 0$, while $\phi(\mathbf{1} \to \emptyset) = \phi(\mathbf{1} + \mathbf{1} + \emptyset) = \phi(\emptyset) = 1$, but $\phi(\mathbf{1}) = 0$. So there *is* no pair of events A and B such that $\phi(A \to B)$, $\phi(A) = 1$, i.e. for which we even need to check whether

$\phi(B) = 1$. Multiplicativity fails, however:

$$\begin{aligned}\phi(\emptyset \cdot \mathbf{1}) &= \phi(\emptyset) = 1 \\ &\neq \phi(\emptyset)\phi(\mathbf{1}) = 1 \cdot 0 = 0.\end{aligned}$$

Neither does MP together with *zero-preservation* guarantee multiplicativity, as demonstrated again by an example. Consider this time the four-element event algebra $\mathfrak{A} = \{\emptyset, A, B, \mathbf{1}\}$, where $B = \mathbf{1} + A$, and the following zero-preserving co-event:

$$\begin{aligned}&\phi(\emptyset) = \phi(B) = \phi(\mathbf{1}) = 0; \\ &\phi(A) = 1.\end{aligned}$$

MP is trivially satisfied by an argument similar to that above, but ϕ is not multiplicative, since

$$\begin{aligned}\phi(A \cdot \mathbf{1}) &= \phi(A) = 1 \\ &\neq \phi(A)\phi(\mathbf{1}) = 1 \cdot 0 = 0.\end{aligned}$$

Having established the relation between multiplicativity of a co-event and the pillar of classical inference—MP—what can be said of the proofs by contradiction? From the proof of Theorem 1 we know that a multiplicative ϕ is unital. It is also C1:

Lemma 3 *If ϕ is a multiplicative co-event then ϕ is zero-preserving and C1.*

Proof Let ϕ be multiplicative. $\phi \neq 1$, so $\exists A \in \mathfrak{A}$ s.t. $\phi(A) = 0$. Thus

$$\phi(\emptyset) = \phi\left(A(\mathbf{1} + A)\right) = \phi(A)\phi(\mathbf{1} + A) = 0\,.$$

Now let $\phi(B) = 1$ for some $B \in \mathfrak{A}$. Then

$$\begin{aligned}0 = \phi(\emptyset) = \phi\left(B(\mathbf{1} + B)\right) = \phi(B)\phi(\mathbf{1} + B) \\ \Rightarrow \phi(\mathbf{1} + B) = 0\,.\end{aligned}$$

□

Corollary 1 *If the co-event ϕ is MP and unital then it is C1.*

It was shown in the previous section that if a co-event ϕ is a homomorphism then it is MP, C1 and C2. Conversely, we can ask: what conditions imply that ϕ is a homomorphism?

Theorem 2 *If co-event ϕ is unital, MP and C2 then it is a homomorphism.*

Proof Let ϕ be unital, MP and C2. By Theorem 1 ϕ is multiplicative, and by Lemma 3 it is C1.

We need to show that ϕ is additive. C1 and C2 imply $\phi(X) + \phi(\mathbf{1} + X) = 1$ for all $X \in \mathfrak{A}$. Let $A, B \in \mathfrak{A}$.

$$
\begin{aligned}
&\phi(A+B) + \phi(\mathbf{1}+A+B) = 1 \quad \text{and} \quad \phi(AB) + \phi(\mathbf{1}+AB) = 1 \\
\Rightarrow\ & [\phi(A+B) + \phi(\mathbf{1}+A+B)]\,[\phi(AB) + \phi(\mathbf{1}+AB)] = 1 \\
\Rightarrow\ & \phi(A+B)\phi(AB) + \phi(A+B)\phi(\mathbf{1}+AB) + \phi(\mathbf{1}+A+B)\phi(AB) \\
&\quad + \phi(\mathbf{1}+A+B)\phi(\mathbf{1}+AB) = 1 \\
\Rightarrow\ & \phi\left((A+B)AB\right) + \phi\left((A+B)(\mathbf{1}+AB)\right) + \phi\left((\mathbf{1}+A+B)AB\right) \\
&\quad + \phi\left((\mathbf{1}+A+B)(\mathbf{1}+AB)\right) = 1 \\
\Rightarrow\ & \phi(\emptyset) + \phi(A+B) + \phi(AB) + \phi(\mathbf{1}+A+B+AB) = 1 \\
\Rightarrow\ & 0 + \phi(A+B) + \phi(AB) + \phi\left((\mathbf{1}+A)(\mathbf{1}+B)\right) = 1 \\
\Rightarrow\ & \phi(A+B) + \phi(AB) + \phi(\mathbf{1}+A)\phi(\mathbf{1}+B) = 1 \\
\Rightarrow\ & \phi(A+B) + \phi(AB) + (1+\phi(A))(1+\phi(B)) = 1 \\
\Rightarrow\ & \phi(A+B) + \phi(AB) + 1 + \phi(A) + \phi(B) + \phi(AB) = 1 \\
\Rightarrow\ & \phi(A+B) = \phi(A) + \phi(B).
\end{aligned}
$$

□

Since zero-preservation and C2 imply unitality we can replace the condition of unitality by that of zero-preservation:

Corollary 2 *If co-event ϕ is zero-preserving, MP and C2 then it is a homomorphism.*

Theorem 2 establishes that, as long as $\phi(\mathbf{1}) = 1$ (the world is affirmed), modus ponens needs the addition of only the rule C2 to lead to classical logic.

5 A Unifying Proposal

5.1 Classical Physics Revisited

We mentioned that when One History Happens, the corresponding co-event is a homomorphism. What about the converse? When the set of spacetime histories Ω is finite, the event algebra $\mathfrak{A}$ is the power set 2^{Ω} of Ω, and in this case the Stone representation theorem tells us that the set of (non-zero) homomorphisms from $\mathfrak{A}$ to $\mathbb{Z}_2$ is isomorphic to Ω. Thus, the axiom that exactly one history from Ω corresponds to the physical world is equivalent—in the finite case—to the assumption that the co-event that corresponds to the physical world is a homomorphism. This is just one of the possible equivalent reformulations of the One History Happens axiom that defines classical physics; we provide a partial list below. Before doing so we must first introduce classical *dynamics* as a rule of inference. The dynamics are encoded in a probability measure μ, a non-negative real function $\mu : \mathfrak{A} \rightarrow \mathbb{R}$ which satisfies the Kolmogorov sum rules and $\mu(\mathbf{1}) = 1$. We call an event in $\mathfrak{A}$ where $\mu(A) = 0$ a *null* event. Classical dynamical law requires that the history that corresponds to the physical world not be an element of any null event: a null event cannot happen. The co-event ϕ that corresponds to the physical world is therefore required to be *preclusive*, where:

Definition 8 A co-event ϕ is *preclusive* if

$$\mu(A) = 0 \Rightarrow \phi(A) = 0, \quad \forall A \in \mathfrak{A}\,. \tag{13}$$

We will also make use of the following definitions:

Definition 9 A filter $F \subseteq \mathfrak{A}$ is *preclusive* if none of its elements are null.

Definition 10 An event $A \in \mathfrak{A}$ is *stymied* if it is a subset of a null event.

The physical world in a classical theory when Ω is finite is then described equivalently by any of the following.

(i) A single history, an element of Ω, which is not an element of any null event.
(ii) A minimal non-empty non-stymied event (ordered by inclusion).
(iii) A preclusive ultrafilter on Ω.
(iv) A maximal preclusive filter (ordered by inclusion).
(v) A preclusive homomorphism $\phi : \mathfrak{A} \rightarrow \mathbb{Z}_2$.
(vi) A preclusive co-event for which all classical, Boolean rules of inference hold.
(vii) A preclusive, unital, MP, C2 co-event.
(viii) A minimal preclusive, multiplicative co-event, where minimality is in the order

$$\phi_1 \preceq \phi_2 \text{ if } \phi_2(A) = 1 \Rightarrow \phi_1(A) = 1\,. \tag{14}$$

(ix) A minimal preclusive, unital, MP co-event, where again minimality is in the order (14).

The equivalence of item (vii) is the import of Theorem 2. The final two items, (viii) and (ix), introduce the concept of *minimality*, which is a finest grainedness condition or a Principle of Maximal Detail: nature affirms as many events as possible without violating preclusion. That the conditions in (viii) imply that ϕ is a homomorphism is proved by Sorkin [6, 7]. That (viii) and (ix) are equivalent is the import of Theorem 1.

In a classical theory one is free to consider any or all of these as corresponding to the physical world, since each is equivalent to a single history $\gamma \in \Omega$.

5.2 *Quantum Measure Theory*

Quantum theories find their place in the framework of GMT at the second level of a countably infinite hierarchy of theories labelled by the amount of interference there is between histories [2]. A quantum measure theory has the threefold structure described in Sect. 2, just as a classical theory does, and it too is based on a set Ω of spacetime histories—the histories summed over in the path integral for the theory. The departure from a classical theory is encoded in the nature of the measure μ which is in general no longer a probability measure. Indeed, given by the path integral, a quantal μ does not satisfy the Kolmogorov sum rule but, rather, a quadratic analogue of it [2–4]. The existence of interference between histories means that there are quantum measure systems for which the union of all the null events is the whole of Ω. Examples are the three-slit experiment [4], the Kochen–Specker antinomy [8, 9, 14, 15] and the inequality-free version of Bell's theorem due to Greenberger et al. [16, 17] and Mermin [11, 18, 19]. The condition of preclusion therefore runs into conflict with the proposal that the physical world corresponds to a single history since if every history is an element of some null event there is no history that can happen: reductio ad absurdum.

Choosing to uphold preclusion as a dynamical law means therefore that, of the above list of nine equivalent descriptions of a classical physical world, (i) fails in the quantal setting, and so do (iii), (v), (vi) and (vii). However, the other four—(ii), (iv), (viii) and (ix)—survive and remain mutually equivalent for a finite quantal measure theory. That (ii), (iv) and (viii) are equivalent can be shown using the fact that a multiplicative co-event ϕ defines and is defined by its *support*, $F(\phi) \in \mathfrak{A}$, the intersection of all those events that are affirmed by ϕ:

$$F(\phi) := \bigcap_{S \in \phi^{-1}(1)} S\,. \tag{15}$$

Adopting (viii) as the axiom for the possible co-events of a theory gives the resulting scheme its name: the multiplicative scheme. The multiplicative scheme is a unifying proposal: whether classical or quantum, the physical world is a minimal preclusive multiplicative co-event [5]. What we have shown here is that it could just as well be dubbed the "modus ponens scheme".

6 Final Words

With hindsight, we can see that the belief that the geometry of physical space was fixed and Euclidean came about because deviations from Euclidean geometry at non-relativistic speeds and small curvatures are difficult to detect. In a similar vein, Sorkin suggests, the need for deviations from classical rules of inference about physical events lay undetected until the discovery of quantum phenomena (see however [6]). That's all very well, but it could seem much harder to wean ourselves off the structure of classical logic than to give up Euclidean geometry. To those who feel that classical rules of inference are essential to science, this reassurance can be offered: in GMT, classical rules of inference are used to reason about the co-events themselves, because a single co-event corresponds to the physical world.

Moreover, in the multiplicative scheme (to the extent that the finite system case is a good guide to the more general, infinite case) each co-event can be equivalently understood in terms of its support—a subset of Ω. In Hartle's Generalised Quantum Mechanics [20, 21], this subset would be called a *coarse-grained history*; the proposal of the multiplicative scheme is to describe the physical world as a single coarse-grained history. The altered rules of inference in the multiplicative scheme for GMT are no more of a conceptual leap than this: the physical world is not as fine grained as it might have been, and there are some details which are missing, ontologically. Furthermore, the results reported here reveal the alteration of logic in the multiplicative scheme to be the mildest possible modification: keeping MP and relinquishing only C2. Relinquishing C2 in physics means allowing the possibility that an electron is not inside a box and not outside it either. Another example is accepting the possibility that a photon in a double slit experiment does not pass through the left slit and does not pass through the right slit. At the level of electrons and photons, such a non-classical state of affairs is not too hard to swallow; indeed, very many, very similar statements are commonly made about the microscopic details of quantum systems. The multiplicative scheme for GMT is a proposal for making precise the nature of Feynman's "logical tightrope" and raises the important question: "Are violations of classical logic confined to the microscopic realm?" Answering this question becomes a matter of calculation within any given theory [7].

Acknowledgements We thank Rafael Sorkin for helpful discussions. Research at Perimeter Institute for Theoretical Physics is supported in part by the Government of Canada through NSERC and by the Province of Ontario through MRI. FD and PW are supported in part by COST Action MP1006. PW was supported in part by EPSRC grant EP/K022717/1. PW acknowledges support from the University of Athens during this work.

References

1. R.P. Feynman, R.B. Leighton, M. Sands, *Lectures on Physics*, vol. iii (Addison-Wesley, 1965)
2. R. Sorkin, Quantum mechanics as quantum measure theory. Mod. Phys. Lett. A **9**(33), 3119–3127 (1994)
3. R. Sorkin, Quantum measure theory and its interpretation, in *Quantum Classical Correspondence: Proceedings of the 4th Drexel Symposium on Quantum Nonintegrability*, ed. by D. Feng, B.-L. Hu (International Press, Cambridge, MA, 1997), pp. 229–251. Preprint gr-qc/9507057v2
4. R. Sorkin, Quantum dynamics without the wavefunction. J. Phys. A: Math. Theor. **40**(12), 3207–3221 (2007)
5. R. Sorkin, An exercise in "anhomomorphic logic". J. Phys.: Conf. Ser. **67**, 012018 (8 pp.) (2007)
6. R. Sorkin, To what type of logic does the "Tetralemma Belong? (2010). Preprint 1003.5735

7. R. Sorkin, Logic is to the quantum as geometry is to gravity, in *Foundations of Space and Time: Reflections on Quantum Gravity*, ed. by G. Ellis, J. Murugan, A. Weltman (Cambridge University Press, Cambridge, 2011). Preprint 1004.1226v1
8. F. Dowker, Y. Ghazi-Tabatabai, The Kochen–Specker theorem revisited in quantum measure theory. J. Phys. A: Math. Theor. **41**(10), 105301-1–105301-17 (2008)
9. F. Dowker, G. Siret, M. Such, A Kochen-Specker system in histories form (2016, in preparation)
10. J. Henson, Causality, Bell's theorem, and Ontic Definiteness (2011). Preprint 1102.2855v1
11. F. Dowker, D. Benincasa, M. Buck, A Greenberger-Horne-Zeilinger system in quantum measure theory (2016, in preparation)
12. L. Carroll, What the tortoise said to Achilles. Mind **4**(14), 278–280 (1895)
13. N. Rescher, R. Brandom, *The Logic of Inconsistency* (Basil Blackwell, Oxford, 1979)
14. S. Kochen, E. Specker, The problem of hidden variables in quantum mechanics. J. Math. Mech. **17**(1), 59–87 (1967)
15. J. Bell, On the problem of hidden variables in quantum mechanics. Rev. Mod. Phys. **38**(3), 447–452 (1966)
16. D. Greenberger, M. Horne, A. Zeilinger, Going beyond Bell's theorem, in *Bell's Theorem, Quantum Theory and Conceptions of the Universe*, vol. 37, ed. by M. Kafatos. Fundamental Theories of Physics (Kluwer, Dordrecht, 1989), pp. 69–72
17. D. Greenberger, M. Horne, A. Shimony, A. Zeilinger, Bell's theorem without inequalities. Am. J. Phys. **58**(12), 1131–1143 (1990)
18. N. Mermin, Quantum mysteries revisited. Am. J. Phys. **58**(8), 731–734 (1990)
19. N. Mermin, What's wrong with these elements of reality? Phys. Today **43**(6), 9–11 (1990)
20. J.B. Hartle, The spacetime approach to quantum mechanics. Vistas Astron. **37**, 569–583 (1993)
21. J. Hartle, Spacetime quantum mechanics and the quantum mechanics of spacetime, in *Gravitation and Quantizations: Proceedings of the 1992 Les Houches Summer School (Les Houches, France, 1992)*, vol. LVII, ed. by B. Julia, J. Zinn-Justin (Elsevier, Amsterdam, 1995). Preprint gr-qc/9304006v2

From Quantum Foundations via Natural Language Meaning to a Theory of Everything

Bob Coecke

Abstract In this paper we argue for a paradigmatic shift from 'reductionism' to 'togetherness'. In particular, we show how interaction between systems in quantum theory naturally carries over to modelling how word meanings interact in natural language. Since meaning in natural language, depending on the subject domain, encompasses discussions within any scientific discipline, we obtain a template for theories such as social interaction, animal behaviour, and many others.

1 ... in the Beginning Was ⊗

No, physicists! ... the symbol ⊗ above does not stand for the operation that turns two Hilbert spaces into the smallest Hilbert space in which the two given ones bilinearly embed. No, category-theoreticians! ... nor does it stand for the composition operation that turns any pair of objects (and morphisms) in a monoidal category into another object that is the symbol ⊗ a horrendous bunch of conditions that guaranty coherence with the remainder of the structure. Instead, this is what it means:

$$\otimes \equiv \text{“togetherness”}$$

More specifically, it represents the togetherness of foo_1 and foo_2 without giving any specification of who/what foo_1 and foo_2 actually are. Differently put, it's the new stuff that emerges when foo_1 and foo_2 get together. If they don't like each other at all, this may be a fight. If they do like each other a lot, this may be a marriage, and a bit later, babies. Note that togetherness is vital for the emergence to actually take place, given that it is quite hard to either have a fight, a wedding, or a baby if there is nobody else around.

It is of course true that in von Neumann's formalisation of quantum theory the *tensor product of Hilbert spaces* (also denoted by ⊗) plays this role [39], giving rise to the emergent phenomenon of *entanglement* [20, 36]. And more generally, in

B. Coecke (✉)
University of Oxford, Oxford, UK
e-mail: coecke@cs.ox.ac.uk

S.B. Cooper, M.I. Soskova (eds.), *The Incomputable*, Theory and Applications of Computability, DOI 10.1007/978-3-319-43669-2_4

category theory one can axiomatise *composition of objects* (again denoted by $\otimes$) within a *symmetric monoidal category* [5], giving rise to elements that don't simply arise by pairing, just like in the case of the Hilbert space tensor product.

However, in the case of von Neumann's formalisation of quantum theory, we are talking about a formalisation which, despite being widely used, its creator himself didn't even like [33]. Moreover, in this formalism, $\otimes$ only arises as a secondary construct, requiring a detailed description of foo_1 and foo_2, whose togetherness it describes. What we are after is a 'foo-less' conception of $\otimes$. The composition operation $\otimes$ in symmetric monoidal categories heads in that direction. However, by making an unnecessary commitment to set theory, it makes things unnecessarily complicated [14]. Moreover, while this operation is general enough to accommodate togetherness, it doesn't really tell us anything about it.

The title of this section is a metaphor aimed at confronting the complete disregard that the concept of togetherness has suffered in the sciences, and especially, in physics, where all of the effort has been on describing the individual, typically by breaking its description down to that of even smaller individuals. While, without any doubt, this has been a useful endeavour, it unfortunately has evolved into a rigid doctrine, leaving no space for anything else. The most extreme manifestation of this dogma is the use of the term 'theory of everything' in particle physics. We will provide an alternative conceptual template for a theory of everything, supported not only by scientific examples, but also by everyday ones.

Biology evolved from chopping up individual animals in laboratories, to considering them in the context of other animals and varying environments. The result is the theory of evolution of species. Similarly, our current (still very poor) understanding of the human brain makes it clear that the human brain should not be studied as something in isolation, but as something that fundamentally requires interaction with other brains [30]. In contemporary audio equipment, music consists of nothing but strings of 0s and 1s. Instead, the entities that truly make up music are pitch, sound, rhythm, chord progression, crescendo, and so on. And in particular, music is not just a bag of these, since their intricate interaction is even more important than these constituents themselves. The same is true for film, where it isn't even that clear what it is made up from, but it does include such things as (easily replaceable) actors, decors, and cameras, which all are part of a soup stirred by a director. But again, in contemporary video equipment, it is nothing but a string of 0s and 1s.

In fact, everything that goes on in pretty much all modern devices is nothing but 0s and 1s. While it was Turing's brilliance to realise that this could in fact be done, and provided a foundation for the theory of computability [38], this is in fact the only place where the 0s and 1s are truly meaningful, in the form of a Turing machine. Elsewhere, it is nothing but a (universal) representation, with no conceptual qualities regarding the subject matter.

2 Formalising Togetherness 1: Not There Yet

So, how does one go about formalising the concept of togetherness? While we don't want an explicit description of the foo involved, we do need some kind of means for identifying foo. Therefore, we simply give each foo a name, say $A, B, C, \ldots$. Then, $A \otimes B$ represents the togetherness of A and B. We also don't want an explicit description of $A \otimes B$, so how can we say anything about $\otimes$ without explicitly describing A, B and $A \otimes B$?

Well, rather than describing these systems themselves, we could describe their relationships. For example, in a certain theory togetherness could obey the following equation:

$$A \otimes A = A$$

That is, togetherness of two copies of something is exactly the same as a single copy, or in simpler terms, one is as good as two. For example, if one is in need of a plumber to fix a pipe, one only needs one. The only thing a second plumber would contribute is a bill for the time he wasted coming to your house. Obviously, this is not the kind of togetherness that we are really interested in, given that this kind adds nothing at all.

A tiny bit more interesting is the case that two is as good as three:

$$A \otimes A \otimes A = A \otimes A,$$

e.g. when something needs to be carried on a staircase, but there really is only space for two people to be involved. Or, when A is female and $\bar{A}$ is male, and the goal is reproduction, we have:

$$A \otimes \bar{A} \otimes \bar{A} = A \otimes \bar{A}$$

(ignoring testosterone-induced scuffles and the benefits of natural selection.)

We really won't get very far in this manner. One way in which things can be improved is by replacing equations by inequalities. For example, while

$$A = B$$

simply means that one of the two is redundant,

$$A \leq B$$

can mean that from A we can produce B, and

$$A \otimes B \leq C$$

can mean that from A and B together we can produce C, and

$$A \otimes C \leq B \otimes C$$

can mean that in the presence of C from A we can produce B, i.e. that C is a catalyst.

What we have now is a so-called *resource theory*, that is, a theory which captures how stuff we care about can be interconverted [19]. Resource theories allow for quantitative analysis, for example, in terms of a *conversion rate*:

$$r(A \to B) := \sup\left\{ \frac{m}{n} \;\middle|\; \underbrace{A \otimes \ldots \otimes A}_{n} \leq \underbrace{B \otimes \ldots \otimes B}_{m} \right\}$$

So evidently we have some genuine substance now.[1]

3 Formalising Togetherness 2: That's Better

But we can still do a lot better. What a resource theory fails to capture (on purpose, in fact) is the actual process that converts one resource into another one. So let's fix that problem, and explicitly account for processes.

In terms of togetherness, this means that we bring the fun foo_1 and foo_2 can have together explicitly in the picture. Let

$$f : A \to B$$

denote some process that transforms A into B. Then, given two such processes f_1 and f_2 we can also consider their togetherness:

$$f_1 \otimes f_2 : A_1 \otimes A_2 \to B_1 \otimes B_2$$

Moreover, some processes can be sequentially chained:

$$g \circ f : A \to B \to C$$

We say 'some', since f has to produce B in order for

$$g : B \to C$$

to take place.

[1]In fact, resource theories are currently a very active area of research in the quantum information and quantum foundations communities, e.g. the resource theories of entanglement [23], symmetry [21], and athermality [7].

Now, here one may end up in a bit of a mess if one isn't clever. In particular, with a bit of thinking one quickly realises that one wants some equations to be obeyed, for example,

$$(f_1 \otimes f_2) \otimes f_3 = f_1 \otimes (f_2 \otimes f_3) \tag{1}$$

$$h \circ (g \circ f) = (h \circ g) \circ f, \tag{2}$$

and a bit more sophisticated:

$$(g_1 \otimes g_2) \circ (f_1 \otimes f_2) = (g_1 \circ f_1) \otimes (g_2 \circ f_2) \tag{3}$$

There may even be some more equations that one wants to have, but which ones? This turns out to be a very difficult problem. Too difficult in the light of our limited existence in this world. The origin of this problem is that we treat $\otimes$, and also $\circ$, as algebraic connectives, and that algebra has its roots in set theory. The larger-than-life problem can be avoided in a manner that is as elegant as it is simple.

To state that things are together, we just write them down together:

$$A \qquad B$$

There really is no reason to adjoin the symbol $\otimes$ between them. Now, this A and B will play the role of an input or an output of processes transforming them. Therefore, it will be useful to represent them by a wire:

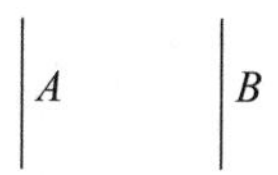

Then, a process transforming A into B can be represented by a box:

Togetherness of processes now becomes

and chaining processes becomes:

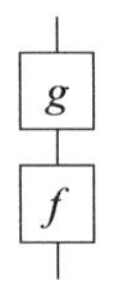

In particular, Eqs. (1)–(3) become:

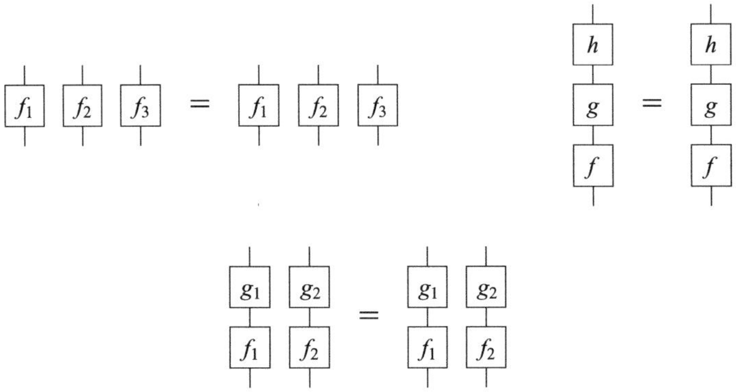

That is, all equations have become tautologies![2]

4 Anti-Cartesian Togetherness

One important kind of processes are *states*:

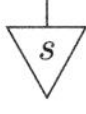

These are depicted without any inputs, where 'no wire' can be read as 'nothing' (or 'no-foo').[3] The opposite notion is that of an *effect*, that is, a process without an output

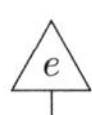

[2]A more extensive discussion of this bit of magic can be found in [12–14].

[3]That we use triangles for these rather than boxes is inspired by the Dirac notation which is used in quantum theory. Please consult [10, 13, 14] for a discussion.

borrowing terminology from quantum theory.[4]

We can now identify those theories in which togetherness *doesn't* yield anything new. Life in such a world is pretty lonely...

Definition 4.1 A theory of togetherness is *Cartesian* if each state

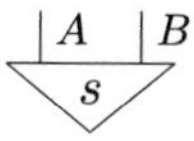

decomposes as follows:

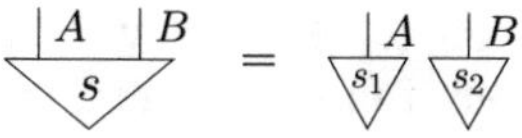

So Cartesianness means that all possible realisations of two foos can be achieved by pairing realisations of the individual foos involved. In short, a whole can be described in term of its parts, rendering togetherness a void concept. So very lonely and indeed... But, wait a minute. Why is it then the case that so much of traditional mathematics follows this Cartesian template, and that even category theory for a long time has followed a strict Cartesian stance? Beats me. Seriously... beats me!

Anyway, an obvious consequence of this is that for those areas where togetherness is a genuinely non-trivial concept, traditional mathematical structures aren't always that useful. That is maybe why social sciences don't make much use of any kind of modern pure mathematics.

And now for something completely different:

Definition 4.2 A theory of togetherness is *anti-Cartesian* if for each A there exists A^*, a special state $\cup$ and a special effect $\cap$,

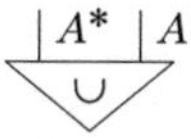

and

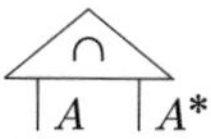

which are such that the following equation holds:

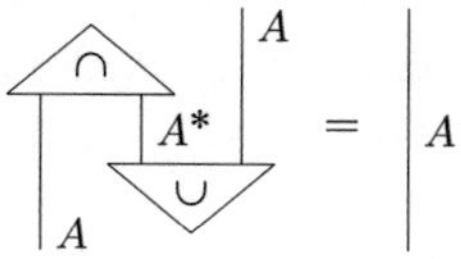

(4)

The reason for 'anti' in the name of this kind of togetherness is the fact that when a theory of togetherness is both cartesian and anti-Cartesian, then it is nothing but

[4]Examples of these include 'tests' [13].

a theory of absolute death, i.e. it describes a world in which nothing ever happens. Indeed, we have:

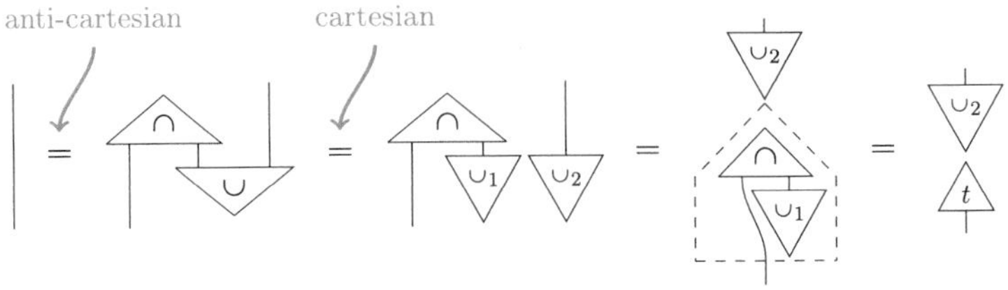

That is, the *identity* is a constant process, always outputting the state $\cup_2$, independently of what the input is. And if that isn't weird enough, any arbitrary process f does the same:

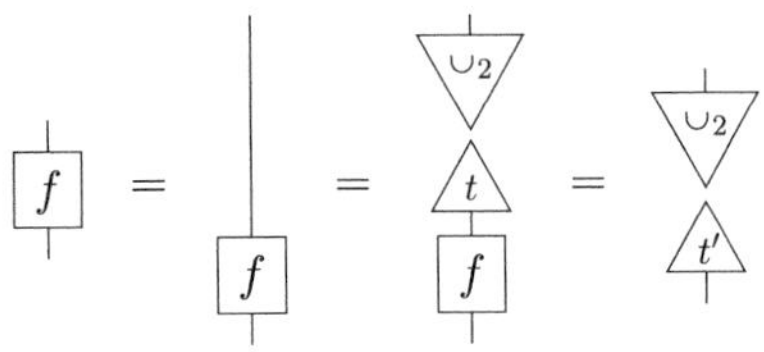

Therefore, any anti-Cartesian theory of togetherness that involves some aspect of change cannot be Cartesian, and hence will have interesting stuff emerging from togetherness.[5]

5 Example 1: Quantum Theory

Anti-Cartesian togetherness is a very particular alternative to Cartesian togetherness (contra any theory that fails to be Cartesian). So one may wonder whether there are any interesting examples. And yes, there are! One example is *quantum entanglement* in quantum theory. That is in fact where the author's interest in anti-Cartesian togetherness started [1, 9, 10].[6] As shown in these papers, Eq. (4) pretty much embodies the phenomenon of quantum teleportation [6]. The full-blown description of quantum teleportation goes as follows [13, 15, 16]:

[5] Many more properties of anti-cartesian togetherness can be found in [13].

[6] Independently, similar insights appeared in [3, 26].

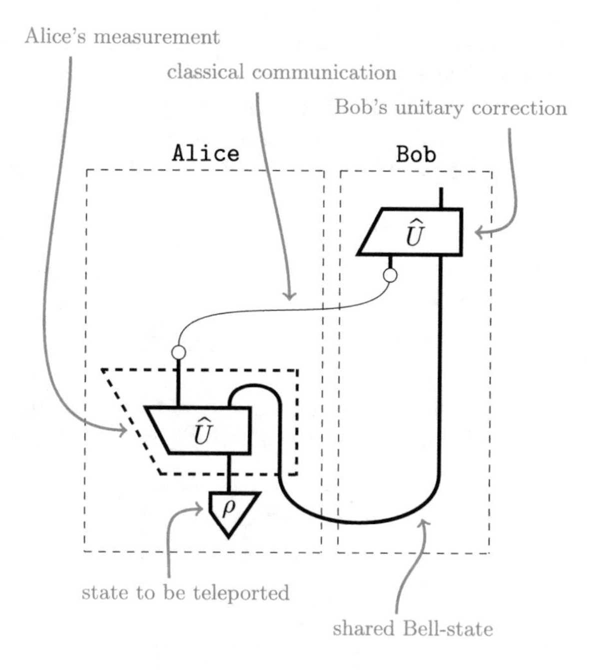

It is not important to fully understand the details here. What is important is to note that the bit of this diagram corresponding to Eq. (4) is the bold wire which zig-zags through it:

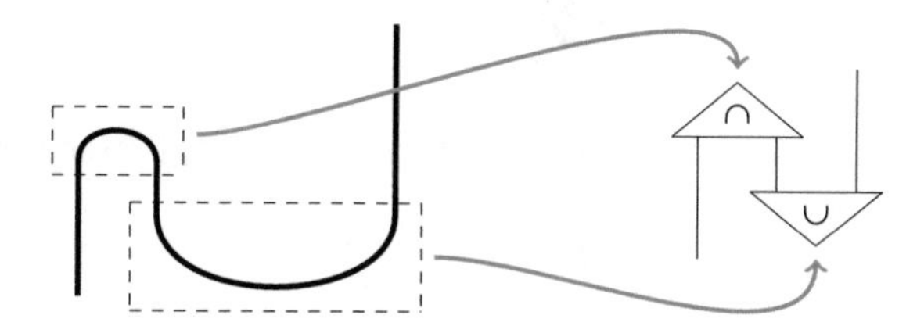

The thin wires and the boxes labelled $\widehat{U}$ are related to the fact that quantum theory is non-deterministic. By conditioning on particular measurement outcomes, teleportation simplifies to [13]:

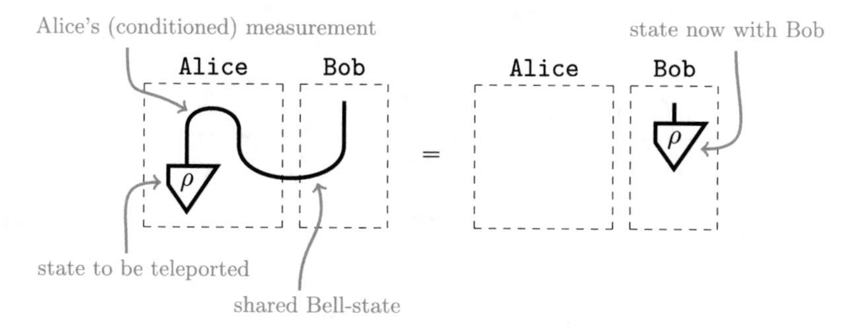

Equality of the left-hand side and of the right-hand side follows directly from Eq. (4). While in this picture we eliminated quantum non-determinism by conditioning on a measurement outcome, there still is something very 'quantum' going on here: Alice's (conditioned) measurement is nothing like a passive observation, but a highly non-trivial intervention that makes Alice's state ρ appear at Bob's side:

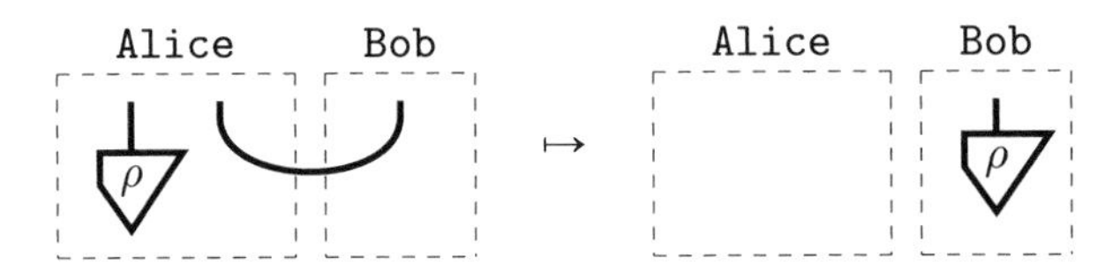

Let's analyse more carefully what's going on here by explicitly distinguishing between the top layer and the bottom layer of this diagram:

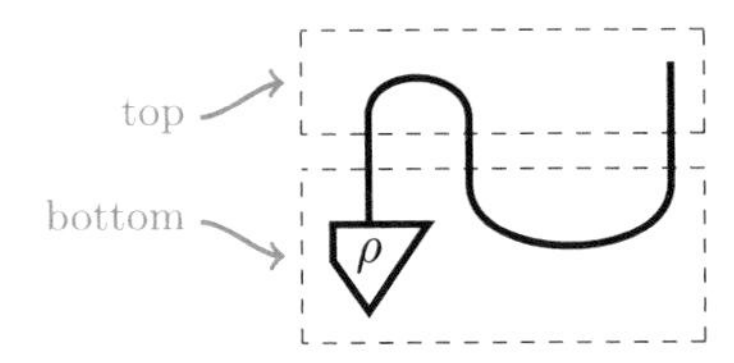

The bottom part,

consists of the state ρ together with a special ∪-*state*, while the top part,

includes the corresponding ∩-*effect*, as well as an output. By making the bottom part and the top part interact, and, in particular, the ∪ and the ∩, the state ρ ends up at the output of the top part.

A more sophisticated variation on the same theme makes it much clearer what is going on here. Using Eq. (4), the diagram:

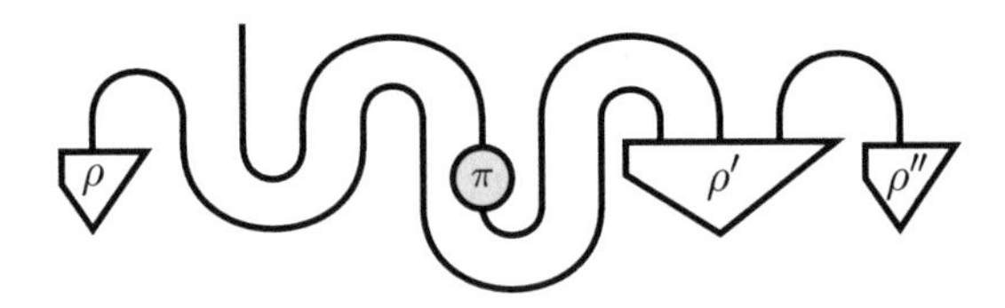

reduces to

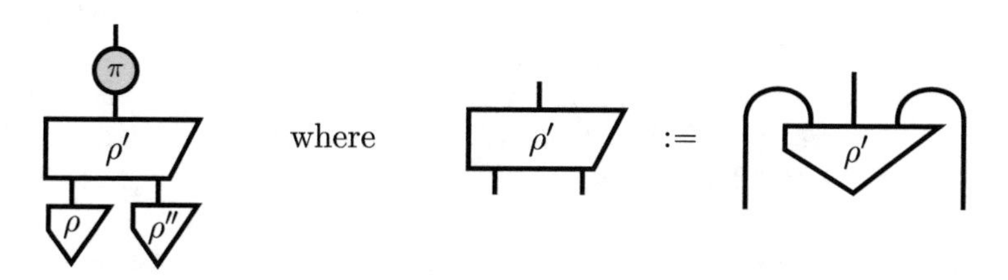

The grey dot labeled π is some (at this point not important) unitary quantum operation [13]. Let us again consider the bottom and top parts:

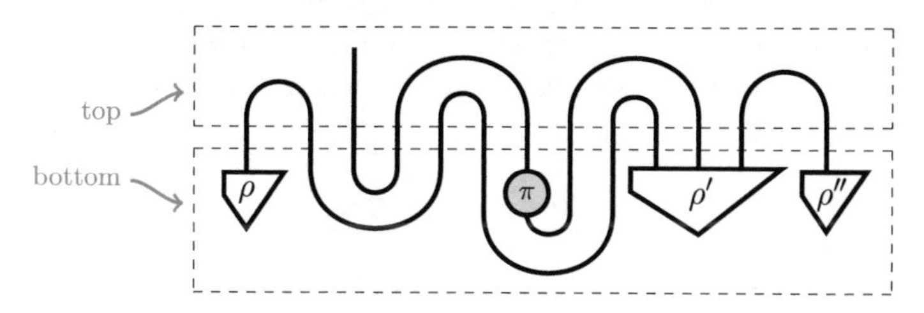

The top part is a far more sophisticated measurement, consisting mainly of ∩s. Also, the bottom part is a lot more sophisticated, involving many ∪s. These now cause a highly non-trivial interaction of the three states ρ, ρ' and ρ''. Why we have chosen this particular example will become clear in the next section. What is important to note is that the overall state and overall effect have to be chosen in a very particular way to create the desired interaction, similarly to an old-fashion telephone switchboard that has to be connected in a very precise manner in order to realise the right connection.

6 Example 2: Natural Language Meaning

Another example of anti-Cartesian togetherness is the manner in which word meanings interact in natural language! Given that logic originated in natural language when Aristotle analysed arguments involving 'and', 'if… then', 'or', etc., anti-Cartesianness can be conceived as some new kind of logic![7] So what are ∪ and ∩ in this context?

In order to understand what ∩ is, we need to understand the mathematics of grammar. The study of the mathematical structure of grammar has indicated that the fundamental things making up sentences are not the words, but some atomic grammatical types, such as the noun-type and the sentence-type [2, 4, 27]. The

[7] A more detailed discussion is in [11].

transitive verb-type is not an atomic grammatical type, but a composite made up of two noun-types and one sentence-type. Hence, particularly interesting here is the fact that atomic doesn't really mean smallest...

On the other hand, just like in particle physics where we have particles and anti-particles, the atomic types include types as well as anti-types. But unlike in particle physics, there are two kinds of anti-types, namely left ones and right ones. This makes language even more non-commutative than quantum theory!

All of this becomes much clearer when considering an example. Let n denote the atomic *noun*-type and let ${}^{-1}n$ and n^{-1} be the corresponding anti-types. Let s denote the atomic *sentence*-type. Then the non-atomic *transitive verb*-type is ${}^{-1}n \cdot s \cdot n^{-1}$. Intuitively, it is easy to understand why. Consider a transitive verb, like 'hate'. Then, simply saying 'hate' doesn't convey any useful information, until, we also specify 'who' hates 'whom'. That's exactly the role of the anti-types: they specify that in order to form a meaningful sentence, a noun is needed on the left, and a noun is needed on the right:

$$\underbrace{Alice}_{n}\ \underbrace{hates}_{{}^{-1}n\cdot s\cdot n^{-1}}\ \underbrace{Bob}_{n}$$

Then, n and ${}^{-1}n$ cancel out, and so do n^{-1} and n. What remains is s, confirming that 'Alice hates Bob' is a grammatically well-typed sentence. We can now depict the cancelations as

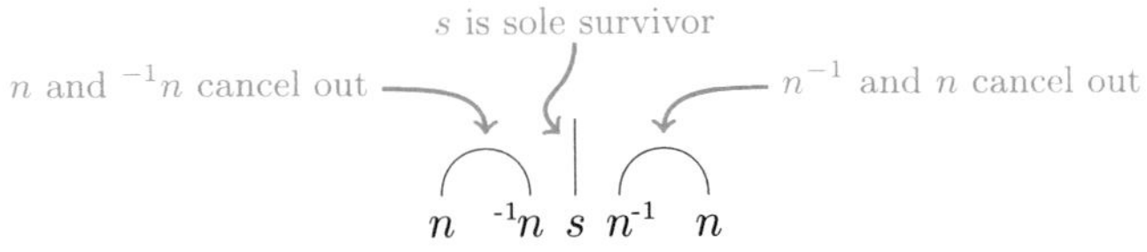

s is sole survivor

n and ${}^{-1}n$ cancel out

n^{-1} and n cancel out

$$n \quad {}^{-1}n \quad s \quad n^{-1} \quad n$$

and bingo, we have found ∩!

While the mathematics of sentence structure has been explored now for some 80 years, the fact that ∩s can account for grammatical structure is merely a 15-year-old idea [28]. So what are the ∪s? That evolves an even more recent story in which we were involved and, in fact, for which we took inspiration from the story of the previous section [8]. While ∩s are about grammar, ∪s are about meaning.

The *distributional* paradigm for natural language meaning states that meaning can be represented by vectors in a vector space [37]. Until recently, grammatical structure was essentially ignored in doing so, and, in particular, there was no theory for how to compute the meaning of a sentence given the meanings of its words. Our new *compositional distributional model of meaning* of [17] does exactly that.[8]

In order to explain how this compositional distributional model of meaning works, let's get back to our example. Since we have grammatical types around,

[8]... and has meanwhile outperformed other attempts in several benchmark natural language processing (NLP) tasks [22, 25].

the meaning vectors should respect grammatical structure, that is, the vectors representing compound types should themselves live in compound vector spaces. So the string of vectors representing the word meanings of our example would look as follows:

Now we want to put forward a new hypothesis:

Grammar is all about how word meanings interact.

Inspired by the previous section, this can be realised as follows:

where the ∩-s are now interpreted in exactly the same manner as in the previous section. And here is a more sophisticated example:

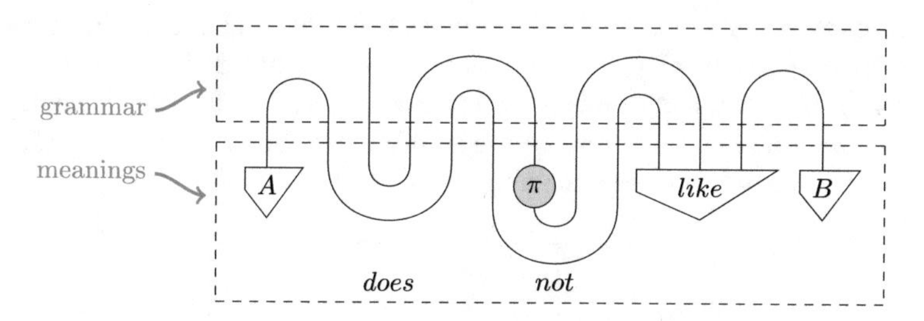

where the π-labeled grey circle should now be conceived as negating meaning [17]. The grammatical structure is here:

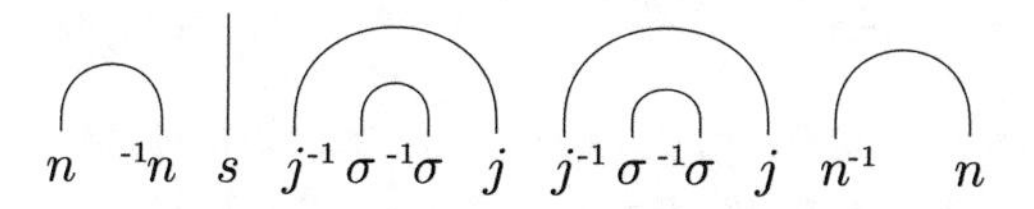

It is simply taken from a textbook such as [29]; the meanings of *Alice*, *likes* and *Bob* can be automatically generated from some corpus, while the meanings of *does* and *not* are just cleverly chosen to be [17, 32]:

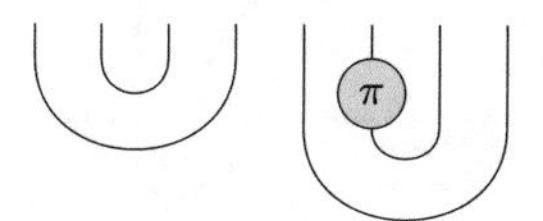

In the previous section we already saw that in this way we obtain:

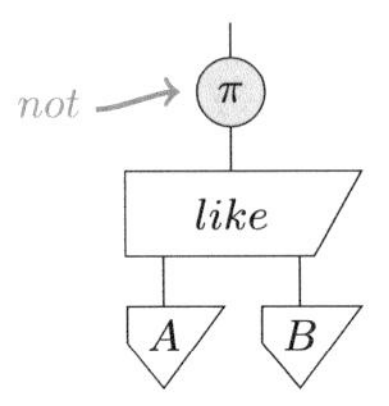

This indeed captures the intended meaning:

$$not\,(like\,(Alice, Bob))$$

where we can think of *like* as being a predicate and *not* as being itself.

So an interesting new aspect of the last example is that some of the meaning vectors of words are simply cleverly chosen and, in particular, involve ∪s. Hence, we genuinely exploit full-blown anti-Cartesianess. What anti-Cartesianess does here is make sure that the transitive verb *likes* 'receives' *Alice* as its object. Note also how *not* does pretty much the same as *does*, guiding word meanings through the sentence, with, of course, one very important additional task: negating the sentence meaning.

The cautious reader must of course have noticed that in the previous section we used thick wires, while here we use thin ones. Also, the dots in the full-blown description of quantum teleportation, which represent classical data operations, have vanished in this section. Meanwhile, thick wires as well as the dots have acquired a vary natural role in a more refined model of natural language meaning. The dots allow us to cleverly choose the meanings of relative pronouns [34, 35]:

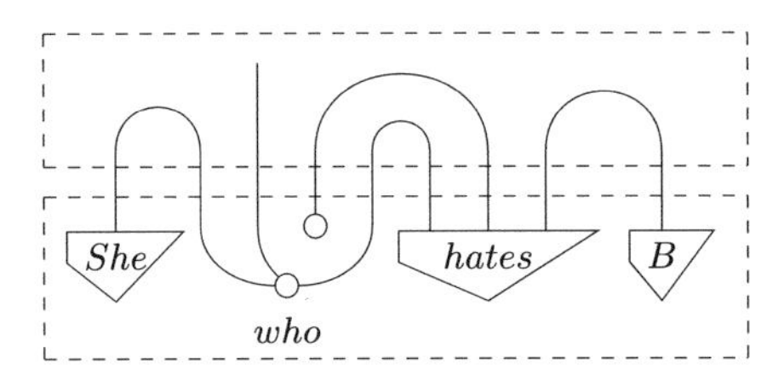

Thick wires (representing density matrices, rather than vectors [13]) allow us to encode word ambiguity as *mixedness* [24, 31]. For example, the different meanings of the word *queen* (a rock band, a person, a bee, a chess piece, or a drag queen). Mixedness vanishes when providing a sufficient string of words that disambiguates that meaning, e.g.,

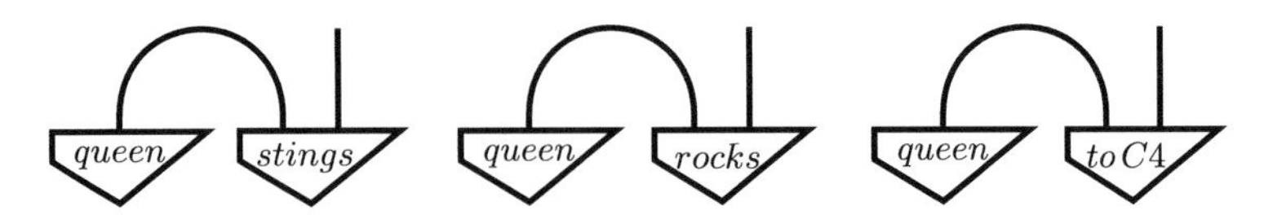

while in the case of

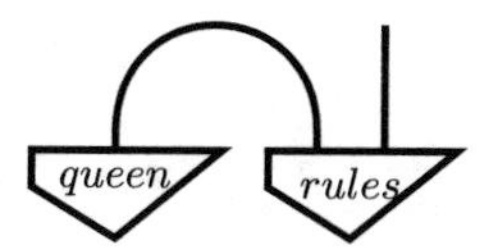

we need more disambiguating words, since *queen* can still refer to a person, a rock band, or a drag queen.

7 Meaning is Everything

The distributional model of meaning [37] is very useful in that it allows for automation, given a substantially large corpus of text. However, from a conceptual point of view it is far from ideal. So one may ask the question:

What is meaning?

One may try to play around with a variety of mathematical structures. The method introduced in [17] doesn't really depend on how one models meaning, as long as we stick to anti-Cartesian togetherness, or something sufficiently closely related [18]. It is an entertaining exercise to play around with the idea of what possibly could be the ultimate mathematical structure that captures meaning in natural language, until one realises that meaning in natural language truly encompasses *everything*. Indeed, we use language to talk about everything, e.g. logic, life, biology, physics, social behaviours, and politics, so the ultimate model of meaning should encompass all of these fields. So, a theory of meaning in natural language is actually a theory of everything! Can we make sense of the template introduced in the previous section for meaning in natural language as one for ... everything?

Let us first investigate whether the general distributional paradigm can be specialised to the variety of subject domains mentioned above. The manner in which the distributional model works is that meanings are assigned relative to a fixed chosen set of context words. The meaning vector of any word then arises by counting the number of occurrences of that word in the close neighbourhood of each of the context words within a large corpus of text. One can think of the context words as attributes, and the relative frequencies as the relevance of an attribute for the word. Simply by specialising the context words and the corpus, one can specialise to a certain subject domain. For example, if one is interested in social behaviours then the corpus could consist of social networking sites, and the context words could be chosen accordingly. This pragmatic approach allows for quantitative analysis, just like the compositional distributional model of [17].

Here's another example:

Here the meaning of "prey" could include specification of the available "prey" and then the meaning of the sentence would capture the survival success of the lion, given the nature of the available prey. All together, the resulting meaning is the result of the interaction between a particular hunter, a particular prey, and the intricacies of the hunting process, which may depend on the particular environment in which it is taking place. It should be clear that again this situation is radically non-Cartesian.

Of course, if we now consider the example of quantum theory from two sections ago, the analogues to grammatical types are system types, i.e. specifications of the kinds (including quantitie) of systems that are involved. So it makes sense to refine the grammatical types according to the subject domain. Just like nouns in physics would involve specification of the kinds of systems involved, in biology, for example, this could involve specification of species, population size, environment, availability of food, etc. Correspondingly, the top part would not just be restricted to grammatical interaction, but also include domain-specific interaction, just like in the case of quantum theory. All together, what we obtain is the following picture:

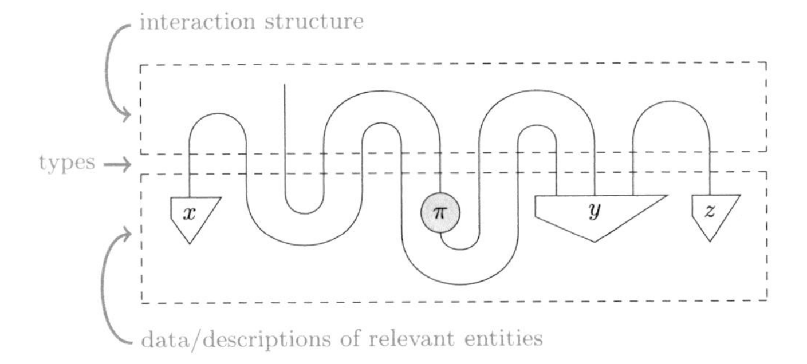

as a (very rough) template for a theory of everything.

Acknowledgements The extrapolation of meaning beyond natural language was prompted by my having to give a course in a workshop on Logics for Social Behaviour, organised by Alexander Kurz and Alessandra Palmigiano at the Lorentz Center in Leiden. The referee provided useful feedback—I learned a new word: 'foo'.

References

1. S. Abramsky, B. Coecke, A categorical semantics of quantum protocols, in *Proceedings of the 19th Annual IEEE Symposium on Logic in Computer Science (LICS)* (2004), pp. 415–425. arXiv:quant-ph/0402130
2. K. Ajdukiewicz, Die syntaktische konnexität. Stud. Philos. **1**, 1–27 (1937)
3. J.C. Baez, Quantum quandaries: a category-theoretic perspective, in *The Structural Foundations of Quantum Gravity*, ed. by D. Rickles, S. French, J.T. Saatsi (Oxford University Press, Oxford, 2006), pp. 240–266. arXiv:quant-ph/0404040
4. Y. Bar-Hillel, A quasiarithmetical notation for syntactic description. Language **29**, 47–58 (1953)
5. J. Benabou, Categories avec multiplication. C. R. Seances Acad. Sci. Paris **256**, 1887–1890 (1963)
6. C.H. Bennett, G. Brassard, C. Crepeau, R., Jozsa, A. Peres, W.K., Wootters, Teleporting an unknown quantum state via dual classical and Einstein-Podolsky-Rosen channels. Phys. Rev. Lett. **70**(13), 1895–1899 (1993)
7. F.G.S.L. Brandão, M. Horodecki, J. Oppenheim, J.M. Renes, R.W. Spekkens, The resource theory of quantum states out of thermal equilibrium. Phys. Rev. Lett. **111**, 250404 (2013)
8. S. Clark, B. Coecke, E. Grefenstette, S. Pulman, M. Sadrzadeh, A quantum teleportation inspired algorithm produces sentence meaning from word meaning and grammatical structure (2013). arXiv:1305.0556
9. B. Coecke, The logic of entanglement. An invitation, Technical Report RR-03-12, Department of Computer Science, Oxford University, 2003
10. B. Coecke, Kindergarten quantum mechanics, in *Quantum Theory: Reconsiderations of the Foundations III*, ed. by A. Khrennikov (AIP Press, New York, 2005), pp. 81–98. arXiv:quant-ph/0510032
11. B. Coecke, The logic of quantum mechanics – take II (2012). arXiv:1204.3458
12. B. Coecke, An alternative Gospel of structure: order, composition, processes, in *Quantum Physics and Linguistics. A Compositional, Diagrammatic Discourse*, ed. by C. Heunen, M. Sadrzadeh, E. Grefenstette (Oxford University Press, Oxford, 2013), pp. 1–22. arXiv:1307.4038
13. B. Coecke, A. Kissinger, *Picturing Quantum Processes. A First Course in Quantum Theory and Diagrammatic Reasoning* (Cambridge University Press, Cambridge, 2016)
14. B. Coecke, É.O. Paquette, Categories for the practicing physicist, in *New Structures for Physics*, ed. by B. Coecke. Lecture Notes in Physics (Springer, New York, 2011), pp. 167–271. arXiv:0905.3010
15. B. Coecke, S. Perdrix, Environment and classical channels in categorical quantum mechanics, in *Proceedings of the 19th EACSL Annual Conference on Computer Science Logic (CSL)*. Lecture Notes in Computer Science, vol. 6247 (2010), pp. 230–244. Extended version: arXiv:1004.1598
16. B. Coecke, É.O. Paquette, D. Pavlović, Classical and quantum structuralism, in *Semantic Techniques in Quantum Computation*, ed. by S. Gay, I. Mackie. (Cambridge University Press, Cambridge, 2010), pp. 29–69. arXiv:0904.1997
17. B. Coecke, M. Sadrzadeh, S. Clark, Mathematical foundations for a compositional distributional model of meaning, in *A Festschrift for Jim Lambek*, vol. 36, ed. by J. van Benthem, M. Moortgat, W. Buszkowski. Linguistic Analysis (2010), pp. 345–384. arxiv:1003.4394
18. B. Coecke, E. Grefenstette, M. Sadrzadeh, Lambek vs. Lambek: functorial vector space semantics and string diagrams for Lambek calculus. Ann. Pure Appl. Log. **164**, 1079–1100 (2013)
19. B. Coecke, T. Fritz, R.W. Spekkens, A mathematical theory of resources. Inf. Comput. (2014, to appear). arXiv:1409.5531
20. A. Einstein, B. Podolsky, N. Rosen, Can quantum-mechanical description of physical reality be considered complete? Phys. Rev. **47**(10), 777 (1935)

21. G. Gour, R.W. Spekkens, The resource theory of quantum reference frames: manipulations and monotones. New J. Phys. **10**, 033023 (2008)
22. E. Grefenstette, M. Sadrzadeh, Experimental support for a categorical compositional distributional model of meaning, in *The 2014 Conference on Empirical Methods on Natural Language Processing* (2011), pp. 1394–1404. arXiv:1106.4058
23. R. Horodecki, P. Horodecki, M. Horodecki, K. Horodecki, Quantum entanglement. Rev. Mod. Phys. **81**, 865–942 (2009). arXiv:quant-ph/0702225
24. D. Kartsaklis, Compositional distributional semantics with compact closed categories and frobenius algebras, Ph.D. thesis, University of Oxford, 2014
25. D. Kartsaklis, M. Sadrzadeh, Prior disambiguation of word tensors for constructing sentence vectors, in *The 2013 Conference on Empirical Methods on Natural Language Processing*, ACL (2013), pp. 1590–1601
26. L.H. Kauffman, Teleportation topology. Opt. Spectrosc. **99**, 227–232 (2005)
27. J. Lambek, The mathematics of sentence structure. Am. Math. Mon. **65**, 154–170 (1958)
28. J. Lambek, Type grammar revisited, in *Logical Aspects of Computational Linguistics*, vol. 1582 (Springer, New York, 1999)
29. J. Lambek, *From Word to Sentence* (Polimetrica, Milan, 2008)
30. M.D. Lieberman, *Social: Why our Brains are Wired to Connect* (Oxford University Press, Oxford, 2013)
31. R. Piedeleu, Ambiguity in categorical models of meaning, Master's thesis, University of Oxford, 2014
32. A. Preller, M. Sadrzadeh, Bell states and negative sentences in the distributed model of meaning. Electron. Notes Theor. Comput. Sci. **270**(2), 141–153 (2011)
33. M. Redei, Why John von Neumann did not like the Hilbert space formalism of quantum mechanics (and what he liked instead). Stud. Hist. Philos. Mod. Phys. **27**(4), 493–510 (1996)
34. M. Sadrzadeh, S. Clark, B. Coecke, The Frobenius anatomy of word meanings I: subject and object relative pronouns. J. Log. Comput. **23**, 1293–1317 (2013). arXiv:1404.5278
35. M. Sadrzadeh, S. Clark, B. Coecke, The Frobenius anatomy of word meanings II: possessive relative pronouns. J. Log. Comput. (2014). doi:10.1093/logcom/exu027
36. E. Schrödinger, Discussion of probability relations between separated systems. Camb. Philos. Soc. **31**, 555–563 (1935)
37. H. Schütze, Automatic word sense discrimination. Comput. Linguist. **24**(1), 97–123 (1998)
38. A.M. Turing, On computable numbers, with an application to the Entscheidungsproblem. Proc. Lond. Math. Soc. **42**, 230–265 (1937)
39. J. von Neumann, *Mathematische Grundlagen der Quantenmechanik* (Springer, Berlin, 1932). Translation, Mathematical Foundations of Quantum Mechanics (Princeton University Press, Princeton, 1955)

Part II
The Search for “Natural” Examples of Incomputable Objects

Some Recent Research Directions in the Computably Enumerable Sets

Peter A. Cholak

Abstract As suggested by the title, this paper is a survey of recent results and questions on the collection of computably enumerable sets under inclusion. This is not a broad survey but one focused on the author's and a few others' current research.

2000 *Mathematics Subject Classification.* Primary 03D25

There are many equivalent ways to define a computably enumerable or c.e. set. The one that we prefer is the domain of a Turing machine or the set of balls accepted by a Turing machine. Perhaps this definition is the main reason that this paper is included in this volume and the corresponding talk in the "Incomputable" conference. The c.e. sets are also the sets which are Σ^0_1 definable in arithmetic.

There is a computable or effective listing, $\{M_e | e \in \omega\}$, of all Turing machines. This gives us a listing of all c.e. sets: x is in W_e at stage s iff M_e with input x accepts by stage s. This enumeration of all c.e. sets is very dynamic. We can think of balls x as flowing from one c.e. set into another. Since they are sets, we can partially order them by inclusion, $\subseteq$, and consider them as models, $\mathcal{E} = \langle \{W_e | e \in \omega\}, \subseteq \rangle$. All sets (not just c.e. sets) are partially ordered by Turing reducibility, where $A \leq_T B$ iff there is a Turing machine that can compute A given an oracle for B.

Broadly, our goal is to study the structure $\mathcal{E}$ and learn what we can about the interplay between definability (in the language of inclusion $\subseteq$), the dynamic properties of c.e. sets and their Turing degrees. A very rich relationship between these three notions has been discovered over the years. We cannot hope to completely cover this history in this short paper. But, we hope that we will cover enough of it to show the reader that the interplay between these three notions on c.e. sets is, and will continue to be, a very interesting subject of research.

We are assuming that the reader has a background in computability theory as found in the first few chapters of [26]. All unknown notation also follows [26].

P.A. Cholak (✉)
Department of Mathematics, University of Notre Dame, Notre Dame, IN 46556-5683, USA
e-mail: Peter.Cholak.1@nd.edu; http://www.nd.edu/~cholak

S.B. Cooper, M.I. Soskova (eds.), *The Incomputable*, Theory and Applications of Computability, DOI 10.1007/978-3-319-43669-2_5

1 Friedberg Splits

The first result in this vein was [15]: every noncomputable c.e. set has a Friedberg split. Let us first understand the result and then explore why we feel it relates to the interplay of definability, Turing degrees and dynamic properties of c.e. sets.

Definition 1.1 $A_0 \sqcup A_1 = A$ is a *Friedberg split* of A iff, for all W (all sets in this paper are always c.e.), if $W - A$ is not a c.e. set, neither are $W - A_i$.

The following definition depends on the chosen enumeration of all c.e. sets. We use the enumeration given to us in the second paragraph of this paper, $x \in W_{e,s}$ iff M_e with input x accepts by stage s, but with the convention that if $x \in W_{e,s}$, then $e, x < s$ and, for all stages s, there is at most one pair e, x where x enters W_e at stage s. Some details on how we can effectively achieve this type of enumeration can be found in [26, Exercise I.3.11]. Moreover, when given a c.e. set, we are given the index of this c.e. set in terms of our enumeration of all c.e. sets. At times we will have to appeal to Kleene's Recursion Theorem to get this index.

Definition 1.2 For c.e. sets $A = W_e$ and $B = W_i$,

$$A \backslash B = \{x | \exists s [x \in (W_{e,s} - W_{i,s})]\}$$

and $A \searrow B = A \backslash B \cap B$.

By the above definition, $A \backslash B$ is a c.e. set. $A \backslash B$ is the set of balls that enter A before they enter B. If $x \in A \backslash B$ then x may or may not enter B, and if x does enter B, it only does so after x enters A (in terms of our enumeration). Since the intersection of two c.e. sets is c.e, $A \searrow B$ is a c.e. set. $A \searrow B$ is the c.e. set of balls x that first enter A and then enter B (under the above enumeration).

Note that $W \backslash A = (W - A) \sqcup (W \searrow A)$ ($\sqcup$ is the disjoint union). Since $W \backslash A$ is a c.e. set, if $W - A$ is not a c.e. set then $W \searrow A$ must be infinite. (This happens for all enumerations.) Hence infinitely many balls from W must flow into A.

Lemma 1.3 (Friedberg) *Assume $A = A_0 \sqcup A_1$, and, for all e, if $W_e \searrow A$ is infinite then both $W_e \searrow A_0$ and $W_e \searrow A_1$ are infinite. Then $A_0 \sqcup A_1$ is a Friedberg split of A. Moreover, if A is not computable, neither are A_0 and A_1.*

Proof Assume that $W - A$ is not a c.e. set but $X = W - A_0$ is a c.e. set. $X - A = W - A$ is not a c.e. set. So $X \searrow A$ is infinite and therefore $X \searrow A_0$ is infinite. Contradiction.

If A_0 is computable then $X = \overline{A_0}$ is a c.e. set and if A is not computable then $X - A$ cannot be a c.e. set. So use the same reasoning as above to show $X \searrow A_0$ is infinite for a contradiction. □

Friedberg more or less invented the priority method to split every c.e. set into two disjoint c.e. sets while meeting the hypothesis of the lemma above. The main idea of Friedberg's construction is to add a ball x to one of A_0 or A_1 when it enters A at stage s, but which set x enters is determined by priority. Let

$$\text{if } W_e \searrow A \text{ is infinite then } |W_e \searrow A_i| \geq k. \qquad \mathcal{P}_{e,i,k}$$

We say x meets $(\mathcal{P}_{e,i,k})$ at stage s if $|W_e \searrow A_i| < k$ by stage $s-1$ and if $|W_e \searrow A_i| \geq k$ by stage s if we add x to A_i. Find the highest $\langle e, i, k\rangle$ that x can meet and add x to A_i at stage s. It is not hard to show that all the $(\mathcal{P}_{e,i,k})$ are met.

It is clear that the existence of a Friedberg split is very dynamic. Let's see why it is also a definable property. But, first, we need to understand what we can say about $\mathcal{E}$ with inclusion. We are not going to go through the details, but we can define union, intersection, disjoint union, the empty set and the whole set. We can say that a set is complemented. A very early result shows that if A and $\overline{A}$ are both c.e. then A is computable. So it is definable if a c.e. set is computable. Inside every computable set, we can repeat the construction of the halting set. So a c.e. set X is finite iff every subset of X is computable. Hence $W - A$ is a c.e. set iff there is a c.e. set X disjoint from A such that $W \cup A = X \sqcup A$. So saying that $A_0 \sqcup A_1 = A$ is a Friedberg split and A is not computable is definable.

Friedberg's result answers a question of Myhill, "Is every non-recursive, recursively enumerable set the union of two disjoint non-recursive, recursively enumerable sets?" The question of Myhill was asked in print in the *Journal of Symbolic Logic* in June 1956, Volume 21, Number 2 on page 215 in the "Problems" section. This question was the eighth problem appearing in this section. The question about the existence of maximal sets, also answered by Friedberg, was ninth. This author does not know how many questions were asked or when this section was dropped. Myhill also reviewed [15] for the AMS, but the review left no clues why he asked the question in the first place.

The big question in computability theory in the 1950s was "Does there exist an incomplete noncomputable c.e. set"? Kleene and Post [20] showed that there are a pair of incomparable Turing degrees below $\mathbf{0}'$. We feel that after Kleene-Post, Myhill's question is very natural. So we can claim that the existence of a Friedberg split for every c.e. set A fits into our theme, the interplay of definability, dynamic properties and Turing degree on the c.e. sets.

2 Recent Work and Questions on Friedberg Splits

Given a c.e. set, one can uniformly find a Friedberg split. It is known that there are other types of splits. One wonders if any of these non-Friedberg splits can be done uniformly. It is also known that for some c.e. sets the only nontrivial splits ($A = A_0 \sqcup A_1$ and the A_0 and A_1 are not computable) are Friedberg. So one cannot hope to get a uniform procedure which always provides a nontrivial non-Friedberg split of every noncomputable c.e. set. But it would be nice to find a computable function $f(e) = \langle e_0, e_1\rangle$ such that, for all e, if W_e is noncomputable then $W_{e_0} \sqcup W_{e_1} = W_e$ is a nontrivial split of W_e and, for every c.e. set A, if A has a nontrivial non-Friedberg split and $A = W_e$ (so W_e is any enumeration of A), then $W_{e_0} \sqcup W_{e_1} = W_e$ is a nontrivial non-Friedberg split. So, if A has a nontrivial non-Friedberg split and W_e is any enumeration of A, f always gives out a nontrivial non-Friedberg split. In work yet to appear, the author has shown that such a computable f cannot exist.

Let $\mathcal{P}$ be a property in $\mathcal{E}$. We say that A is *hemi-*$\mathcal{P}$ iff there are c.e. sets B and C such that $A \sqcup B = C$ and C has $\mathcal{P}$. We can also define *Friedberg-*$\mathcal{P}$ iff there are c.e. sets B and C such that $A \sqcup B = C$ is a Friedberg split and C has $\mathcal{P}$. If $\mathcal{P}$ is definable then *hemi-*$\mathcal{P}$ and *Friedberg-*$\mathcal{P}$ are also definable. One can get lots of mileage from the *hemi-*$\mathcal{P}$; see [10] and [11]. Most of these results are about properties $\mathcal{P}$ where every nontrivial split of a set with $\mathcal{P}$ is Friedberg. We feel that one should be using *Friedberg-*$\mathcal{P}$ rather than *hemi-*$\mathcal{P}$. To that end, we ask the following:

Question 2.4 Is there a definable $\mathcal{P}$ such that the Friedberg splits are a proper subclass of the nontrivial splits?

We feel that the Friedberg splits are very special and they should not be able to always cover all the nontrivial splits of every definable property.

3 All Orbits Nice? No!

As we mentioned earlier, Friedberg also constructed a maximal set answering another question of Myhill. A maximal set, M, is a c.e. set such that for every superset X either $X =^* M$ ($=^*$ is equal modulo finite) or $W =^* \omega$. Being maximal is definable. Friedberg's construction of a maximal set is very dynamic. Martin [23] showed that all maximal sets must be high. A further result of [23] shows that a c.e. degree is high iff it contains a maximal set. A remarkable result of [24] shows that the maximal sets form an orbit, even an orbit under automorphisms computable from $\mathbf{0}''$ or Δ^0_3-automorphisms.

The result of Soare gives rise to the question, are all orbits as nice as the orbit of the maximal sets? We can go more into the formality of the question, but that was dealt with already in another survey paper [6]. To tell if two c.e. sets, A and B, are in the same orbit, it is enough to show if there is an automorphism Φ of $\mathcal{E}$ taking the one to the other, $\Phi(A) = B$ (we write this as A is *automorphic* to B). Hence it is Σ^1_1 to tell whether two sets are in the same orbit. The following theorem says that is the best that we can do and hence not all orbits are as nice as the orbits of maximal sets. The theorem has a number of interesting corollaries.

Theorem 3.1 (Cholak et al. [7]) *There is a c.e. set A such that the index set $\{i : W_i \approx A\}$ is Σ^1_1-complete.*

Corollary 3.2 (Cholak et al. [7]) *Not all orbits are elementarily definable; there is no arithmetic description of all orbits of $\mathcal{E}$.*

Corollary 3.3 (Cholak et al. [7]) *The Scott rank of $\mathcal{E}$ is $\omega_1^{\mathrm{CK}} + 1$.*

Theorem 3.4 (Cholak et al. [7]) *For all finite $\alpha > 8$, there is a properly Δ^0_α orbit.*

These results were completely explored in the survey [6]. So we will focus on some more recent work. In the work leading to the above theorems, Cholak and Harrington also showed that:

Theorem 3.5 (Cholak and Harrington [4]) *Two simple sets are automorphic iff they are Δ^0_6 automorphic. A set A is* simple *iff for every (c.e.) set B, if $A \cap B$ is empty then B is finite.*

Recently Harrington improved this result to show:

Theorem 3.6 (Harrington 2012, Private Email) *The complexity of the $\mathcal{L}_{\omega_1,\omega}$ formula describing the orbit of any simple set is very low (close to 6).*

That leads us to make the following conjecture:

Conjecture 3.7 We can build the above orbits in Theorem 3.4 to have complexity close to α in terms of the $\mathcal{L}_{\omega_1,\omega}$ formula describing the orbit.

4 Complete Sets

Perhaps the biggest questions on the c.e. sets are the following:

Question 4.1 (Completeness) Which c.e. sets are automorphic to complete sets?

The Motivation for this question dates back to Post. Post was trying to use properties of the complement of a c.e. set to show that the set was not complete. In the structure $\mathcal{E}$, all the sets in the same orbit have the same definable properties.

By Harrington and Robert [16–18], we know that not every c.e. set is automorphic to a complete set and, furthermore, there is a dichotomy between the "prompt" sets and the "tardy" (nonprompt) sets, with the "prompt" sets being automorphic to complete sets. We will explore this dichotomy in more detail, but more definitions are needed:

Definition 4.2 $X = (W_{e_1} - W_{e_2}) \cup (W_{e_3} - W_{e_4}) \cup \ldots (W_{e_{2n-1}} - W_{e_{2n}})$ iff X is $2n$-c.e., and X is $2n+1$-c.e. iff $X = Y \cup W_e$, where Y is $2n$-c.e.

Definition 4.3 Let X^n_e be the eth n-c.e. set. A is *almost prompt* iff there is a computable nondecreasing function $p(s)$ such that, for all e and n, if $X^n_e = \overline{A}$ then $(\exists x)(\exists s)[x \in X^n_{e,s}$ and $x \in A_{p(s)}]$.

Theorem 4.4 (Harrington and Soare [17]) *Each almost prompt set is are automorphic to some complete set.*

Definition 4.5 D is *2-tardy* iff for every computable nondecreasing function $p(s)$ there is an e such that $X^2_e = \overline{D}$ and $(\forall x)(\forall s)[\text{if } x \in X^2_{e,s} \text{ then } x \notin D_{p(s)}]$

Theorem 4.6 (Harrington and Soare [16]) *There are $\mathcal{E}$ definable properties $Q(D)$ and $P(D, C)$ such that*

(1) *$Q(D)$ implies that D is 2-tardy and hence the orbit of D does not contain a complete set;*
(2) *for D, if there is a C such that $P(D, C)$ and D is 2-tardy, then $Q(D)$ (and D is high).*

The 2-tardy sets are not almost prompt and the fact they are not almost prompt is witnessed by $e = 2$. It would be nice if the above theorem implied that being 2-tardy was definable. But it says with an extra definable condition that being 2-tardy is definable.

Harrington and Soare [16] ask if each 3-tardy set is computable by some 2-tardy set. They also ask if all low$_2$ simple sets are almost prompt (this is the case if A is low). With Gerdes and Lange, Cholak answered these negatively:

Theorem 4.7 (Cholak et al. [8]) *There exists a properly 3-tardy B such that there is no 2-tardy A such that $B \leq_T A$. Moreover, B can be built below any prompt degree.*

Theorem 4.8 (Cholak et al. [8]) *There is a low$_2$, simple 2-tardy set.*

Moreover, with Gerdes and Lange, Cholak showed that there are definable (first-order) properties $Q_n(A)$ such that if $Q_n(A)$, then A is n-tardy and there is a properly n-tardy set A such that $Q_n(A)$ holds. Thus the collection of all c.e. sets not automorphic to a complete set breaks up into infinitely many orbits.

But, even with the work above, the main question about completeness and a few others remain open. These open questions are of a more degree-theoretic flavor. The main questions still open are:

Question 4.9 (Completeness) Which c.e. sets are automorphic to complete sets?

Question 4.10 (Cone Avoidance) Given an incomplete c.e. degree $\mathbf{d}$ and an incomplete c.e. set A, is there an $\hat{A}$ automorphic to A such that $\mathbf{d} \not\leq_T \hat{A}$?

It is unclear whether these questions have concrete answers. Thus the following seems reasonable.

Question 4.11 Are these arithmetical questions?

Let us consider how we might approach these questions. One possible attempt would be to modify the proof of Theorem 3.1 to add degree-theoretic concerns. Since the coding comes from how A interacts with the sets disjoint from it, we should have reasonable degree-theoretic control over A. The best we have been able to do so far is alter Theorem 3.1 so that the set constructed has hemimaximal degree and everything in its orbit also has hemimaximal degree. However, what is open is whether the orbit of any set constructed via Theorem 3.1 must contain a representative of every hemimaximal degree or only have hemimaximal degrees. If the infinite join of hemimaximal degrees is hemimaximal then the degrees of the sets in these orbits only contain the hemimaximal degrees. But, it is open whether the infinite join of hemimaximal degrees is hemimaximal.

5 Tardy Sets

As mentioned above, there are some recent results on n-tardy and very tardy sets (a set is very tardy iff it is not almost prompt). But there are several open questions related to this work. For example, is there a (first-order) property Q_∞ such that if $Q_\infty(A)$ holds, then A is very tardy (or n-tardy, for some n)? Could we define Q_∞ such that $Q_n(A) \implies Q_\infty(A)$? How do hemi-$Q$ and Q_3 compare? But the big open questions here are the following:

Question 5.12 Is the set B constructed in Theorem 4.7 automorphic to a complete set? If not, does $Q_3(B)$ hold?

It would be very interesting if both of the above questions have a negative answer.

Not a lot about the degree-theoretic properties of the n-tardies is known. The main question here is whether Theorem 4.7 can be improved to n other than 2.

Question 5.13 For which n are there $(n + 1)$-tardies which are not computed by n-tardies?

But there are many other approachable questions. For example, how do the following sets of degrees compare:

- the hemimaximal degrees?
- the tardy degrees?
- for each n, $\{\mathbf{d} : \text{there is an } n\text{-tardy } D \text{ such that } \mathbf{d} \leq_T D\}$?
- $\{\mathbf{d} : \text{there is a 2-tardy } D \text{ such that } Q(D) \text{ and } \mathbf{d} \leq_T D\}$?
- $\{\mathbf{d} : \text{there is an } A \in \mathbf{d} \text{ which is not automorphic to a complete set}\}$?

Does every almost prompt set compute a 3-tardy? Or a very tardy? Harrington and Soare [18] show there is a maximal 2-tardy set. So there are 2-tardy sets which are automorphic to complete sets. Is there a non-high, nonhemimaximal, 2-tardy set which is automorphic to a complete set?

6 Cone Avoidance, Question 4.10

The above prompt vs. tardy dichotomy gives rise to a reasonable way to address Question 4.10. An old result of [1] and, independently [17] says that every c.e. set is automorphic to a high set. Hence, a positive answer to both the following questions would answer the cone avoidance question but not the completeness question. These questions seem reasonable as we know how to work with high degrees and automorphisms; see [1],

Question 6.14 Let A be incomplete. If the orbit of A contains a set of high prompt degree, must the orbit of A contain a set from all high prompt degrees?

Question 6.15 If the orbit of A contains a set of high tardy degree, must the orbit of A contain a set from all high tardy degrees?

Similarly, we know how to work with prompt degrees and automorphisms; see [5] and [17]. We should be able to combine the two. No one has yet explored how to work with automorphisms and tardy degrees.

7 $\mathcal{D}$-Maximal Sets

In the above sections we have mentioned maximal and hemimaximal sets several times. It turns out that maximal and hemimaximal sets are both $\mathcal{D}$-maximal.

Definition 7.1 $\mathcal{D}(A) = \{B : \exists W(B \subseteq A \cup W \text{ and } W \cap A = \emptyset)\}$ under inclusion. Let $\mathcal{E}_{\mathcal{D}(A)}$ be $\mathcal{E}$ modulo $\mathcal{D}(A)$.

$\mathcal{D}(A)$ is the ideal of c.e. sets of the form $\tilde{A} \sqcup \tilde{D}$, where $\tilde{A} \subseteq A$ and $\tilde{D} \cap A = \emptyset$.

Definition 7.2 A is *$\mathcal{D}$-hhsimple* iff $\mathcal{E}_{\mathcal{D}(A)}$ is a Σ^0_3 Boolean algebra. A is *$\mathcal{D}$-maximal* iff $\mathcal{E}_{\mathcal{D}(A)}$ is the trivial Boolean algebra iff for all c.e. sets B, there is a c.e. set D disjoint from A such that either $B \subset A \cup D$ or $B \cup D \cup A = \omega$.

Maximal sets and hemimaximal sets are $\mathcal{D}$-maximal. Plus, there are many other examples of $\mathcal{D}$-maximal sets. In fact, with the exception of the creative sets, all known elementary definable orbits are orbits of $\mathcal{D}$-maximal sets. In the lead up to Theorem 3.1, Cholak and Harrington were able to show:

Theorem 7.3 (Cholak and Harrington [3]) *If A is $\mathcal{D}$-hhsimple and A and $\hat{A}$ are in the same orbit, then $\mathcal{E}_{\mathcal{D}(A)} \cong_{\Delta^0_3} \mathcal{E}_{\mathcal{D}(\hat{A})}$.*

So it is an arithmetic question to ask whether the orbit of a $\mathcal{D}$-maximal set contains a complete set. But the question remains, does the orbit of every $\mathcal{D}$-maximal set contain a complete set? It was hoped that the structural properties of $\mathcal{D}$-maximal sets would be sufficient to allow us to answer this question.

Cholak et al. [9] have completed a classification of all $\mathcal{D}$-maximal sets. The idea is to look at how $\mathcal{D}(A)$ is generated. For example, for a hemimaximal set A_0, $\mathcal{D}(A_0)$ is generated by A_1, where $A_0 \sqcup A_1$ is maximal. There are ten different ways that $\mathcal{D}(A)$ can be generated. Seven were previously known and all these orbits contain complete and incomplete sets. Work from Herrmann and Kummer [19] shows that these seven types are not enough to provide a characterization of all $\mathcal{D}$-maximal sets. Cholak, Gerdes, and Lange construct three more types and show that these ten types provide a characterization of all $\mathcal{D}$-maximal sets. We have constructed three new types of $\mathcal{D}$-maximal sets; for example, a $\mathcal{D}$-maximal set where $\mathcal{D}(A)$ is generated by infinitely many not disjoint c.e sets. We show these three types, plus another, split into infinitely many different orbits. We can build examples of these sets which are incomplete or complete. But, it is open whether each such orbit contains a complete set. So, the structural properties of $\mathcal{D}$-maximal sets were not enough to determine if each $\mathcal{D}$-maximal set is automorphic to a complete set.

It is possible that one could provide a similar characterization of the $\mathcal{D}$-hhsimple sets. One should fix a Σ^0_3 Boolean algebra, $\mathcal{B}$, and characterize the $\mathcal{D}$-hhsimple sets, A, where $\mathcal{E}_{\mathcal{D}(A)} \cong \mathcal{B}$. It would be surprising if, for some $\mathcal{B}$, the characterization would allow us to determine if every orbit of these sets contains a complete set.

8 Lowness

Following his result that the maximal sets form an orbit [25] showed that the low sets resemble computable sets. A set A is low$_n$ iff $\mathbf{0}^{(n)} \equiv_T A^{(n)}$. We know that noncomputable low sets cannot have a computable set in their orbit, so the best that Soare was able to do is the following:

Definition 8.1 $\mathcal{L}(A)$ are the c.e. supersets of A under inclusion. $\mathcal{F}$ is the filter of finite sets. $\mathcal{L}^*(A)$ is $\mathcal{L}(A)$ modulo $\mathcal{F}$.

Theorem 8.2 (Soare [25]) *If A is low then $\mathcal{L}^*(A) \approx \mathcal{L}^*(\emptyset)$.*

In 1990, Soare conjectured that this can be improved to low$_2$. Since then there have been a number of related results but this conjecture remains open. To move forward, some definitions are needed:

Definition 8.3 A is *semilow* iff $\{i | W_i \cap A \neq \emptyset\}$ is computable from $\mathbf{0}'$. A is *semilow*$_{1.5}$ iff $\{i | W_i \cap A \text{ is finite}\} \leq_1 \mathbf{0}''$. A is *semilow*$_2$ iff $\{i | W_i \cap A \text{ is finite}\}$ is computable from $\mathbf{0}''$.

Semilow implies semilow$_{1.5}$ implies semilow$_2$, if A is low then $\overline{A}$ is semilow, and low$_2$ implies semilow$_2$ (details can be found in [22] and [1]). Soare [25] actually showed that if $\overline{A}$ is semilow then $\mathcal{L}^*(A) \approx \mathcal{L}^*(\emptyset)$. Maass [22] improved this to when $\overline{A}$ is semilow$_{1.5}$.

In Maass proof, semilow$_{1.5}$ness is used in two ways: A c.e. set, W, is *well-resided outside* A iff $W \cap \overline{A}$ is infinite. Semilow$_{1.5}$ makes determining which sets are well-resided outside A a Π^0_2 question. The second use of semilow$_{1.5}$ was to capture finitely many elements of $W \cap \overline{A}$. For that, Maass showed that semilow$_{1.5}$ implies the *outer splitting property*:

Definition 8.4 A has the *outer splitting property* iff there are computable functions f, h such that, for all e, $W_e = W_{f(e)} \sqcup W_{h(e)}$, $W_{f(e)} \cap \overline{A}$ is finite, and if $W_e \cap \overline{A}$ is infinite then $W_{f(e)} \cap \overline{A}$ is nonempty.

Cholak used these ideas to show the following:

Theorem 8.5 (Cholak [1]) *If A has the outer splitting property and $\overline{A}$ is semilow$_2$, then $\mathcal{L}^*(A) \approx \mathcal{L}^*(\emptyset)$.*

It is known that there is a low$_2$ set which does not have the outer splitting property; see [12, Theorem 4.6]. So to prove that if A is low$_2$ then $\mathcal{L}^*(A) \approx \mathcal{L}^*(\emptyset)$

will require a different technique. However [21] showed that every low_2 set has a maximal superset using the technique of *true stages*. Perhaps the true stages technique can be used to show Soare's conjecture.

Recently, there has been a result by Epstein.

Theorem 8.6 (Epstein [13] and [14]) *There is a properly* low_2 *degree* $\mathbf{d}$ *such that if* $A \leq_T \mathbf{d}$ *then* A *is automorphic to a low set.*

Epstein's result shows that there is no collection of c.e. sets which is closed under automorphisms and contains at least one set of every nonlow degree. Related results were discussed in [2].

This theorem does have a nice yet unmentioned corollary: The collection of all sets A such that $\overline{A}$ is semilow (these sets are called *speedable*) is not definable. By Downey et al. [12, Theorem 4.5], every nonlow c.e. degree contains a set A such that $\overline{A}$ is not $\mathrm{semilow}_{1.5}$ and hence not semilow. So there is such a set A in $\mathbf{d}$. A is automorphic to a low set $\hat{A}$. Since $\hat{A}$ is low, $\overline{\hat{A}}$ is semilow.

Esptein's result leads us wonder if the above results can be improved as follows:

Conjecture 8.7 (Soare) Every semilow set is (effectively) automorphic to a low set.

Conjecture 8.8 (Cholak and Epstein) Every set A such that A has the outer splitting property and $\overline{A}$ is $\mathrm{semilow}_2$ is automorphic to a low_2 set.

Cholak and Epstein are currently working on a proof of the latter conjecture and some related results. Hopefully, a draft will be available soon.

References

1. P. Cholak, Automorphisms of the lattice of recursively enumerable sets. Mem. Am. Math. Soc. **113**(541), viii+151 (1995). ISSN: 0065-9266
2. P. Cholak, L.A. Harrington, Definable encodings in the computably enumerable sets. Bull. Symb. Log. **6**(2), 185–196 (2000). ISSN: 1079-8986
3. P. Cholak, L.A. Harrington, Isomorphisms of splits of computably enumerable sets. J. Symb. Log. **68**(3), 1044–1064 (2003). ISSN: 0022-4812
4. P. Cholak, L.A. Harrington, Extension theorems, orbits, and automorphisms of the computably enumerable sets. Trans. Am. Math. Soc. **360**(4), 1759–1791 (2008). ISSN: 0002-9947, math.LO/0408279
5. P. Cholak, R. Downey, M. Stob, Automorphisms of the lattice of recursively enumerable sets: promptly simple sets. Trans. Am. Math. Soc. **332**(2), 555–570 (1992). ISSN: 0002-9947
6. P. Cholak, R. Downey, L.A. Harrington, The complexity of orbits of computably enumerable sets. Bull. Symb. Log. **14**(1), 69–87 (2008). ISSN: 1079-8986
7. P. Cholak, R. Downey, L.A. Harrington, On the orbits of computably enumerable sets. J. Am. Math. Soc. **21**(4), 1105–1135 (2008). ISSN: 0894-0347
8. P. Cholak, P.M. Gerdes, K. Lange, On n-tardy sets. Ann. Pure Appl. Log. **163**(9), 1252–1270 (2012). ISSN: 0168-0072, doi:10.1016/j.apal.2012.02.001. http://dx.doi.org/10.1016/j.apal.2012.02.001
9. P. Cholak, P. Gerdes, K. Lange, The $\mathcal{D}$-maximal sets. J. Symbolic Logic **80**(4), 1182–1210 (2015). doi:10.1017/jsl.2015.3

10. R.G. Downey, M. Stob, Automorphisms of the lattice of recursively enumerable sets: orbits. Adv. Math. **92**, 237–265 (1992)
11. R. Downey, M. Stob, Splitting theorems in recursion theory. Ann. Pure Appl. Log. **65**(1), 106 pp. (1993). ISSN: 0168-0072
12. R.G. Downey, C.G. Jockusch Jr., P.E. Schupp, Asymptotic density and computably enumerable sets. arXiv.org, June 2013
13. R. Epstein, Invariance and automorphisms of the computably enumerable sets, Ph.D. thesis, University of Chicago, 2010
14. R. Epstein, The nonlow computably enumerable degrees are not invariant in $\mathcal{E}$. Trans. Am. Math. Soc. **365**(3), 1305–1345 (2013). ISSN: 0002-9947, doi:10.1090/S0002-9947-2012-05600-5, http://dx.doi.org/10.1090/S0002-9947-2012-05600-5
15. R.M. Friedberg, Three theorems on recursive enumeration. I. Decomposition. II. Maximal set. III. Enumeration without duplication. J. Symb. Log. **23**, 309–316 (1958)
16. L.A. Harrington, R.I. Soare, Post's program and incomplete recursively enumerable sets. Proc. Natl. Acad. Sci. U.S.A. **88**, 10242–10246 (1991)
17. L.A. Harrington, R.I. Soare, The Δ_3^0-automorphism method and noninvariant classes of degrees. J. Am. Math. Soc. **9**(3), 617–666 (1996). ISSN: 0894-0347
18. L. Harrington, R.I. Soare, Codable sets and orbits of computably enumerable sets. J. Symb. Log. **63**(1), 1–28 (1998). ISSN: 0022-4812
19. E. Herrmann, M. Kummer, Diagonals and $\mathcal{D}$-maximal sets. J. Symb. Log. **59**(1), 60–72 (1994). ISSN: 0022-4812, doi:10.2307/2275249, http://dx.doi.org/10.2307/2275249
20. S.C. Kleene, E.L. Post, The upper semi-lattice of degrees of recursive unsolvability. Ann. Math. (2) **59**, 379–407 (1954)
21. A.H. Lachlan, Degrees of recursively enumerable sets which have no maximal supersets. J. Symb. Log. **33**, 431–443 (1968)
22. W. Maass, Characterization of recursively enumerable sets with supersets effectively isomorphic to all recursively enumerable sets. Trans. Am. Math. Soc. **279**, 311–336 (1983)
23. D.A. Martin, Classes of recursively enumerable sets and degrees of unsolvability. Z. Math. Logik Grundlag. Math. **12**, 295–310 (1966)
24. R.I. Soare, Automorphisms of the lattice of recursively enumerable sets I: maximal sets. Ann. Math. (2) **100**, 80–120 (1974)
25. R.I. Soare, Automorphisms of the lattice of recursively enumerable sets II: low sets. Ann. Math. Log. **22**, 69–107 (1982)
26. R.I. Soare, *Recursively Enumerable Sets and Degrees*. Perspectives in Mathematical Logic, Omega Series (Springer, Heidelberg, 1987)

Uncomputability and Physical Law

Seth Lloyd

Abstract This paper investigates the role that uncomputability plays in the laws of physics. While uncomputability might seem like an abstract mathematical concept, many questions in physics involve uncomputability. In particular, the form of the energy spectrum of quantum systems capable of universal computation is uncomputable: to answer whether a Hamiltonian system is gapped or gapless in a particular sector requires one to solve the Halting problem. Finally, the problem of determining the most concise expression of physical laws requires one to determine the algorithmic complexity of those laws, and so, the answer to this problem is uncomputable.

Computers can do so much that it's easy to forget that they were invented for what they could not do. In his 1937 paper, "On Computable Numbers, with an Application to the *Entscheidungsproblem,*" Alan Turing defined the notion of a universal digital computer (a Turing machine), which became the conceptual basis for the contemporary electronic computer [1]. Turing's goal was to show that there were tasks that even the most powerful computing machine could not perform. In particular, Turing showed that no Turing machine could solve the problem of whether a given Turing machine would halt and give an output when programmed with a given input. In creating a computational analogue to Gödel's incompleteness theorem [2], Turing introduced the concept of uncomputability.

Although Turing's machine was an abstract mathematical construct, his ideas soon found implementation as physical devices [3]. At the end of the 1930s, Konrad Zuse in Germany began building digital computers, first mechanical (the Z-1), and then electrical (the Z-3), and Claude Shannon's 1937 Master's thesis showed how digital computers could be constructed using electronic switching circuits [4], a development that presaged the construction of the British code-breaking electronic computer, the Colossus, and the American Mark I. Since computers

S. Lloyd (✉)
Mechanical Engineering, Massachusetts Institute of Technology,
Cambridge, MA, USA
e-mail: slloyd@mit.edu

S.B. Cooper, M.I. Soskova (eds.), *The Incomputable*, Theory and Applications of Computability, DOI 10.1007/978-3-319-43669-2_6

are physical systems, Turing's uncomputability results show that there are well-formulated questions that can be asked about physical systems whose answers are uncomputable.

That is, there are questions that we can ask about the physical behavior of the universe whose answers are not resolvable by any finite computation. In this chapter, I will investigate the question of how such questions permeate the fabric of physical law. I will show that answers to some of the most basic questions concerning physical law are in fact uncomputable. In particular, one of the primary driving forces of science in general and of physics in particular is to create the most succinct formulation of the natural laws. However, straightforward results from algorithmic information theory imply that the most concise formulation of the fundamental laws of nature is in fact uncomputable.

Many physical systems are capable of universal computation: indeed, it is difficult to find an extended system with nonlinear interactions that is *not* capable of universal computation given proper initial conditions and inputs [5–8]. For such systems, there are always questions that one can ask about their physical dynamics whose answers are uncomputable. One might hope that such questions are sufficiently esoteric that they do not overlap with the usual questions that physicists ask of their systems. I will show that even the answers to such basic questions are often uncomputable. In particular, consider the commonly asked question of whether a physical system has an energy gap—so that its energy spectrum is discrete in the vicinity of its ground state—or whether it is gapless—so that the spectrum is continuous. If the system is capable of universal computation, then the answer to the question of whether it has an energy gap or not is uncomputable.

The uncomputability of the answers to common and basic questions of physics might seem to be a serious hindrance to doing science. To the contrary, the underlying uncomputability of physical law simply puts physicists in the same position that mathematicians have occupied for some time: many quantities of interest can be computed, but not all. For example, even though the most concise formulation of the underlying laws of physics is uncomputable, short and elegant formulations of physical laws certainly exist. Not knowing in advance whether or not the quantity that one is trying to compute is uncomputable reflects the shared experience of all scientists: one never knows when the path of one's research will become impassible. The underlying uncomputability of physical law simply adds zest and danger to an already exciting quest.

1 The Halting Problem

We start by reviewing the origins of uncomputability. Like Gödel's incompleteness theorem [2], Turing's halting problem has its origins in the paradoxes of self-reference. Gödel's theorem can be thought of as a mathematization of the Cretan liar paradox. The sixth-century BC Cretan philosopher Epimenides is said to have declared that all Cretans are liars. (As Saint Paul says of the Cretans in his letter to Titus 1:12: 'One of themselves, even a prophet of their own, said, The Cretans are alway liars, evil beasts, slow bellies. This witness is true.') More concisely, the fourth-century BC Miletian philosopher Eubulides, stated the paradox as 'A man says, "What I am saying is a lie."' The paradox arises because if the man is telling the truth, then he is lying, while if he is lying, then he is telling the truth.

Potential resolutions of this paradox had been discussed for more than two millenia by the time that Kurt Gödel constructed his mathematical elaboration of the paradox in 1931. Gödel developed a method for assigning a natural number or 'Gödel number' to each well-formed statement or formula in a mathematical theory or formal language. He then constructed theories that contained statements that were versions of the liar paradox ('this statement cannot be proved to be true') and used Gödel numbering to prove that such a theory must be either inconsistent (false statements can be proved to be true) or incomplete: there exist statements that are true but that cannot be proved to be true within the theory.

Gödel's original incompleteness theorems were based on formal language theory. Turing's contribution was to re-express these theorems in terms of a mechanistic theory of computation. The Turing machine is an abstract computing machine that is based on how a mathematician calculates. Turing noted that mathematicians think, write down formulae on sheets of paper, return to earlier sheets to look at and potentially change the formulae there, and pick up new sheets of paper on which to write. A Turing machine consists of a 'head,' a system with a finite number of discrete states that is analogues to the mathematician, and a 'tape,' a set of squares each of which can either be blank or contain one of finite number of symbols, analogous to the mathematician's sheets of paper. The tape originally contains a finite number of non-blank squares, the program, which specify the computation to be performed. The head is prepared in a special 'start' state and placed at a specified square on the tape. Then as a function of the symbol on the square and of its own internal state, the head can alter the symbol, change its own internal state, and move to the square on the left or right. The machine starts out, and like a mathematician, the head reads symbols, changes them, moves on to different squares/sheets of paper, writes and erases symbols there, and moves on. The computation halts when the head enters a special 'stop' state, at which point the tape contains the output to the computation. If the head never enters the stop state, then computation never halts.

Turing's definition of an abstract computing machine is both simple and subtle. It is simple because the components (head and squares of tape) are all discrete and finite, and the computational dynamics are finitely specified. In fact, a Turing machine whose head has only two states and whose squares have only three states can be universal in the sense that it can simulate any other Turing machine. The definition is subtle because the computation can potentially continue forever, so that the number of squares of tape covered with symbols can increase without bound. One of the simplest questions that one can ask about a Turing machine is whether the machine, given a particular input, will ever halt and give an output, or whether it will continue to compute forever. Turing showed that no Turing machine can compute the answer to this question for all Turing machines and for all inputs: the answer to the question of whether a given Turing machine will halt on a given input is uncomputable.

Like the proof of Gödel's incompleteness theorems, Turing's proof of the halting problem relies on the capacity of Turing machines for self-reference. Because of their finite nature, one can construct a Gödel numbering for the set of Turing machines. Similarly, each input for a Turing machine can be mapped to a natural number. Turing showed that a Turing machine that computes whether a generic Turing machine halts given a particular input cannot exist. The self-referential part of the proof consists of examining what happens when this putative Turing machine evaluates what happens when a Turing machine is given its own description as input. If such a Turing machine exists, then we can define another Turing machine that halts only if it doesn't halt, and fails to halt only if it halts, a computational analogue to the Cretan liar paradox (Saint Paul: 'damned slow-bellied Turing machines').

The only assumption that led to this self-contradiction is the existence of a Turing machine that calculates whether Turing machines halt or not. Accordingly, such a Turing machine does not exist: the question of whether a Turing machine halts or not given a particular input is uncomputable. This doesn't mean that one can't compute some of the time whether Turing machines halt or not: indeed, you can just let a particular Turing machine run, and see what happens. Sometimes it will halt and sometimes it won't. The fact that it hasn't halted yet even after a very long time makes it unlikely to halt, but it doesn't mean that it won't halt someday. To paraphrase Abraham Lincoln, you can tell whether some Turing machines halt all the time, and you can tell whether all Turing machines halt some of the time, but you can't tell whether all Turing machines halt all of the time.

2 The Halting Problem and the Energy Spectrum of Physical Systems

At first it might seem that the halting problem, because of its abstract and paradoxical nature, might find little application in the description of physical systems. I will now show, to the contrary, that the halting problem arises in the computation of basic features of many physical systems. The key point to keep in mind is that even quite simple physical systems can be capable of universal digital computation. For example, the Ising model is perhaps the simplest model of a set of coupled spins, atoms, quantum dots, or general two-level quantum systems (qubits). Arbitrary logical circuits can effectively be written into the ground state of the inhomogeneous Ising model, showing that the Ising model is capable of universal computation in the limit that the number of spins goes to infinity [7]. When time-varying fields such as the electromagnetic field are added, the number of systems capable of universal computation expands significantly: almost any set of interacting quantum systems is capable of universal computation when subject to a global time-varying field [8].

The ubiquity of computing systems means that the answers to many physical questions are in fact uncomputable. One of the most useful questions that one can ask of a physical system is what are the possible values for the system's energy, i.e., what is the system's spectrum. A system's spectrum determines many if not most of the features of the system's dynamics and thermodynamics. In particular, an important question to ask is whether a system's spectrum is discrete, consisting of well-separated values for the energy, or continuous.

For example, if there is a gap between the lowest or 'ground state' energy and the next lowest or 'first excited state' energy, then the spectrum is discrete and the system is said to possess an energy gap. By contrast, if the spectrum is continuous in the vicinity of the ground state, then the system is said to be gapless. Gapped and gapless systems exhibit significantly different dynamic and thermodynamic behavior. In a gapped system, for example, the fundamental excitations or particles are massive, while in a gapless system they are massless. In a gapped system, the entropy goes to zero as the temperature goes to zero, while for a gapless system, the entropy remains finite, leading to significantly different thermodynamic behavior in the low temperature limit.

As will now be seen, if a physical system is capable of universal computation, the answer to the question of whether a particular part of its spectrum is discrete or continuous is uncomputable. In 1986, the Richard Feynman exhibited a simple quantum system whose whose dynamics encodes universal computation [9]. Feynman's system consists of a set of two-level quantum systems (qubits) coupled to a clock. The computation is encoded as a sequence of interactions between qubits. Every time an interaction is performed, the clock 'ticks' or increments by 1. In a computation that halts, the computer starts in the initial program state and then explores a finite set of states. In a computation that doesn't halt, the clock keeps on ticking forever, and the computer explores an infinite set of states. In 1992, I showed that this feature of Feynman's quantum computer implied that halting programs correspond to discrete sectors of the system's energy spectrum, while non-halting programs correspond to continuous sectors of the system's spectrum [10]. In particular, when the system has been programmed to perform a particular computation, the question of whether its spectrum has a gap or not is equivalent to the question of whether the computer halts or not.

More precisely, a non-halting computer whose clock keeps ticking forever corresponds to a physical system that goes through a countably infinite sequence of distinguishable states. Such a system by necessity has a continuum of energy eigenstates. Qualitatively, the derivation of this continuum comes because the energy eigenstates are essentially the Fourier transforms of the clock states. But the Fourier transform of a function over an infinite discrete set (labels of different clock states explored by the computation) is a function over a continuous, bounded set (energy eigenvalues). So for a non-halting program, the spectrum is continuous. Similarly, a halting computer that only explores a finite set of distinguishable states has a discrete spectrum, because the Fourier transform of a function over a finite, discrete set (labels of the clock states) is also a function over a finite discrete set (energy eigenvalues). So for a halting program, the spectrum is discrete. The reader is referred to [10, 11] for the details of the derivation.

Although the derivation in [10] was given specifically for Feynman's quantum computer, the generic nature of the argument implies that the answer to the question of whether *any* quantum system capable of universal computation is gapped or gapless is generically uncomputable. If the computation never halts, then the system goes through a countably infinite sequence of computational states and the spectrum is continuous. If the computation halts, then the system goes through a finite sequence of computational states and the spectrum is discrete. But since many if not most infinite quantum systems that evolve according to nonlinear interactions are capable of universal computation, this implies that uncomputability is ubiquitous in physical law.

3 The Theory of Everything is Uncomputable

As just shown, relatively prosaic aspects of physical systems, such as whether a system has an energy gap or not, are uncomputable. As will now be seen, grander aspirations of physics are also uncomputable. In particular, one of the long-term goals of elementary particle physics is to construct 'the theory of everything'—the most concise unified theoretical description of all the laws of physics, including the interactions between elementary particles and gravitational interactions. But the most concise description of any set of physical laws is in general uncomputable.

The reason stems from the theory of algorithmic information [12–14]. The algorithmic information content of a string of bits is the length of the shortest computer program that can produce that string as output. In other words, the algorithmic information content of a bit string is the most concise description of that string that can be written in a particular computer language. The idea of algorithmic information was first defined by Solomonoff [12] in order to construct a computational version of Ockham's razor, which urges us to find the most parsimonious explanation for a given phenomenon. (William of Ockham: *Numquam ponenda est pluralitas sine necessitate*—plurality should never be posited without necessity—and *Frustra fit per plura quod potest fieri per pauciora*—it is futile to do with more things what can be done with fewer.) Kolmogorov [13] and Chaitin [14] independently arrived at the concept of algorithmic information.

Algorithmic information is an elegant concept which makes precise the notion of the most concise description. In aiming for a theory of everything, physicists are trying to find the most concise description of the set of laws that govern our observed universe. The problem is that this most concise description is uncomputable. The uncomputability of algorithmic information stems from Berry's paradox. Like all the paradoxes discussed here, Berry's paradox arises from the capacity for self-reference.

The English language can be used to specify numbers, e.g., 'the smallest natural number that can expressed as the sum of two distinct prime numbers.' Any natural number can be defined in English, and amongst all English specifications for a given number, there is some specification that has the smallest number of words. Berry's paradox can be expressed as the following phrase: 'the smallest natural number that requires more than twelve words to specify.' Does this phrase specify a number? If it does, then that number can be specified in twelve words, in contradiction to the statement that it cannot be specified in twelve words or less.

The theory of computation makes Berry's paradox precise. The shortest description of a number is given by the shortest program on a given universal Turing machine that produces that number as output. Suppose that algorithmic information is computable. Then there is some program for the same Turing machine which, given a number as input, outputs the length of the shortest program that produces that number. Suppose that this program that computes algorithmic information content has length ℓ. Now look at the algorithmic version of Berry's phrase: 'The smallest number whose algorithmic information content is greater than ℓ plus a billion.' Does this phrase specify a number? If algorithmic information is computable, then the answer is Yes. A program can compute that number by going through all natural numbers in ascending order, and computing their algorithmic information content. When the program reaches the smallest number whose algorithmic information content is greater than ℓ plus a billion, it outputs this number and halts. The length of this program is the length of subroutine that computes algorithmic information content, i.e., ℓ, plus the length of the additional code needed to check whether the algorithmic information content of a number is greater than ℓ plus a billion, and if not, increment the number and check again. But the length of this additional code is far less than a billion symbols, and so the length of the program to produce the number in the algorithmic version of Berry's phrase is far less than ℓ plus a billion. So the algorithmic information content of the number in Berry's phase is also less than ℓ plus a billion, in contradiction to the phrase itself. Paradox!

As in the halting problem, the apparent paradox can be resolved by the concept of uncomputability. The only assumption that went into the algorithmic version of Berry's argument was that algorithmic information content is computable. This assumption lead to a contradiction. By the principle of reductio ad absurdum, the only conclusion is that algorithmic information content is uncomputable. The shortest description of a natural number or bit string cannot in general be computed.

The uncomputability of shortest descriptions holds for any bit string, including the hypothetical bit string that describes the physical theory of everything. Physical laws are mathematical specifications of the behavior of physical systems: they provide formulae that tell how measurable quantities change over time. Such quantities include the position and momentum of a particle falling under the force of gravity, the strength of the electromagnetic field in the vicinity of a capacitor, the state of an electron in a hydrogen atom, etc. Physical laws consist of equations that govern the behavior of physical systems, and those equations in turn can be expressed algorithmically.

There is a subtlety here. Many physical laws describe the behavior of continuous quantities. Algorithmic representations of the equations governing continuous quantities necessarily involve a discretization: for example, using a finite number of bits to represent floating-point variables. The concept of computability can be extended to the computability of continuous quantities as well [15]. It is important to verify that the number of bits needed to approximate continuous behavior to a desired degree of accuracy remains finite. For the known laws of physics, such discrete approximations of continuous behavior seem to be adequate in the regimes to which those laws apply. The places where discrete approximation to continuous laws break down—e.g. the singularities in the center of black holes—are also the places where the laws themselves are thought to break down.

The uncomputability of algorithmic information content implies that the most concise expression of Maxwell's equations or the standard model for elementary particles is uncomputable. Since Maxwell's equations are already concise—they are frequently seen on the back of a tee shirt—few scientists are working on making them even more terse. (Note, however, that terser the expression of Maxwell's equations, the more difficult they are to 'unpack': in all fairness, to assess the algorithmic information content of a physical law, one should also count the length of the extra computer code needed to calculate the predictions of Maxwell's equations.) In cases where the underlying physical law that characterizes some phenomenon is unknown, however, as is currently the case for high T_C superconductivity, uncomputability can be a thorny problem: finding even one concise theory, let alone the most concise one, could be uncomputable.

Uncomputability afflicts all sciences, not just physics. A simplified but useful picture of the goal of scientific research is that scientists obtain large amounts of data about the world via observation and experiment, and then try to find regularities and patterns in that data. But a regularity or pattern is nothing more or less than a method for *compressing* the data: if a particular pattern shows up in many places in a data set, then we can create a compressed version of the data by describing the pattern only once, and then specifying the different places that the pattern shows up. The most compressed version of the data is in some sense the ultimate scientific description. There is a sense in which the goal of all science is finding theories that provide ever more concise descriptions of data.

4 Computational Complexity and Physical Law

What makes a good physical law? Being concise and easily expressed is only one criterion. A second criterion is that the predictions of the law be readily evaluated. If a law is concisely expressed but its predictions can only be revealed by a computation that takes the age of the universe, then the law is not very useful. Phrased in the language of computational complexity, if a physical law is expressed in terms of equations that predict the future given a description of the past, ideally those predictions can be obtained in time polynomial in the description of the past state. For example, the laws of classical mechanics and field theory are described in terms of ordinary and partial differential equations that are readily encoded and evaluated on a digital computer.

There is no guarantee that easily evaluated laws exist for all phenomena. For example, no classical computer algorithm currently exists that can predict the future behavior of a complex quantum system. The obstacles to efficient classical simulation of quantum systems are the counterintuitive aspects of quantum mechanics, such as quantum superposition and entanglement, which evidently require exponential amounts of memory space to represent on a classical computer. It seems to be as hard for classical computers to simulate quantum weirdness as it is for human beings to comprehend it.

Embryonic quantum computers exist, however, and are capable of simulating complex quantum systems in principle [16]. Indeed, the largest-scale quantum computations to date have been performed by specialized quantum information processors that simulate different aspects of quantum systems. In [17] Cory simulated the propagation of spin waves in a sample of $O(10^{18})$ fluorine spins. More recently, the D-Wave adiabatic quantum computer has found the ground state of Ising models using 512 superconducting quantum bits [18]. If we expand the set of computational devices that we use to define computational complexity, then the consequences of the known laws of physics can in principle be elaborated in polynomical time on classical and quantum computers.

If we restrict our attention to laws whose consequences can be evaluated in polynomial time, then the problem of finding concise expressions of physical laws is no longer uncomputable. The obstacle to finding the shortest program that produces a particular data string is essentially the halting problem. If we could tell that a program never halts, then we could find the shortest program to produce the data by going through all potential programs in ascending order, eliminating as we go along all non-halting programs from the set of programs that could potentially reproduce the data. That is, a halting oracle would allow us to find the shortest program in finite time.

In the case where we restrict our attention to programs that halt in time bounded by some polynomial in the length of the inputs, by contrast, if the program hasn't halted and reproduced the data by the prescribed time, we eliminate it from consideration and move on to the next potential program. This construction shows that the problem of finding the shortest easily evaluated program is in the computational complexity class NP: one can check in polynomial time whether a given program will reproduce the data. Indeed, the problem of finding the shortest program to produce the data is NP-complete: it is a variant of the NP-complete problem of finding if there is *any* program within a specified subset of programs that gives a particular output. If we restrict our attention to physical laws whose predictions can be readily evaluated, then the problem of finding the most concise law to explain a particular data set is no longer uncomputable; it is merely NP complete. Probably exponentially hard is better than probably impossible, however.

Similarly, in the case of evaluating the energy gap of a physical system capable of computation, if the system runs for time n, then the gap is no greater than $1/n^2$. Accordingly, the problem of finding out whether the gap is smaller than some bound ϵ is also NP-hard.

5 Discussion

This chapter reviewed how uncomputability impacts the laws of physics. Although uncomputability as in the halting problem arises from seemingly esoteric logical paradoxes, I showed that common and basic questions in physics have answers that are uncomputable. Many physical systems are capable of universal computation; to solve the question of whether such a system has a discrete or continuous spectrum in a particular regime, or whether it is gapped or gapless, requires one to solve the halting problem. At a more general level, one can think of all the scientific laws as providing concise, easily unpacked descriptions of observational and experimental data. The problem of finding the most concise description of a data set is uncomputable in general, and the problem of finding the most concise description whose predictions are easily evaluated is NP-complete. Science is hard, and sometimes impossible. But that doesn't mean we shouldn't do it.

Acknowledgements The author would like to thank Scott Aaronson for helpful discussions.

References

1. A.M. Turing, Proc. Lond. Math. Soc. **242**, 230–265, **243**, 544–546 (1937)
2. K. Gödel, Monatsschr. Math. Phys. **38**, 173–198 (1931)
3. M. Cunningham, *The History of Computation* (AtlantiSoft, New York, 1997)
4. C. Shannon, A symbolic analysis of relay and switching circuits. M.S. thesis, MIT, 1937
5. S. Lloyd, Phys. Rev. Lett. **75**, 346–349 (1995)
6. D. Deutsch, A. Barenco, A. Ekert, Proc. R. Soc. Lond. A **8**, 669–677 (1995)
7. F. Barahona, J. Phys. A **15**, 3241 (1982)
8. S. Lloyd, Science **261**, 1569–1571 (1993)
9. R.P. Feynman, Found. Phys. **16**, 507–531 (1986)
10. S. Lloyd, Phys. Rev. Lett. **71**, 943–946 (1993); J. Mod. Opt. **41**, 2503–2520 (1994)
11. T. Cubitt et al. (in preparation)
12. R.J. Solomonoff, Inf. Control. **7**, 224–254 (1964)
13. A.N. Kolmogorov, Probl. Inf. Transm. **1**, 3–11 (1965)
14. G.J. Chaitin, J. Assoc. Comput. Mach. **13**, 547–569 (1966)
15. L. Blum, M. Shub, S. Smale, Bull. Am. Math. Soc. **21**, 1–46 (1989)
16. S. Lloyd, Science **273**, 1073–1078 (1996)
17. W. Zhang, D. Cory, Phys. Rev. Lett. **80**, 1324–1347 (1998)
18. M.W. Johnson et al., Nature **473**, 194–198 (2011)

Algorithmic Economics: Incomputability, Undecidability and Unsolvability in Economics

K. Vela Velupillai

*Dedicated to **Martin Davis** for introducing me to 'the pleasures of the illicit'. Preface to Davis [6].*

Abstract Economic theory, like any theoretical subject where formalised deduction is routinely practised, is subject to *recursive*, or *algorithmic*, *incomputabilities*, *undecidabilities* and unsolvabilities. Some examples from core areas of economic theory, micro, macro and game theory are discussed in this paper. Some background of the nature of the problems that arise in economic decision-theoretic contexts is given before the presentation of results, albeit largely in a sketchy form.

> Being unfamiliar with what counts as a problem in economics I am astounded that *decidability and other foundational problems* should turn up here. Quantum mechanics is pretty OK; there are problems, true, but there are also lots of very precise predictions. I have not noticed anyone in the field went into the foundations of mathematics to get things going. *Why do these matters turn up in economics*?
> **Paul Feyerabend**, *Letter to the author*, '02/04/92' (4th February, 1992); italics added.

1 By Way of a Prologue on *Theses*[1]

> The last of the original three [i.e., *λ-definability*, *General Recursiveness* and Turing Machines] equivalent *exact* definitions of *effective calculability* is *computability by a Turing Machine*. I assume my readers are familiar with the concept of a Turing Machine ....
> In a conversation at San Juan on October 31, 1979, [Martin] Davis expressed to me the opinion that the equivalence between Gödel's definition of general recursiveness and

[1]The paradigmatic example of which is, of course, the Church-Turing *Thesis*, but there are two others I should mention in this prologue: *Brattka's Thesis* [4] and what I have called the *Kolmogorov-Chaitin-Solomonoff Thesis* [41]. I should have added the name of Per Martin-Löf to the latter trio.

K.V. Velupillai (✉)
Madras School of Economics, Chennai, India
e-mail: kvelupillai@gmail.com

S.B. Cooper, M.I. Soskova (eds.), *The Incomputable*, Theory and Applications of Computability, DOI 10.1007/978-3-319-43669-2_7

mine ... and my normal form theorem, were considerations that combined with Turing's arguments to convince Gödel of the *Church-Turing thesis*.

Kleene [17], pp. 61–62; italics added.

'Why do these matters' of (in)computability, (un)decidability and (un)solvability 'turn up' in economics, as Paul Feyerabend posed it in his letter to me.[2] I think, however, Feyerabend is not quite correct about 'not... anyone in the field [of quantum mechanics going] into the foundations of mathematics to get things going.' Svozil [32], with copious references to the pre-1992 results, Finkelstein [11] and Penrose [20, 21] are obvious counterexamples to this view. I would like to add, further, notwithstanding the general focus of this paper, 'there are lots of very precise predictions' in economics, even if not founded on orthodox economic theory.

Today, or 2 decades later, if it is possible to give a guarded, more than even unequivocal, answer, it is because the concepts and foundations of economic theory have increasingly been given a computational basis, founded on computability theory, constructive mathematics and even intuitionistic, logic-based smooth infinitesimal analysis (thus dispensing with any use of the *tertium non datur* in proofs involving uncountable infinities). Even classic economic problems invoke *computer aided proofs*—for example, in the travelling salesperson's problem or in the use of the Poincaré-Bendixson theorem for proving the existence of fluctuations in macrodynamical models, algorithmic procedures are routinely invoked. Even in the domain of classical game theory, as in Euwe's [9] demonstration of the 'existence' of a min-max solution to a two-person game and Rabin's pioneering work on arithmetical games [23], foundational issues of a mathematical nature arise in economic settings.

In other writings, for example, Velupillai [43], p. 34, I am on record as saying:

> The three 'crown jewels' of the mathematical economics of the second half of the twentieth century are undoubtedly the proof of the *existence of a Walrasian Exchange Equilibrium* and the mathematically rigorous demonstration of the validity of the *two fundamental theorems of welfare economics*.

Unfortunately, in orthodox mathematical economics these important theorems are proved non-computationally.[3] For example, in Brainard and Scarf [3], p. 58, we read:

> But we know of no argument for the existence of equilibrium prices in this restricted model that does not require the full use of *Brouwer's fixed point theorem*. Of course fixed point theorems were not available to Fisher... .

This claim was made as late as only about 10 years ago, despite Smale's important point made 30 years before Brained and Scarf (op. cit.) in Smale [29],

[2]Cited above, on the title page.

[3]By 'computational' I mean either computable or constructable, i.e., algorithmic; hence, 'non-computationally' is supposed to mean 'non-algorithmically'. The above statement applies also to all standard results of orthodox game theory, despite occasional assertions to the contrary. Euwe, as mentioned above, is the exception.

p. 290 (italics added):

> The existence theory of the static approach is deeply rooted to the use of mathematics of fixed point theory. Thus one step in the liberation from the static point of view would be to *use a mathematics of a different kind.*
>
> ...
>
> I think it is fair to say that for the main existence problem in the theory of economic equilibria, one can now bypass the fixed point approach and attack the equations directly to give *the existence of solutions, with a simpler kind of mathematics and even mathematics with dynamic and algorithmic overtones.*

As for the two fundamental theorems of welfare economics non-constructive and non-algorithmic versions of the Hahn-Banach Theorem(s) are invoked in the proofs (particularly of the more important second fundamental theorem of welfare economics).

I need to add here[4] that very little work in economics in the mathematical mode is done with models of computable reals; nor are—to the best of my knowledge—algorithms in economics specified in terms of *interval analysis*. Of course, neither *smooth infinitesimal analysis* nor constructive formalisms are routine in economics (Bridges [5] is a noble exception), despite claims to the contrary (e.g., [18]).

In this paper there is an attempt to use a 'simpler kind of mathematics and even with dynamic and algorithmic overtones', even if the results appear to be 'negative' solutions; it must be remembered that there are obvious positive aspects to negative solutions—one doesn't attempt to analyse the impossible, or construct the infeasible, and so on (hence the importance of theses, which is, after all, what the second 'law'[5] of thermodynamics is).

Computable General Equilibrium[6] theory, *computational* economics, agent-based *computational* models, and *algorithmic* game theory are some of the frontier topics in core areas of economic theory and applied economics. There is even a journal with the title *Computational Economics*.[7]

---

[4]I am greatly indebted to an anonymous referee for raising the relevance of this point here. I have dealt with the issue in many of my writings on algorithmic economics in the last decade or so. The trouble in economics is that reals or approximations are used in blind and *ad hoc* ways. Even the approximations involved in either discretising continuous dynamical systems—however elementary, though nonlinear—or computing approximate Nash or Arrow-Debreu-Walrasian equilibria are done carelessly.

[5]It must be remembered that Emil Post referred to 'theses' as 'natural laws'.

[6]The foundations on which the much vaunted recursive competitive equilibrium (RCE) is based, from which, via, real business cycle (RBC) theory, the frontier, fashionable models of dynamic stochastic general equilibrium (DSGE) framework is carved out. None of these models are computable or recursive in any of the formal senses of computability theory, or in any version of constructive mathematics. In Velupillai [44] I discuss the non-algorithmic aspects of these problems in greater detail, from either a computable or a constructive point of view.

[7]One of whose associate editors once wrote me—when inviting me to organize a session in the annual event sponsored by the journal—that 'we compute the uncomputable'. He was, with some seriousness, referring to the fact that the overwhelming majority of computational economists are blissfully ignorant of the computability theory underpinnings of whatever it is they compute.

However, very few in economics seem to pay attention to the notion of a *thesis*—such as the Church-Turing *Thesis*—and therefore do not pay sufficient attention to the importance of tying a concept developed by *human intuition* to a *formal* notion that is claimed to encapsulate that intuitive notion *exactly*.[8] That the Church-Turing *Thesis* is a result of this kind of identification, between the intuitive notion of *effective calculability* and the formal notions of—independently developed—*general recursiveness*, *λ-definability* and *Turing Machines* is not fully appreciated by the mathematical economics community.

As a result of this particular kind of disinterest, economic theory in its mathematical mode forces economic concepts to conform to independently developed mathematical notions, such as continuity, compactness, and so on. It is—to the best of my knowledge—never acknowledged that there are intuitive notions of *continuity* which cannot be encapsulated by, for example, the concept of a *topological space* ([13], p. 73).

A *thesis* is not a *theorem*. The Church-Turing Thesis came about, as suggested above, as a result of trying to find a *formal* encapsulation of the *intuitive* notion of *effective calculability*. What is the difference between a *Thesis* and a *Theorem*? Perhaps one illuminating way to try to answer this question is to reflect on 'an imaginary interview between a modern mathematician [Professor X] and... Descartes', as devised by Rosser [24], pp. 2–3, to illustrate the importance of the open-ended nature of any claimed *exact* equivalence between an intuitive concept and a formal notion:

> ... Descartes raised one difficulty which Professor X[9] had not foreseen. Descartes put it as follows:
>
> 'I have here an important concept which I call continuity. At present my notion of it is rather vague, not sufficiently vague that I cannot decide which curves are continuous, but too vague to permit careful proofs. You are proposing a precise definition[10] of this same notion. However, since my definition is too vague to be the basis for a careful proof, how are we going to verify that my vague definition and your precise definition are definitions of the same thing?
>
> If by 'verify' Descartes meant 'prove,' it obviously could not be done, since his definition was too vague for proof. If by 'verify' Descartes meant 'decide,' then it might be done, since his definition was not too vague for purposes of coming to decisions. Actually, Descartes and Professor X did finally decide that the two definitions were equivalent, and arrived at the decision as follows. Descartes had drawn a large number of curves and classified them into continuous and discontinuous, using his vague definition of continuity. He and Professor X checked through all these curves and classified them into continuous

---

Nor are they seriously interested in the link between dynamical systems, numerical analysis—sometimes referred to as 'scientific computing'—and computability theory (cf. [31]).

[8] I use this word in the precise sense in which it is invoked by Kleene, in the quote above.

[9] Rosser's explanation for this particular 'christening' of the 'modern mathematician' was (*ibid*, p. 1):

> [I]n the classic tradition of mathematics we shall refer to him as Professor X.

[10] The 'proposal' by Professor X was the familiar 'ε-δ' definition of continuity.

> and discontinuous using the ε-δ definition of continuity. Both definitions gave the same classification. As these were all the interesting curves that either of them had been able to think of, the evidence seemed 'conclusive' that the two definitions were equivalent.

How did they come to this conclusion? By comparing the classifications into continuous and discontinuous all those 'interesting curves' both of them could 'think of', using their own respective definitions—the intuitive and the (so-called) precise—and finding they resulted in identical characterisations. Thus, 'the evidence seemed "conclusive" that the two definitions were equivalent' (*ibid*, p. 3). The *evidence for equivalence can only 'seem' conclusive*.

Any and every computation that is implementable by a Turing Machine answers *all* such questions of the 'equivalence' between 'intuitive' notions of effective calculability and formal definitions of computability unambiguously: every model of computation thus far formally defined (going beyond the triple noted by Kleene, above)—Turing Machines, Post's Machine, Church's λ-Calculus, General Recursiveness, the Shepherdson-Sturgis Register Machines, etc.,—is formally equivalent to any other.[11] But this kind of categorical assertion requires me to assume a framework in which the Church-Turing Thesis is assumed. This is not so, for example, in Brouwerian constructive mathematics, where, nevertheless, *all* functions are continuous; *ditto* for *smooth infinitesimal analysis*, which is founded upon a kind of *intuitionistic logic*.

As summarised by the classic and original definition of this concept by Kleene [16], pp. 300–301:

- Any general recursive function (predicate) is effectively calculable.
- Every effectively calculable function (effectively decidable predicate) is general recursive.
- The Church-Turing Thesis is also implicit in the conception of a computing machine formulated by Turing and Post.

And Kleene went on (*ibid*, pp. 317–318; italics added):

> Since our original notion of effective calculability of a function (or of effective decidability of a predicate) is a somewhat *vague intuitive* one, the thesis cannot be proved.
>
> The *intuitive notion* however is real, in that it vouchsafes as effectively calculable many particular functions, ... and on the other hand *enables us to recognize that our knowledge about many other functions is insufficient to place them in the category of effectively calculable functions*.[12]

[11] It is, of course, this that was stressed by Gödel when he finally accepted the content of the Church-Turing Thesis ([14], p. 84; italics added):

> It seems to me that [the] importance [of Turing's computability] is largely due to the fact that with this concept one has for the first time succeeded in giving an *absolute definition* of an interesting epistemological notion, i.e., one *not depending on the formalism chosen*.

[12] It is, surely, the method adopted by Ramanujan, a further step down the line of mathematical reasoning:

Once economic theoretical concepts are underpinned by computability or constructive theoretical formalisations, and once computation is itself considered on a reasonably equivalent footing with traditional analytical methods, it is *inevitable* that decidability,[13] computability and solvability issues—almost all in a recursive sense—will rise to the forefront.

The computable approach to the mathematisation of economics *enables us to recognize that our knowledge about relevant functions is insufficient to place them in the category of effectively calculable functions*. This 'insufficiency' and its formal 'recognition' is what enables one to derive undecidable, incomputable and unsolvable problems in economics—but also to find ways to decide the undecidable, compute the incomputable and solve the unsolvable.

2 A Menu of Undecidable, Uncomputable and Unsolvable Problems in Economics

> Indeed virtually any 'interesting' question about dynamical systems is – in general – *undecidable*.
>
> This does not imply that it cannot be answered for a specific system of interest: ... [I]t does demonstrate the *absence of any general formal technique*.
>
> Stewart [30], p. 664; italics added.

2.1 The Undecidability of the Excess Demand Function

On 21st July 1986, Arrow [1] wrote as follows to Alain Lewis (Arrow Papers, Box 23; italics added):

> [T]he claim the *excess demands are not computable* is a much profounder question for economics than the claim that equilibria are not computable. The former challenges economic theory itself; if we assume that human beings have calculating capacities not exceeding those of Turing machines, then the *non-computability of optimal demands is a serious challenge to the theory* that individuals choose demands optimally.

> [I]f a significant piece of *reasoning* occurred somewhere, and the total mixture of evidence and intuition gave [Ramanujan] *certainty*, he looked no further.
>
> Hardy [15], p. 147; italics in the original.

[13]The 'duality' between effective calculability—i.e., computability—and effective undecidability, made explicit by the Church-Turing Thesis is described with characteristic and concise elegance by Rózsa Péter ([22], p. 254; italics in the original):

> One of the most important applications of the [Church-Turing] thesis, making precise the concept of effectivity, is the proof of the *effective undecidability of certain problems*.

That "the excess demands are not computable"—in the dual, *undecidable*, form suggested by Rózsa Péter (*op. cit.*)—can be shown using *one* of Turing's enduring results—the *Unsolvability of the Halting Problem for Turing Machines*.[14] On the other hand, it can also be proved without appealing to this result (although the phrase 'not exceeding those of Turing machines' is not unambiguous).

To prove that the excess demand functions are effectively undecidable, it seems easiest to start from one half of the celebrated *Uzawa Equivalence Theorem* [40]. This half of the theorem shows that the *Walrasian Equilibrium Existence Theorem* (**WEET**) implies the *Brouwer fix point theorem,* and the finesse in the proof is to show the feasibility of devising a continuous excess demand function, $\boldsymbol{X}(\boldsymbol{p})$, satisfying *Walras' Law* (and *homogeneity*) from an arbitrary continuous function, say $\boldsymbol{f}(.)$: $\boldsymbol{S} \rightarrow \boldsymbol{S}$, such that the equilibrium price vector implied by $\boldsymbol{X}(\boldsymbol{p})$ is also the fix point for $\boldsymbol{f}(.)$, from which it is 'constructed'. The key step in proceeding from a given, arbitrary $\boldsymbol{f}(.)$: $\boldsymbol{S} \rightarrow \boldsymbol{S}$ to an excess demand function $\boldsymbol{X}(\boldsymbol{p})$ is the definition of an appropriate scalar:

$$\mu(p) = \frac{\sum_{i=1}^{n} p_i f_i \left[\frac{p}{\lambda(p)} \right]}{\sum_{i=1}^{n} p_i^2} = \frac{p \cdot f(p)}{|p|^2} \tag{1}$$

where:

$$\lambda(p) = \sum_{i=1}^{n} p_i \tag{2}$$

From (1) and (2), the following excess demand function, $X(p)$, is defined:

$$x_i(p) = f_i \left(\frac{p}{\lambda(p)} \right) - p_i \mu(p) \tag{3}$$

[14]The anonymous referee has pointed out, most perceptively, that "there are many 'specific' Turing machines for which the associated Halting problem 'is' [algorithmically] decidable." Absolutely true—a result known for at least the past 4 or 5 decades. It is fairly 'easy' to establish *criteria for the solvability of the Halting problem,* thereby also showing that the relevant Turing Machine is *not* universal. For simplicity, in the above discussion I shall work only with Turing Machines that do not satisfy any of the known criteria for the 'solvability of the Halting problem.' I have to add two caveats to this in the context of the problem of the computability of the *excess demand function.* First of all, the claims in orthodox theory for the validity of the excess demand function are 'universal', essentially, that one is working with Universal Turing Machines. Secondly, the kinds of functions that should be used as the *excess demand function* should be constructive or, at least, of the *smooth infinitesimal* type.

i.e.,

$$X(p) = f(p) - \mu(p)p \tag{4}$$

It is simple to show that (3) [or (4)] satisfies:

(i) $\boldsymbol{X}(\boldsymbol{p})$ is continuous for all prices, $\boldsymbol{p} \in \boldsymbol{S}$;
(ii) $\boldsymbol{X}(\boldsymbol{p})$ is homogeneous of degree 0;
(iii) $\boldsymbol{p} \bullet \boldsymbol{X}(\boldsymbol{p}) = 0$, $\forall \boldsymbol{p} \in \mathrm{S}$, i.e., Walras' Law holds:

$$\Sigma p_i x_i(p) = \boldsymbol{0}, \quad \forall p \in S \,\&\, \forall i = 1 \ldots \mathrm{n} \tag{5}$$

Hence, $\exists \boldsymbol{p}^*$ s.t. $\boldsymbol{X}(\boldsymbol{p}^*) \leq 0$ (with equality unless $\boldsymbol{p}^* = 0$). Elementary logic and economics then imply that $\boldsymbol{f}(\boldsymbol{p}^*) = \boldsymbol{p}^*$. I claim that the procedure that leads to the definition of (3) [or, equivalently, (4)] to determine $\boldsymbol{p}^*$ is provably *undecidable*. In other words, the crucial scalar in (1) cannot be defined recursion theoretically to *effectivize* a sequence of projections that would ensure convergence to the equilibrium price vector.

Theorem 1 *$\boldsymbol{X}(\boldsymbol{p}^*)$, as defined in (3) [or (4)] above, is undecidable, i.e., cannot be determined algorithmically.*

Proof Suppose, contrariwise, there is an algorithm which, given an arbitrary $\boldsymbol{f}(.)$: $\boldsymbol{S} \rightarrow \boldsymbol{S}$, determines $\boldsymbol{X}(\boldsymbol{p}^*)$. This means, therefore, in view of (i)–(iii) above, that the given algorithm determines the equilibrium $\boldsymbol{p}^*$ implied by WEET. In other words, given the arbitrary initial conditions $\boldsymbol{p} \in \boldsymbol{S}$ and $\boldsymbol{f}(.)$: $\boldsymbol{S} \rightarrow \boldsymbol{S}$, the assumption of the existence of an algorithm to determine $\boldsymbol{X}(\boldsymbol{p}^*)$ implies that its halting configurations are decidable. But this violates the **undecidability of the Halting Problem for Turing Machines**. Hence, the assumption that there exists an algorithm to determine—i.e., to construct—$\boldsymbol{X}(\boldsymbol{p}^*)$ is untenable.

Remark 1 Alas, the proof is uncompromisingly non-constructive—i.e., the proposition is itself established by means of appeals to noneffective methods. This is, I believe, an accurate reflection of a perceptive observation made by Davis, Matiyasevic and Robinson [8], p. 340:

> [T]hese [*reductio ad absurdum* and other noneffective methods] may well be the best available to us because *universal methods for giving all solutions* do not exist.

This is a strengthening of the point made by Stewart (*op.cit*) on the absence of 'general formal techniques', that to find *all* solutions to any given problem (or any solution to *all* given problems of a given type) leads to undecidabilities, incompatibilities or unsolvabilities. Modesty and humility when posing solvable problems, decidable issues and computable entities are virtues.

2.2 Uncomputability of Rational Expectations Equilibria

There are two crucial aspects to the notion of rational expectations equilibrium (henceforth, *REE*): an individual optimization problem, subject to perceived constraints, and a system-wide, autonomous set of constraints imposing consistency across the collection of the perceived constraints of the individuals. The latter would be, in a most general sense, the accounting constraint, generated autonomously, by the logic of the macroeconomic system. In a representative agent framework the determination of *REE*s entails the solution of a general fix point problem. Suppose the representative agent's perceived law of motion of the macroeconomic system (as a function of state variables and exogenous 'disturbances') as a whole is given by $\boldsymbol{H}$.[15]

The system-wide autonomous set of constraints, implied, partially at least, by the optimal decisions based on perceived constraints by the agents, on the other hand, implies an actual law of motion given by, say, $\boldsymbol{H}^0$. The search for fixed points of a mapping, $\boldsymbol{T}$, linking the individually perceived macroeconomic law of motion, $\boldsymbol{H}$, and the actual law of motion, $\boldsymbol{H}^0$, is assumed to be given by a general functional relationship subject to the standard mathematical assumptions:

$$\boldsymbol{H}^0 = \boldsymbol{T}(\boldsymbol{H}). \tag{6}$$

Thus, the fixed-points of $\boldsymbol{H}^*$ of $\boldsymbol{T}$, in a space of functions, determine *REE*s:

$$\boldsymbol{H}^* = \boldsymbol{T}\left(\boldsymbol{H}^*\right). \tag{7}$$

Suppose $\boldsymbol{T}: \boldsymbol{H} \rightarrow \boldsymbol{H}$ is a recursive operator (or a recursive program $\boldsymbol{\Gamma}$). Then there is a computable function $\boldsymbol{f}_t$ that is a *least fixed point* of $\boldsymbol{T}$:

$$\mathrm{T}(f_t) = f_t. \tag{8}$$

$$\textit{If } \boldsymbol{T}(\boldsymbol{g}) = \boldsymbol{g}, \textit{ then } \boldsymbol{f}_t \sqsubseteq \boldsymbol{g}. \tag{9}$$

This result can be used directly to show that there is a (recursive) program that, under any input, outputs exactly itself. It is this program that acts as the relevant reaction or response function, $\boldsymbol{T}$, for an economy in REE. However, finding this particular recursive program by specifying a dynamical system leads to the noneffectivity of REE in a dynamic context.[16] Hence, the need for learning processes to find

[15]Readers familiar with the literature will recognise that the notation $\boldsymbol{H}$ reflects the fact that, in the underlying optimisation problem, a ***Hamiltonian*** function has to be formed.

[16]Proving the noncomputability of a *static* ***REE*** is as trivial as proving the uncomputability of a *Walrasian* or a *Nash* equilibrium.

this program unless the theorem is utilized in its constructive version. Thus, the uncomputability of ***REE.***

Theorem 2 *No dynamical system can effectively generate the recursive program* $\boldsymbol{\Gamma}$.

Proof A trivial application of **Rice's Theorem**—in the sense that any nontrivial property of a dynamical system—viewed algorithmically—is undecidable. The intended interpretation is that only the properties of the empty set and the universal set can be effectively generated.

Remark 2 Given the way Rice's Theorem is proved, the same remarks as above, in Remark 1, apply.

2.3 Algorithmically Unsolvable Problems in Economics

No mathematical method can be *useful* for *any* problem *if it involves much calculation.*
Turing [38], p. 9; italics added.

Although it is easy to illustrate unsolvability of economic problems in the same vein as in the previous two subsections, I shall not take that approach. Instead, this subsection will be an introduction to Herbert Simon's vision on *Human Problem Solving* [19], in the light of Turing's approach to *Solvable and Unsolvable Problems* [38].

Turing's fundamental work on *Solvable and Unsolvable Problems* [38], *Intelligent Machinery* [35] and *Computing Machinery and Intelligence* [36] had a profound effect on the work of Herbert Simon, the only man to win both the ACM Turing Prize and the Nobel Memorial Prize in Economics, particularly in defining *boundedly rational* economic agents as *information processing systems* (IPSs) solving decision problems.[17]

A comparison of Turing's classic formulation of *Solvable and Unsolvable Problems* and Simon's variation on that theme, as *Human Problem Solving* [19], would be an interesting exercise, but it must be left for a different occasion. This is partly because the *human problem solver* in the world of Simon needs to be defined in the same way Turing's approach to *Solvable and Unsolvable Problems* was built on the foundations he had established in his classic of 1936–1937.

It is little realised that four of what I call the *Six Turing Classics—On Computable Numbers* [33], *Systems of Logic Based on Ordinals* [34]; *Proposal for Development in the Mathematics Division of an Automatic Computing Engine* (ACE) of 1946, published in Turing [39]; *Computing Machinery and Intelligence* [36]; *The Chemical Basis of Morphogenesis* [37]; and *Solvable and Unsolvable*

[17]In the precise sense in which this is given content in mathematical logic, metamathematics, computability theory and model theory.

Problems [38]—should be read together to glean *Turing's Philosophy*[18] *of Mind*. Simon, as one of the acknowledged founding fathers of computational cognitive science was deeply indebted to Turing in the way he tried to fashion what I have called *Computable Economics* [41]. It was not for nothing that Simon warmly acknowledged—and admonished—Turing in his essay in a volume 'memorializing Turing' [27], p. 81, titled *Machine as Mind*[19]:

> If we hurry, we can catch up to Turing on the path he pointed out to us so many years ago.

Simon was on that path for almost the whole of his research life.

Building a Brain, in the context of economic decision making, meant building a mechanism for encapsulating human intelligence, underpinned by rational behaviour in economic contexts. This was successfully achieved by Herbert Simon's lifelong research program on computational behavioural economics.[20]

From the early 1950s Simon had empirically investigated evidence on human problem solving and had organised that evidence within an explicit framework of a theory of sequential information processing by a Turing Machine. This resulted in ([26], p. x; italics added):

> [A] general theory of human cognition, not limited to problem solving, [and] a methodology for expressing *theories of cognition as programs* [for digital computers] and for using [digital] computers [in general, Turing Machines] to simulate *human thinking*.

This was the first step in replacing the traditional *Rational Economic Man* with the computationally underpinned *Thinking*, i.e., Intelligent Man. The next step was to stress two empirical facts (ibid, p. x; italics added):

i. "There exists *a basic repertory of mechanisms and processes* that Thinking Man uses in all the domains in which he exhibits *intelligent behaviour*."
ii. "The models we build initially for the several domains must all be assembled from this same basic repertory, and common principles of architecture must be followed throughout."

---

[18]Remembering Feferman's [10], p. 79 cautionary note that 'Turing never tried to develop an overall *philosophy of mathematics*...', but not forgetting that his above-mentioned works were decisive in the resurrection of a particular vein of research in the philosophy of the mind, particularly in its cognitive, neuroscientific versions pioneered by Simon.

[19]To which he added the caveat (*ibid*, p. 81):

> I speak of 'mind' and not 'brain'. By mind I mean a system [a mechanism] that produces thought... .

I have always interpreted this notion of 'mechanism' with Gandy's *Principles for Mechanisms* [12] in *mind* [sic].

[20]I refer to this variant of behavioural economics, which is underpinned by a basis in *computational complexity theory*, as *classical* behavioural economics, to distinguish it from currently orthodox behavioural economics, sometimes referred to as *modern* behavioural economics, which has no computational basis whatsoever.

It is easy to substantiate the claim that the *basic repertory of mechanisms and processes* are those that define, in the limit, a *Turing Machine* formalisation of the Intelligent Man when placed in the decision-making, problem-solving context of economics (cf. [42]).

However, the unsolvability of a problem, implied in any Turing Machine formalization of decision processes, did not really stop people from looking for a solution for it, particularly not Herbert Simon. Sipser (1997) admirably summarises the *pros* and *cons* of proving the unsolvability of a problem, and then coming to terms with it:

> After all, showing that a problem is *unsolvable* doesn't appear to be any use if you have to solve it. You need to study this phenomenon for two reasons. First, knowing when a problem is *algorithmically unsolvable* is useful because then you realize that the problem must be simplified or altered before you can find an algorithmic solution. Like any tool, computers have capabilities and limitations that must be appreciated if they are to be used well. The second reason is *cultural*. Even if you deal with problems that clearly are solvable, a glimpse of the unsolvable can stimulate your imagination and help you gain an important perspective on computation.
>
> Sipser [28], p. 151, italics added.

This was quintessentially the vision and method adopted by Simon in framing decision problems to be solved by boundedly rational agents, satisficing in the face of computationally complex search spaces.

The formal decision problem framework for a boundedly rational information processing system can be constructed in one of the following ways: systems of linear Diophantine inequalities, systems of linear equations in non-negative integer variables, integer programming. Solving the aforementioned three problems are equivalent in the sense that the method of solving one problem provides a method to solve the other two as well. The Integer Linear Programming (**ILP**) problem and **SAT** can be translated both ways, i.e., one can be transformed into another.

In Velupillai [42], it is demonstrated that the **SAT** problem is the meeting ground of Diophantine problem solving and satisficing search; in turn, this connection leads to the conclusion that bounded rationality is the superset of orthodox rationality, which had been Simon's point all along.

Quite apart from the positive aspects of taking a Simonian approach to *Human Problem Solving*, in its analogy with Turing's way of discussing *Solvable and Unsolvable Problems*, the felicity of being able to show that orthodox optimization, by what Simon referred to as the *Olympian* rational agent, is a *special case*[21] of boundedly rational agents, *satisficing* in a computationally underpinned behavioural decision-making context, is, surely, a vindication of the computable approach to economics!

[21]In the same sense in which the reals are 'constructed' from the integers and, then, the rational numbers.

3 Beyond Computable Economics

> Most mathematicians are so deeply immersed in the classical tradition that the idea of abandoning classical logic and mathematics for a constructive approach seems very strange. Those of us interested in approaching mathematics constructively usually need to spend a great deal of time and energy justifying our interest, and we are often dismissed as cranks. ...
>
> ...
>
> [T]he crux of the insistence by intuitionists (and, presumably, by other strict constructivists as well) that nonconstructive classical mathematics is meaningless is their view that it is not meaningful to talk about completing a computation process requiring an infinite number of steps.
>
> Seldin [25]; p. 105.

Those of us who take a computable approach to economic theory face the same problems, The mind boggles at the thought of one who is not only a computable economist, but who also endorses a constructive and non-standard analytic approach to the mathematisation of economic theory.

What, 'exactly', is a computation? The lucid, elementary, answer to this question, entirely in terms of computability theory and Turing Machines, was given by Martin Davis in a masterly exposition almost three and a half decades ago [7]. Here I am interested in an answer that links the triad of *computation*, *simulation* and *dynamics* in an *epistemological* way. This is because I believe simulation provides the epistemological cement between a computation and the dynamical system that is implemented during a computation—by means of an algorithm, for example.

I shall assume the following *theses*, in the spirit of the *Church-Turing thesis* with which I began this paper, so that I can answer the question 'What is a computation?' in an epistemologically meaningful sense.

Thesis 1 Every computation is a dynamical system

Thesis 2 Every simulation of a dynamical system is a computation

Analogously, then: What can we know about, what must we do with and what can we hope for from a computation that is, by the above claim, a dynamical system? This, in turn, means what can we know about, what must we do with and what can we hope for from studying the behaviour of a dynamical process during a computation? Since, however, not everything can be computed, it follows that not every question about a dynamical system can be answered unambiguously. But by the second of the above claims, I have expressed an 'identity' between a simulation and a computation, via the intermediation of a dynamical system, which implies that *not everything can be learned* about the behaviour of a dynamical system by simulating it on (even) an ideal device that can compute anything that is theoretically computable (i.e., a Turing Machine, assuming the Church-Turing Thesis). Above all, we cannot distinguish, in any meaningful sense—i.e., in an algorithmically decidable sense—between what can be known or learned and that which lies 'beyond' this undecidable indefinable, border, on one side of which we live our scientific lives.

It is humility in the face of this epistemological barrier that one learns to cultivate when approaching the decision problems of economics on the basis of computability theory.

With this as a starting point, the next step beyond Computable Economics is towards Constructive Economics and Intuitionistic Logic—towards an *Intuitionistically Constructive Economics*,[22] to untie oneself from the bondage of classical logic, too.

It is, therefore, appropriate to end this exercise with Jeremy Avigad's perceptive, yet pungent, reflection ([2], pp. 64–65; italics added):

> [The] adoption of the infinitary, nonconstructive, set theoretic, algebraic, and structural methods that are characteristic to modern mathematics [...] were controversial, however. At issue was not just whether they are consistent, but, more pointedly, whether they are meaningful and appropriate to mathematics. After all, if one views mathematics as an essentially computational science, then arguments without computational content, whatever their heuristic value, are not properly mathematical... [At] the bare minimum, we wish to know that the universal assertions we derive in the system will not be contradicted by our experiences, and the existential predictions will be borne out by calculation. This is exactly what Hilbert's program was designed to do.

And it is precisely in that particular aim Hilbert's Program failed; yet mathematical economics, in all its orthodox modes, adheres to it with unreflective passion.

Acknowledgment I am greatly indebted to the late Barry Cooper, dear friend and a fellow Turing *aficionado* for inviting me, an outsider and an outlier *par excellence*, to contribute to the *Proceedings of the Turing Centennial* conference, held at *Chicheley Hall*, in June 2012. The conference was both a pleasure and an honour to attend as an invited speaker. It also gave me a chance, finally, to meet and get to know personally some of the giants of twentieth century computability theory, in particular, *Martin Davis*, *Yuri Matiyasevich* and *Robert Soare*. I have benefited greatly from the friendly suggestions for improvement by an anonymous referee and the incredibly good-humoured patience shown by Mariya Soskova. Since June 2012, when a first draft of this paper was composed, my knowledge of computability theory, constructive mathematics and Alan Turing's writings has deepened; I hope that is reflected in this revision.

References

1. K.J. Arrow, *Letter to Alain Lewis*, July 21, deposited in: Kenneth Arrow Papers, Perkins Library, Duke University (1986)
2. J. Avigad, The metamathematics of ergodic theory. Ann. Pure. Appl. Log. **157**, 64–76 (2009)
3. W.C Brainard, H.E. Scarf, How to compute equilibrium prices in 1891. Am. J. Econ. Sociol.: Special Invited Issue: Celebrating Irving Fisher: The Legacy of a Great Economist, **64**(1), 57–83 (2005)
4. V. Brattka, Plottable real number functions and the computable graph theorem. SIAM J. Comput. **38**(1), 303–328 (2008)

[22]Or, alternatively, in terms of *Smooth Infinitesimal Analysis* (**SIA**), and thus an economic theory in mathematical modes that is based on *category theory* and *topos theory*, all three of which have computational underpinnings.

5. D. Bridges, Constructive methods in mathematical economics. J. Econ. Suppl. **8**, 1–21 (1999)
6. M. Davis, *Applied Nonstandard Analysis* (Wiley, New York, 1977)
7. M. Davis, What is a computation? in *Mathematics Today—Twelve Informal Essays*, ed. by L.A. Steen (New York, Springer, 1978), pp. 241–267
8. M. Davis, Y. Matijasevic, J. Robinson, Hilbert's tenth problem. Diophantine equations: positive aspects of a negative solution, in *Mathematical Developments Arising from Hilbert Problems*, ed. by F.E. Browder (American Mathematical Association, Providence, 1986)
9. M. Euwe, *Mengentheoretische Betrachtungen über des Schachspiel* (Communicated by Prof. R. Weitzenböck), translation forthcoming in: New Mathematics and Natural Computation, 2016 (1929, 2016)
10. S. Feferman, Preface to: systems of logic based on ordinals, in *Collected Works of A.M. Turing – Mathematical Logic*, ed. by R.O. Gandy, C.E.M Yates (North-Holland, Amsterdam, 1992)
11. D. Finkelstein, Finite Physics, in *The Universal Turing Machine – A Half-Century Survey (Second Edition)*, ed. by R. Herken (Springer, Wien/New York, 1994), pp. 323–347
12. R. Gandy, Church's thesis and principles for mechanisms, in *The Kleene Symposium*, ed. by J. Barwise, H.J. Keisler, K. Kunen (North-Holland, Amsterdam, 1980)
13. R. Gandy, The confluence of ideas in 1936, in *The Universal Turing Machine – A Half-Century Survey (Second Edition)*, ed. by R. Herken (Springer, Wien/New York, 1994), pp. 51–102
14. K. Gödel, Remarks before the Princeton bicentennial conference on problems in mathematics, in *The Undecidable—Basic Papers on Undecidable Propositions, Unsolvable Problems and Computable Functions*, ed. by M. Davis (Raven Press, New York, (1946, [1965]), pp. 84–88
15. G.H. Hardy, The Indian mathematician Ramanujan. Am. Math. Mon. **44**(3), 137–155 (1937)
16. S.C. Kleene, *Introduction to Metamathematics* (North-Holland Publishing Company, Amsterdam, 1952)
17. S.C. Kleene, Origins of recursive function theory. Ann. Hist. Comput. **3**(1), 52–67 (1981)
18. R.R. Mantel, in *Toward a Constructive Proof of the Existence of Equilibrium in a Competitive Economy*, Yale Economic Essays, vol. 8, # 1 (Spring, 1968), pp. 155–196
19. A. Newell, H.A. Simon, *Human Problem Solving* (Prentice-Hall Inc., Englewood Cliffs, 1972)
20. R. Penrose, *The Emperor's New Mind: Concerning Computers, Minds, and the Laws of Physics* (Oxford University Press, Oxford, 1987)
21. R. Penrose, *Shadows of the Mind: A Search for the Missing Science of Consciousness* (Oxford University Press, Oxford, 1994)
22. R. Péter, *Recursive Functions (Third Revised Edition)*, translated from the German by I. Földes (Academic Press, New York, 1967)
23. M.O. Rabin, Effective computability of winning strategies, in *Contributions to the Theory of Games, Vol. III*, ed. by M. Dresher, A.W. Tucker, P. Wolfe (Princeton University Press, Princeton, 1957), pp. 147–157
24. B.J. Rosser, *Logic for Mathematicians* (McGraw-Hill Book Company, Inc., New York, 1953)
25. J.P. Seldin, A constructive approach to classical mathematics, in Constructive Mathematics – *Proceedings of the New Mexico State University Conference Held at Las Cruces, New Mexico*, August 11–15, 1980, ed. by F. Richman (Springer, Berlin/New York, 1981), pp. 105–110
26. H.A. Simon, *Models of Thought*, vol 1 (Yale University Press, New Haven, 1979)
27. H.A. Simon, Machine as mind, in *Machines and Thought - The Legacy of Alan Turing, Volume 1*, ed. by P. Macmillan, A. Clark (Oxford University Press, Oxford, 1996), pp. 81–101
28. M. Sipser, *Introduction to the Theory of Computation* (PWS Publishing Company, Boston, 1997)
29. S. Smale, Dynamics in general equilibrium theory. Am. Econ. Rev. **66**(2), 288–294 (1976)
30. I. Stewart, Deciding the undecidable. Nature **352**, 664–5 (1992)
31. A. Stuart, A.R. Humphries, *Dynamical Systems and Numerical Analysis* (Cambridge University Press, Cambridge, 1998)
32. K. Svozil, *Randomness & Undecidability in Physics* (World Scientific, Singapore, 1993)
33. A.M. Turing, On computable numbers with an application to the Escheidungsproblem. in *Proceedings of the London Mathematical Society*, Series 2, vol. 42 (1936–1937), pp. 230–265

34. A.M. Turing, *Systems of logic based on ordinals,* in *Proceedings of the London Mathematical Society*, Series 2, vol. 45 (1939)
35. A.M. Turing, *Intelligent machinery*, Report, National Physical Laboratory, in *Machine Intelligence*, ed. by B. Meltzer, D. Michie, vol. 5 (Edinburgh University Press, Edinburgh), pp. 3–23; reprinted in *Mechanical Intelligence - Collected Works of Alan Turing*, ed. by D.C. Ince (North-Holland, Amsterdam, (1948, 1969, 1992)), pp. 107–127
36. A.M. Turing, Computing machinery and intelligence. Mind **50**, 433–60 (1950)
37. A.M. Turing, in *The Chemical Basis of Morphogenesis*, Philosophical Transactions of the Royal Society, Series B, vol. 237 (1952), pp. 37–72
38. A.M. Turing, Solvable and unsolvable problems. Sci News (31), 7–23 (1954)
39. A. Turing, in *A. M. Turing's ACE Report of 1946 and Other Papers*, ed. by B.E. Carpenter, R.W. Doran (The MIT Press, Cambridge, 1986)
40. H. Uzawa, Walras' existence theorem and Brouwer's fixed point theorem. Econ. Stud. Q. **8**(1), 59–62 (1962)
41. K. Velupillai, *Computable Economics* (Oxford University Press, Oxford, 2000)
42. K.V. Velupillai, *Computable Foundations for Economics* (Routledge, London, 2010)
43. K.V. Velupillai, Constructive and computable Hahn–Banach theorems for the (second) fundamental theorem of welfare economics. J Math Econ **54**, 34–37 (2014)
44. K.V. Velupillai, *Seven kinds of computable and constructive infelicities,* in Forthcoming in: *New Mathematics and Natural Computation* (2016)

Part III
Mind, Matter and Computation

Is Quantum Physics Relevant for Life?

Vlatko Vedral

Abstract In this paper I explore the possibility that quantum physics might play a fundamental role in living systems. The first part explains how thermodynamical insights may be used to derive a characterization of life as a physical phenomenon. In this case, quantum mechanics is used to describe the state of every atom on earth, providing an estimate of the entropy required for life to exist. In the second part, I argue that quantum mechanics, according to Per-Olov Löwdin, can lead to an explanation of DNA replication in terms of energy consumption. The third part describes two modern developments, photosynthesis and magneto-reception, both of which rely on quantum physics to explain their efficiency. The fourth part asks whether quantum computers can be used to simulate life in a way that goes beyond the classical approach. Finally, the fifth part suggests a new approach to physics in terms of possible and impossible processes that are based on observations of the natural world.

> ... living matter, while not eluding the 'laws of physics' as established up to date, is likely to involve 'other laws of physics' hitherto unknown, which however, once they have been revealed, will form just as integral a part of science as the former.—Erwin Schrödinger (quote from *What is Life?*)

V. Vedral (✉)
Department of Physics, Clarendon Laboratory, University of Oxford, Parks Road, Oxford OXI 3PU, UK

Centre for Quantum Technologies, National University of Singapore, Singapore 117543, Singapore

Department of Physics, National University of Singapore, 2 Science Drive 3, Singapore 117551, Singapore

Center for Quantum Information, Institute for Interdisciplinary Information Sciences, Tsinghua University, Beijing 100084, China
e-mail: v.vedral1@physics.ox.ac.uk

S.B. Cooper, M.I. Soskova (eds.), *The Incomputable*, Theory and Applications of Computability, DOI 10.1007/978-3-319-43669-2_8

1 Review of Schrödinger

After revolutionizing physics, Erwin Schrödinger turned to biology. In 1944 he wrote a highly influential book, called *What is Life?* [1], that discussed the physical basis of biological processes. From the present perspective he got an astonishing number of things right. He basically anticipated the stable encoding of biological genetic information, and he guessed that crystals form the basis for encoding (Watson and Crick later showed DNA has a periodic crystalline-like structure). He also understood the thermodynamical underpinnings of living processes that have since become the bread and butter of biology.

Fundamentally, however, in his treatment of biology using the laws of physics, Schrödinger, follows Ludwig Boltzmann. Here the first and the second laws of thermodynamics drive biological processes. The first law stipulates that the overall energy has to be conserved in all processes, though it can transform from heat (a disordered form of energy) to work (a useful form of energy). The second law says that the overall disorder (as quantified by entropy) has to increase in a closed system, but of course this does not prohibit one part of the system from becoming more ordered at the expense of the rest, which becomes even more disordered. The key is how to get as much useful energy (work) as possible within the constraints of the second law, namely that the overall disorder has to increase. The trick is to make the rest even more disordered, i.e. of higher entropy, and then to exploit the entropy difference. This is the trick that life pulls.

Boltzmann expressed the thermodynamically driven logic beautifully in the late nineteenth century when he said: "The general struggle for existence of living beings is therefore not a fight for energy, which is plentiful in the form of heat, unfortunately untransformably, in every body. Rather, it is a struggle for entropy that becomes available through the flow of energy from the hot Sun to the cold Earth. To make the fullest use of this energy, the plants spread out the immeasurable areas of their leaves and harness the Sun's energy by a process as yet unexplored, before it sinks down to the temperature level of our Earth, to drive chemical syntheses of which one has no inkling as yet in our laboratories [2]."

According to this view all living systems are actually Maxwell's demons. This is not in the original sense that Maxwell meant, namely that demons can violate the second law, but in the sense that we all try to maximize the information about energy to utilize it to minimize entropy and therefore extract useful work. This view of life is beautifully articulated by the French biologist and Nobel Laureate Jacques Monod in his classic *Chance and Necessity* [3]. Monod even goes on to say: "... it is legitimate to view the irreversibility of evolution as an expression of the second law in the biosphere". In animals, the key energy generating processes take place in the mitochondria, which converts food into useful energy; plants rely on photosynthesis instead of food. We can almost take the fact that living systems "strive" to convert heat into useful work as a defining feature that discriminates life from inanimate matter.[1]

This is why biologists tend to think about all living processes as driven by entropic forces. This force is fictitious (i.e. not fundamental, like the electromagnetic one) and it captures the fact that life maintains itself at an entropy value lower than that of the environment. The entropy gradient that drives all life is, of course, given by the temperature difference between the Sun and the Earth.

Another physicist, Leon Brillouin, has called this the neg-entropy ("neg" is short for negative) principle [4]. Life operates far from equilibrium with its environment, which is characterized by the maximum entropy. To maintain itself far from equilibrium life needs to import negative entropy from the environment. This is why we eat food which is highly structured, either in the form of plants (which use the Earth-Sun neg-entropy to develop their own structure through processes involving carbon dioxide and water) or animals (which eat plants to maintain low entropy). We utilize the chemical energy stored in the bonds between atoms that make up our food.

The neg-entropy provided by the Sun-Earth temperature difference can be estimated to be 10 to the power of 37 times the Boltzmann constant per second. This is a huge amount. How much neg-entropy is required to create life? We will assume that to turn lifeless Earth into present Earth, it is necessary to pluck every atom required for the biomass from the atmosphere and place it into its exact present-day quantum state. These assumptions maximize the entropy of dead-Earth and minimize that of Earth, so the difference between the two entropies grossly overestimates the entropy reduction needed for life. A simple calculation shows that the entropy difference required is about 10 to the power of 44 times the Boltzmann constant. This suggests that, in principle, about an hour (10 to the power of 44–37) of Sun-Earth motion should give us enough entropy to create all life!

Of course it took much longer for life to evolve because the conditions were not right and the whole process is random. Living systems are by no means perfect work engines, and so on, but the bottom line is that the Sun-Earth temperature difference is enough to maintain life. On top of this, we have all the entropy generated by life, such as culture, industry, big cities and so on.[2] Anything that would require more than 10–53 units of entropy (a unit of entropy is k times T where k is Boltzmann's constant and T is the temperature) would take longer than the age of universe.[3]

[1]Though, of course, there are always grey areas. Man-made machines, like cars, also convert energy (fuel) into work, but they don't really strive to do it independently of us. On the other hand, we ourselves are not independent of the external factors either, so the whole issue regarding how to define life is not that easy.

[2]Culture, industry and other human inventions can of course help us handle energy more efficiently.

[3]Note that the particular way in which the organisms on Earth extract energy puts more severe limitations on the sustainability of life. For instance, it's only plants that utilize the Sun's energy directly. Herbivores obtain it by eating plants and carnivores by eating herbivores (and plants). The more removed an organism is from direct sunlight utilization, the less efficient is its extraction. Isaac Asimov in "Life and Energy" has estimated that the Sun-Earth system can sustain at best 1.5

That said, the interesting point is that Schrödinger goes beyond all this to anticipate that we might need other laws of physics to fully understand life. By this he most likely would have meant quantum physics (though at the end of the article we will explore other possibilities). This would mean that thermodynamics, designed to be oblivious to the underlying macroscopic theory, is not sufficient to explain how life operates, or indeed, how it starts.

On the one hand, finding quantum effects in biology might not be thought too surprising. After all, quantum physics has been successful for over 100 years now and no experimental deviation has ever been observed. Moreover, quantum physics is important in understanding not only atomic and subatomic particles, but also molecules, as well as macroscopic systems such as various solids and systems as large as neutron stars. On the other hand, however, living systems are warm and wet and we know that these two conditions are usually adverse to quantum physics. Large systems that strongly interact with the environment are typically well described with classical physics. So how much evidence is there really for quantum effects in biology?

2 Löwdin and Quantum DNA

The first person to take Schrödinger's suggestion seriously was a Swedish condensed matter physicist, Per-Olov Löwdin [5]. In the abstract on his 1964 paper on the mechanisms behind DNA mutations, he even coins the phrase "quantum biology". The main quantum contribution he had in mind was proton tunneling and its role in DNA replication and genetic mutations.[4] If tunneling is really fundamental to DNA replication and mutations, then of course quantum physics is crucial for life (as tunneling cannot occur in classical physics). The key in DNA replication is the matching between different base pairs on two different strands of the DNA and the position of protons required for base pairing. If protons tunnel to a different spatial location, then a mismatch in the pairs can occur, which effectively constitutes a genetic mutation.

The main idea behind biological quantum tunneling is that a certain process can still take place in quantum physics even if there is not enough energy for it under classical physics. So, even if the energy due to temperature is not sufficient to make certain molecules pair up, quantum physics tells us that there is a chance a process can still take place. Fifty years have passed since Löwdin's proposal, but no conclusive evidence is available to prove that protons tunnel in DNA and that this has an impact on DNA replication. But that it not a reason to despair, as

trillion humans (eating algae directly!). We are still far away from this, but the bound assumes that we have the relevant technology to do this too.

[4]It was already known that the explanation of chemical bonding itself lies in quantum physics. The tunneling is an additional feature.

we have other evidence that quantum physics might be at work in other biological phenomena.

3 Recent Quantum Biology Experiments

Photosynthesis is the name of the mechanism by which plants absorb, store, and use this light energy from the Sun. Here, basically the solar energy is used to utilize carbon dioxide and water to build various plant structures such as roots, branches and leaves. Recent fascinating experiments by the group of Graham Fleming at the University of Berkeley, California, suggest that quantum effects might matter in photosynthesis [7]. Furthermore, they point out a close connection between the photosynthetic energy transfer and certain types of quantum computations. In other words, plants are so much more efficient than what we expected that there must be some underlying quantum information processing.

Plants are fiendishly efficient (90–99 %) at channeling the light absorbed by antennas in their leaves to energy storage sites. The best man-made photocells barely achieve a 20 % efficiency.

So there is an enormous difference, but how do plants do it? The complete answer is not entirely clear but the overall picture is this. When sunlight hits a surface that is not designed to carefully absorb and store it, the energy is usually dissipated to heat within the surface. Either way, it is lost as far as any subsequent useful work is concerned. The dissipation within the surface happens because each atom in the surface acts independently of other atoms.

When radiation is absorbed in this incoherent way, then all its useful properties vanish. What is needed is that atoms and molecules in the surface act in unison. And this is a feat that all green plants manage to achieve. In order to understand how this happens, it is helpful to think of each molecule as a small vibrating string. All molecules vibrate as they interact with one another, transferring energy between them. When they are hit by light, they change their vibration and dynamics and end up in the most stable configuration. The crux is that if the vibrations are not quantum, then they cannot find the reliable configuration as efficiently (at best with 50 % efficiency).

Fleming's experiments were initially performed at low temperature (77 K, while plants normally operate at 300 K) but the subsequent studies indicated this behavior persists at room temperature (though the experiments have never been done using sunlight). Therefore, it is not entirely clear if any quantum effects can survive under fully realistic conditions. However, even the fact that there exists a real possibility that a quantum computation[5] has been implemented by living systems has made for a very exciting and growing area of research.

Magneto-reception is the other instance where animals might be utilizing quantum physics. European robins are crafty little creatures. Each year they make a round trip from the cold Scandinavian Peninsula to the warm equatorial planes of Africa, a hazardous trip of about 4000 miles each way. Armed with only their internal sense of direction, these diligent birds regularly make the journey without any fuss.

When faced with a similar challenge, what do we humans do? Ferdinand Magellan (the late fifteenth-century Portuguese explorer), for instance, had the same problem—and solved it. He worked out what a useful contribution a compass could make to his journey: he showed how the Earth's magnetic field—to which the compass is sensitive—could be used as a stable reference system to circumnavigate the Earth. So now, when we start in Europe and use a compass to follow the southward direction of the Earth's magnetic field, we are confident we will eventually find ourselves in Africa. But while a compass may guide us humans, it's not at all clear how robins find their way so unerringly and consistently. Do they also have a kind of inbuilt compass? Evidence suggests that there is some kind of internal guidance mechanism, but not of the type Magellan used.

Wolfgang Wiltschko, a biologist at the University of Frankfurt, came up with the first evidence of this guidance mechanism in the early 1970s [6]. He caught robins on their flight path from Scandinavia to Africa and put them in different artificial magnetic fields to test their behavior.[6] One of Wiltschko's key insights was to interchange the direction of the North and South and then observe how the robins reacted to this. Much to his surprise, nothing happened! Robins simply did not notice the reversal of the magnetic field. This is very telling: if you did this swap with a compass, its needle would follow the external field, make a U-turn, and point in the opposite direction to its original one. The human traveler would be totally confused. But somehow, the robins proved to be impermeable to the change.

Wiltschko's experiments went on to show that although robins cannot tell magnetic North from magnetic South, they are able to estimate the angle the magnetic field makes with Earth's surface. And this is really all they needed in order to navigate themselves. In a separate experiment, Wiltschko covered robins' eyes and discovered that they were unable to detect the magnetic field at all. He concluded that, without light, the robins cannot even 'see' the magnetic field, whereas of course a compass works perfectly well in the dark. This was a significant breakthrough in our understanding of the birds' navigation mechanism. The now

[5]Quantum computation is a more powerful form of computation than the current one that fully relies on the laws of quantum physics, both in order to encode information as well as to process it. In the case of photosynthesis, the information that is conveyed is simply the energy of photons, and the vibrations are a form of quantum computation that transfers this information to the appropriate reaction centre where chemistry takes over to generate energy.

[6]Note to reader: no robins were injured or suffered any obvious side effects in the making of this experiment!

widely accepted explanation to Wiltschko's result was proposed by Klaus Schulten of Maryland and developed by Thorsten Ritz of the University of Southern Florida.

The main idea behind the proposal is that light excites electrons in the molecules of the retina of the robin. The key, however, is that the excitation also causes the electron to become 'super correlated' with another electron in the same molecule. This super-correlation, which is a purely quantum mechanical effect, manifests itself in the form that whatever is happening to one electron somehow affects the other—they become inseparable 'twins'. Given that each of these twinned electrons is under the influence of the earth's magnetic field, the field can be adjusted to affect the relative degree of 'super-correlation'. So by picking up on the relative degree of 'super-correlation' (and relating this to the variation in the magnetic field) the birds somehow form an image of the magnetic field in their mind, and then use this to orient and navigate themselves. As a physicist, Ritz already knew a great deal about this super-correlation phenomenon: it had been proven many times in quantum physics under the name of 'quantum entanglement', or just 'entanglement' for short. It is this same entanglement that scientists are trying to exploit to build a new type of superfast quantum computer.

Our very simple model suggests that the computation performed by these robins is as powerful (in the sense that entanglement lasts longer) as any similar quantum computation we can currently perform![7] If this is corroborated by further evidence, its implications would be truly remarkable. For one, this would make quantum computation yet another technology discovered by Nature long before any of us humans thought it possible. While Nature continues to humble us, it also brings new hope: the realization of a large-scale usable quantum computer is possibly not as distant as we once thought. All we need to do is perhaps find a way of better replicating what already exists out there in the natural world.

4 Simulating Life with Quantum Computers

A fascinating possibility is now emerging in which we try to reverse-engineer living processes and simulate them on computers. This is not just like in John Conway's "game of life" where a cellular automata algorithm generates replicating patterns on the computer screen based on a simple underlying algorithm (whose code is only a few lines long). Instead, we can now start to think about simulating biological processes, such as energy flow, in greater detail using the new quantum technology. After all, if living systems are those that strive to get useful work out of heat, then the game of life misses the point: none of its structures can even be thought of or become alive! Do we need to breathe in the "striving" component into inanimate

[7]More specifically, robins can keep electrons entangled up to 100 μs, while we humans can manage just about the same (at room temperature).

structures or will a faithful simulation of life automatically be living? This is a deep question and we are no nearer to addressing it than Schrödinger was.

We are now already using quantum computers of up to 20 qubits to simulate and understand properties of solids. But solids are inanimate. Could it be that as our simulations become more and more intricate we could faithfully simulate living processes with quantum computers to recreate life? And could this artificial, computational simulation really become alive? Oxford University's newest research college, the Oxford Martin School, is currently funding the first steps towards such a program and aims to develop future quantum technologies inspired by the design of natural systems [8].

5 Conclusions

Let us briefly return to Schrödinger. When he mentions "other laws of physics", could that have meant that we may need to go beyond quantum physics to understand life? Could indeed biology teach us something about the underlying laws of physics? This idea is radical and most physicists would probably reject it. After all, we think of physics as underlying chemistry and chemistry as underlying biology. The arrow of causation works from small (physical systems) to large (biological) systems. But could biology still tell us something about physics?

My colleague at Oxford, David Deutsch, is not inspired by biology, but he still thinks we should take things much further as far as physics is concerned. His ideas, though, curiously seem to resonate with Schrödinger's. Not only should any future physics conform to thermodynamics, but the whole of physics should be constructed in its image. The idea is to generalise the logic of the second law as it was stringently formulated by the German-Greek mathematician Constantino Caratheodory in 1909: in the vicinity of any state of a physical system, there are other states that cannot physically be reached if we forbid any exchange of heat with the environment.

As an illustration, we can think of the experiments of James Joule, a Mancunian brewer whose name lives on in the standard unit of energy. In the nineteenth century, he studied the properties of thermally isolated beer. The beer was sealed in a tub containing a paddle-wheel that was connected to weights falling under gravity outside the tub. The rotation of the paddle warmed the beer, increasing the disorder of its molecules and therefore its entropy. But however hard we might try, we simply cannot use Joule's set-up to decrease the beer's temperature, even by a tiny fraction. Cooler beer is, in this instance, a state beyond the reach of physics.

The question is whether we can express the whole of physics simply by stipulating which processes are possible and which are not. This is very different from how physics is usually phrased, in both the classical and quantum regimes, in terms of states of systems and equations that describe how those states change in time. The blind alleys down which this approach can lead are easiest to understand in classical physics, where the dynamical equations we derive allow a whole host of processes that patently do not occur—the ones we have to conjure up the laws of thermodynamics expressly to forbid. But this is even more so in quantum physics, where modifications of the basic laws lead to even greater contradictions. The most famous example is perhaps the American Physics Nobel Laureate Steven Weinberg's modification of the basic quantum equation of motion, the Schrödinger equation, which was subsequently shown by the Swiss physicist Nicolas Gisin to lead to communications faster than the speed of light (impossible).

Reversing the logic allows us again to let our observations of the natural world take the lead in deriving our theories. We observe the prohibitions that nature puts in place—be it decreasing entropy, getting energy from nothing, travelling faster than light or whatever. The ultimately "correct" theory of physics—the logically tightest—would be the one from which the smallest deviation would give us something that breaks those commandments. If this is indeed how physics is to proceed—and this is a big "if"—then there might indeed be something in the existence of biological systems and their processes that can teach us about the basic fundamental principles of physics.[8]

References

1. E. Schrödinger, *What is Life?* (Cambridge University Press, Cambridge, 1946)
2. E. Broda, *Ludwig Boltzmann, Man—Physicist—Philosopher* (Ox Bow Press, Woodbridge, 1983). translated from the German
3. J. Monod, *Chance and Necessity* (Vintage, New York, 1971)
4. L. Brillouin, Negentropy principle of information. J. Appl. Phys. **24**(1152) (1953)
5. P.-O. Löwdin, Rev. Mod. Phys. **35** (1963)
6. R. Wiltschko, W. Wiltschko, Magnetic compass of European robins. Science **176**, 62 (1972)
7. G.R. Fleming, G.D. Scholes, Physical chemistry: quantum mechanics for plants. Nature **431**, 256 (2004)
8. R. Dorner, J. Goold, V. Vedral, Interface Focus **2**(4), 522 (2012)
9. K. Sigmund, *Games of Life* (Oxford University Press, Oxford, 1993)
10. C.E. Shannon, W. Weaver, *Mathematical Theory of Communications* (University of Illinois Press, Urbana, 1948)
11. V. Vedral, *Decoding Reality* (Oxford University Press, Oxford, 2010)
12. A. Pross, *What is Life? How Chemistry Becomes Biology* (Oxford University Press, Oxford, 2012)
13. H. Price, *Time's Arrow and Archimedes' Point* (Oxford University Press, Oxford, 1996)
14. G. Chaitin, in *Collection of Essays* (World Scientific, 2007)
15. R. Solomonoff, A formal theory of inductive inference. Inf. Control **7**, 1 (1964)

[8]For further reading consult [11–15].

Trouble with Computation: A Refutation of Digital Ontology

J. Mark Bishop

Abstract One of the questions that have defined and simultaneously divided philosophy is the question of the absolute nature of reality. Whether we have the right to ask the very question; whether we can know reality or merely be content with the epistemic conditions that make its experience possible. One response to this question, currently enjoying something of a renaissance, can be found in so-called 'digital philosophy'—the view that nature is ultimately discrete, it can be modelled digitally, its evolution is the computable output of an elegant algorithmic process, and its laws are deterministic. However, if digital philosophy presents an accurate description of the universe, then it follows that the ultimate nature of all phenomena exhibited in and by the universe must ultimately at their root be both digital and explicable in digital terms; clearly, under this view, the final explanation of consciousness must also be digital. Digital ontology so defined thus has resonance with those versed in science and science fiction who periodically reignite the hope that a computer system will one day be conscious purely by virtue of its execution of an appropriate program. In this paper I highlight a contrasting argument which reveals a *trouble with computation* whereby computational explanations of mind, and digital ontology, lead inexorably to panpsychism. (Panpsychism: the belief that the physical universe is composed of elements each of which is conscious.)

1 Digital Philosophy

In 'The Philosophy of Information' [8] Luciano Floridi characterises 'digital philosophy' as a strand of thought, proposing the following four theses: (1) nature is ultimately discrete; (2) it can be modelled digitally; (3) its evolution is the computable output of an elegant algorithmic process, and (4) its laws are deterministic. By insisting that 'reality can be decomposed into ultimate, discrete *indivisibilia*' the first two of these theses give away the underlying neo-Pythagorean nature of digital philosophy, because for Pythagoras ontology itself was defined as

J.M. Bishop (✉)
Goldsmiths, University of London, London, UK
e-mail: m.bishop@gold.ac.uk

S.B. Cooper, M.I. Soskova (eds.), *The Incomputable*, Theory and Applications of Computability, DOI 10.1007/978-3-319-43669-2_9

"that which is countable'; as Robert Lawlor, in 'Pythagorean number as form, color and light', writes [10]:

> Plato's monadic ontology implies that every number presupposes a definite and discrete unit taken from a limitless, homogeneous field. Contemplation of it thus provides access to the contemplation not only a limit, but also of the limitless. These extremes are the fundamental tension and Pythagorean thought, rather than the terms "Being" and "Non-Being". In Pythagoreanism, Being is limited and countable (perceivable); Non-Being is limitless and uncountable; Being and Non-Being can only be linguistically considered in terms of the limited and the limitless, and thus take on a particular conceptual tonality. Non-Being (the limitless) is not an opposite to Being (limit), but is only other than Being. In numerical practise, the relationship between the limited and the limitless, or between Being and Non-Being, is not fixed.[1]

Floridi [7, 8] suggests that the third thesis effectively 'interprets the neo-Pythagorean ontology in computational terms: the ultimate, discrete *indivisibilia* are actually computable digits, while elegance and Ockham's razor inclines digital ontologists to favour an algorithmic theory as simple as possible' and 'the position that unifies most supporters of digital ontology is summarised in what is known as Zuse's Thesis':

> .. the universe is being deterministically computed on some sort of giant but discrete computer [21].

Thus, for Floridi (ibid), most digital ontologists tend to subscribe to a form of pancomputationalism:[2]

> The overall perspective, emerging from digital ontology, is one of a metaphysical monism: ultimately, the physical universe is a gigantic digital computer. It is fundamentally composed of digits, instead of matter or energy, with material objects as a complex secondary manifestation, while dynamic processes are some kind of computational state transitions. There are no digitally irreducible infinities, infinitesimals, continuities, or locally determined random variables. In short, the ultimate nature of reality is not smooth and random but grainy and deterministic.

And in the context of such digital ontology, two fundamental questions arise (ibid):

1. whether the physical universe might be adequately modelled digitally and computationally, independently of whether it is actually digital and computational in itself; and

[1]Lawlor (ibid) subsequently unfolds, "*Pythagorean ontology begins with a homogeneous field of imperishable monadic units* [my emphasis] analogous to the cosmic ether of Western Science and to the mythic image of creation arising from a vast undifferentiated vibratory ocean. This primary beginning has also been alluded to as a universal consciousness, that is, as pure Unity infinitely replicating itself as monadic units. The epistemology, or knowledge of this world, however, is dependent form (*eidos*). The word *eidos* also translates as *idea*. Creation only becomes manifest through both *ontos* and *episteme*, being and knowledge of being, and knowledge of being arises from the contemplation of the laws of organisation which we experience as form."

[2]Albeit digital ontology and pancomputationalism are distinct positions, for example Wheeler [20] supports the former but not the latter; processes in Wheeler's universe may not be reducible to mere *computational* state transitions.

2. whether the ultimate nature of the physical universe might be actually digital and computational in itself, independently of how it can be effectively or adequately modelled.

Considering these two questions, Floridi (ibid) claims that the first is an 'empirico-mathematical' question that, so far, remains unsettled whilst the second is a metaphysical question that, by an ingenious thought experiment, he endeavours to show is ill posed and hence, when answered, misapplied:

> The first step [of the thought experiment] consists in arguing that, even assuming that reality in itself is indeed digital or analogue, an epistemic agent, confronted by what appears to be an (at least partly) analogue world of experience, could not establish whether its source (that is, reality in itself as the source of the agent's experience or knowledge) is digital (or analogue).
> One could object, however, that this first, epistemological step is merely negative, for it establishes, at best, only the unknowability of the intrinsically digital (or analogue) nature of reality lying behind the world of experience, not that reality in itself is not digital (or analogue). Independently of the epistemic access enjoyed by an agent, the objection continues, logic dictates that reality must be assumed to be digital/discrete (grainy) or continuous/analogue (smooth).
> So the second step is more positive and ontological. It consists in showing that the initial concession, made in the first step, can be withdrawn: the intrinsic nature of reality does not have to be digital or analogue because the dichotomy might well be misapplied. Reality is experienced, conceptualised and known as digital or analogue depending on the level of abstraction (LoA) assumed by the epistemic agent when interacting with it. Digital and analogue are features of the LoA modelling the system, not of the modelled system in itself.

In a recent paper [14] Chryssa Sdrolia and I argue that while we sympathise with Floridi's criticism 'against' digital philosophy, we express doubt about the validity of the thought experiment he introduces against it; alternatively, in this paper I revisit an argument I first framed in 2002 [1] in the context of [so-called] 'Strong Artificial Intelligence'[3] revealing *trouble at the heart of computation*—a trouble that prompts the rejection of both Zuse (pancomputationalist) digital ontology and any purely computational explanation of mind.

2 Dancing with Pixies

Many people hold the view that 'there is a crucial barrier between computer models of minds and real minds: the barrier of consciousness' and thus that computational connectionist simulations of mind (e.g. the huge, hi-fidelity simulation of the brain

[3]The notion of a 'strong computational explanation of mind' derives from Searle, who in [15] famously defined 'Strong Artificial Intelligence' as follows: "*the computer is not merely a tool in the study of the mind; rather the appropriately programmed computer really is a mind, in the sense that computers given the right programs can be literally said to understand and have other cognitive states.*" In this 'strong view', a computational explanation of mind explains **all** aspects of mind, including phenomenal consciousness; for discussion of Searle's position on Strong AI and the 'Chinese room', see [12].

currently being instantiated in Henry Markram's EU-funded €1.19 billion 'Human Brain Project') and 'phenomenal [conscious] experiences' are conceptually distinct (e.g. [2, 12, 15, 18]).

But is consciousness a prerequisite for genuine cognition and the realisation of mental states? Certainly Searle believes so, "the study of the mind is the study of consciousness, in much the same sense that biology is the study of life" [17], and this observation leads him to postulate a 'connection principle' whereby, "...any mental state must be, at least in principle, capable of being brought to conscious awareness". Hence, if computational machines are not capable of enjoying consciousness, they are incapable of carrying genuine mental states and computational connectionist projects must ultimately fail as an adequate model for cognition.

In the following I briefly review a simple reductio ad absurdum argument that suggests there may be problems in granting phenomenal (conscious) experience to any computational system purely by virtue of its execution of a particular program; if correct the argument suggests either that strong computational accounts of consciousness (and a fortiori all computational connectionist accounts) must fail *or* that panpsychism is true.

The argument—the *dancing with pixies reductio*—derives from ideas originally outlined by Hilary Putnam [13], Tim Maudlin [11], and John Searle [16] and subsequently criticised by David Chalmers [4], Colin Klein [9] and Ron Chrisley [5, 6] amongst others.[4]

In what follows, instead of seeking to justify Putnam's claim that "every open system[5] implements every Finite State Automaton" (FSA) and hence that "psychological states of the brain cannot be functional states of a computer", I will simply establish the weaker result that, over a finite time window, every open physical system (OPS) implements the execution trace—the series of state transitions - defined by FSA Q acting on specified input (I).

That this result leads to panpsychism is clear as, equating FSA Q (I) to a specific computational system that is claimed to instantiate phenomenal states as it executes, and following Putnam's procedure, identical computational (and *exhypothesi* phenomenal) state transitions can be found (dancing) in every open physical system.

---

[4]For comprehensive early discussion of these themes see Stevan Harnad (ed) '*What is Computation?*'; the November 1994 special issue of 'Minds and Machines' journal (volume 4: 4); ISSN: 0924-6495.

[5]In the context of this discussion an open system is a system which can exchange energy [and/or matter] with its surroundings.

2.1 *The Dancing with Pixies (DwP) Reductio...*

In his 1950 paper, 'Computing Machinery and Intelligence', Turing [19] defined Discrete State Machines (DSMs) as "machines that move in sudden jumps or clicks from one quite definite state to another", and explained that modern digital computers fall within the class of them. An example DSM from Turing is one that cycles through three computational states (Q_1, Q_2, Q_3) at discrete clock clicks. Such a device, which cycles through a linear series of state transitions 'like clockwork', may be implemented by a simple wheel-machine that revolves through 120^0 intervals (see Fig. 1).

By labelling the three discrete positions of the wheel (A, B, C) we can map computational states of the DSM (Q_1, Q_2, Q_3) to the physical positions of the wheel (A, B, C) such that, for example, ($A \Rightarrow Q_1; B \Rightarrow Q_2; C \Rightarrow Q_3$). Clearly this mapping is observer relative: position A could map to Q_2 or Q_3 and, with other states appropriately assigned, the machine's function would be unchanged. In general, we can generate the behaviour of any k-state (input-less) DSM, $f(Q) \Rightarrow Q'$, by a k-state wheel-machine (e.g. a suitably large digital counter) and a function that maps each 'wheel-machine/counter' state C_n to each computational state Q_n as required.

In addition, Turing's wheel-machine may be configured such that whenever it enters a particular position [computational state] a lamp can be made to switch on and furthermore, the machine may be paused in any of its k [discrete] positions by the application of a brake mechanism. Input to the machine is thus the state of the brake ($I = ON|OFF$) and its output (Z) the state of the lamp. The operation of Turing's DSM with input is thus described by a series of contingent branching state transitions, which map from current state to next state $f(Q, I) \Rightarrow Q'$, and define output (in the Moore form) $f(Q') \Rightarrow Z$.

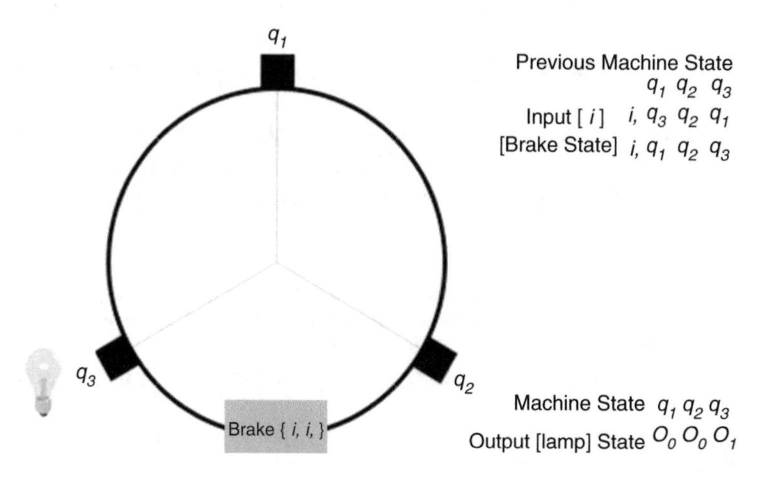

Fig. 1 Turing's 'Discrete State Wheel Machine'

However **over any finite time interval** fully specifying the input to the device[6] entails that such contingent behaviour reverts to a mere linear-series of modal state transitions; $f(Q) \Rightarrow Q'$.

For example, if Turing's three state DSM starts in Q_1 and the brake is OFF for two consecutive clock-ticks, its behaviour, (execution trace), is fully described by the sequence of state transitions ($Q_1 \Rightarrow Q_2 \Rightarrow Q_3$); alternatively, if the machine is initially in state Q_3 and the brake is OFF for one clock-tick and ON for the next, its will transit the following states ($Q_3 \Rightarrow Q_1 \Rightarrow Q_1$).

Hence, over a finite time window, if the input to a DSM is specified, we can trivially map from each wheel-machine (digital counter) state C_n to each computational state Q_n, as required. Furthermore, in Bishop [3] I demonstrate, following Putnam [13], how to map the computational state transition sequence of any such DSM onto the [non-repeating] internal states instantiated by any *open physical system* (such as a rock):

> Discussed in a brief appendix to Hilary Putnam's 1988 book 'Representation and Reality' is a short argument that endeavours to prove that every open physical system is a realisation of every abstract Finite State Automaton and hence that functionalism fails to provide an adequate foundation for the study of the mind.
> Central to Putnam's argument is the observation that every open physical system, S, is in different 'maximal' states[7] at every discrete instant and hence can be characterised by a discrete series of non-cyclic natural state transitions, $[s_1, s_2 .. s_t .. s_n]$. Putnam argues for this on the basis that every such open system, S, is continually exposed to electromagnetic and gravitational signals from, say, a natural clock. Hence by quantizing these natural states appropriately, every open physical system can by considered as a generator of discrete non-repeating modal state sequences, $[s_1, s_2 .. s_\infty]$.[8]

Thus, considering Turing's input-less three state DSM Q and an open physical system S, (where $S = \{s_1, s_2, s_3, s_4, s_5, s_6\}$ over time interval $[t_1 .. t_6]$), using Putnam's mapping it is trivial to observe that if we map DSM state $[Q_1]$ to the disjunction of open physical system states, $[s_1 \text{ v } s_4]$, DSM state $[Q_2]$ to the disjunction of OPS states, $[s_2 \text{ v } s_5]$ and DSM state $[Q_3]$ to the disjunction of OPS states, $[s_3 \text{ v } s_6]$, then the open physical system, S, will fully implement Q as the OPS transits states $[s_1 .. s_6]$ over the interval $[t_1 .. t_6]$.

Furthermore, given any OPS state, s, ($s \in \{s_1, s_2, s_3, s_4, s_5, s_6\}$), say $s = s_4$ at time $[t]$, (in which case by application of Putnam's mapping we note that the DSM is in computational state $[Q_1]$), we can predict that the OPS will modally enter state

[6] As might happen if we present exactly the same input to a robot on repeated experimental trials; in which case, ceteris paribus, the computationalist is committed to saying the 'experience' of the robot will remain the same during each experimental trial.

[7] A 'maximal' state is a total state of the system, specifying the system's physical makeup in absolute detail.

[8] Chalmers [4] observes, "*Even if it [the claim that 'every open physical system is a realisation of every abstract Finite State Automaton'] does not hold across the board (arguably, signals from a number of sources might cancel each other's effects, leading to a cycle in behaviour), the more limited result that every non-cyclic system implements every finite-state automaton would still be a strong one.*"

$s = s_5$ and hence, by a further application of Putnam's mapping, observe that the DSM entered computational state $[Q_2]$ at time $[t + 1]$ etc.

Now, returning to a putative conscious robot: at the heart of such a beast there is a computational system—typically a microprocessor, with finite memory and peripherals; such a system **is** a DSM. Thus, with input to the robot fully specified over some finite time interval, we can map its execution trace onto the state evolution of any digital counter or, after Putnam, any open physical system.

Thus, if the control program of a robot genuinely instantiates phenomenal experience as it executes, then so must the state evolution of any OPS, in which case we are inexorably led to a panpsychist worldview whereby disembodied phenomenal consciousnesses [aka 'pixies'] are found in every open physical system.

NB. For detailed analysis of the above DwP reductio, including discussion of the key objections to it, see Bishop [2, 3].

3 Conclusion: Refuting Digital Ontology

In Sect. 1 I highlighted two fundamental questions related to digital ontology: the first of these asks if, irrespective of its ultimate nature, the universe can be adequately modelled digitally and computationally[9] and the second enquires if the ultimate nature of reality is digital, and the universe is a computational system equivalent to a Turing Machine. Computationally, sensu stricto, it is apparent that the latter ontological case—if digital ontology (so defined) is true—implies the former epistemic ('empirico-mathematical') case; as Floridi [7, 8] observes, "*the empirico-mathematical and the metaphysical position with respect to digital ontology are compatible and complimentary*". Furthermore, *exhypothesi*, as the existence of humans and human consciousness is a fundamental component of our world, if true, both questions imply that there must exist at least one Turing machine computation that will instantiate human [phenomenal] consciousness.

Conversely, the DwP reductio ad absurdum implies that if phenomenal consciousness is instantiated by the mere execution of a computer program, then a vicious form of panpsychism holds, whereby every possible conscious experience can be found in every open physical system. However, against the backdrop of our immense scientific knowledge of the closed physical world, and the corresponding widespread desire to explain everything ultimately in physical terms, such panpsychism has come to seem an implausible worldview; hence I conclude that we should reject both digital ontology and strong computational explanations of mind.

Acknowledgements The central argument presented herein was developed under the aegis of the Templeton project 21853, *Cognition as Communication and Interaction*. Some *Troubles with*

[9]By adequate here I imply that **all** relevant properties and functions of the universe can be so instantiated which, *exhypothesi*, includes human phenomenal consciousness.

computation were first aired at the Newton Institute workshop on 'The Incomputable', held at the Kavli Royal Society International Centre at Chicheley Hall (12–15 June 2012). Some elements of this paper have been extracted from Sdrolia, C. & Bishop, J.M., (2014), *Rethinking Construction. On Luciano Floridi's 'Against digital Ontology'*, Minds and Machines 24(1), and Bishop, J.M., (2014), *History and Philosophy of Neural Networks*, in Ishibuchi, H., (ed), *Computational Intelligence*, UNESCO Encyclopaedia of Life Support Systems (EOLSS).

References

1. J.M. Bishop, Dancing with pixies: strong artificial intelligence and panpyschism, in *Views into the Chinese Room: New Essays on Searle and Artificial Intelligence*, ed. by J. Preston, J.M. Bishop (Oxford University Press, Oxford, 2002)
2. J.M. Bishop, A cognitive computation fallacy? Cognition, computations and panpsychism. Cogn. Comput. **1**(3), 221–233 (2009)
3. J.M. Bishop, Why computers can't feel pain. Mind. Mach. **19**(4), 507–516 (2009)
4. D.J. Chalmers, *The Conscious Mind: In Search of a Fundamental Theory* (Oxford University Press, Oxford, 1996)
5. R. Chrisley, Why everything doesn't realize every computation. Mind. Mach. **4**, 403–420 (1995)
6. R. Chrisley, Counterfactual computational vehicles of consciousness. Toward a Science of Consciousness, University of Arizona, Tucson (2006)
7. L. Floridi, Against digital ontology. Synthese **161**, 219–253 (2009)
8. L. Floridi, *The Philosophy of Information* (Oxford University Press, Oxford, 2011)
9. C. Klein, Maudlin on computation. Working paper (2004)
10. R. Lawlor, Pythagorean number as form, color and light, in *Homage to Pythagoras: Rediscovering Sacred Science*, ed. by C. Bamford (Lindisfarne Press, Felton, 1994)
11. T. Maudlin, Computation and consciousness. J. Philos. **86**, 407–432 (1989)
12. J. Preston, J.M. Bishop (eds.), *Views into the Chinese Room: New Essays on Searle and Artificial Intelligence* (Oxford University Press, Oxford, 2002)
13. H. Putnam, *Representation and Reality* (Bradford Books, Cambridge, MA, 1988)
14. C. Sdrolia, J.M. Bishop, Rethinking construction. On Luciano Floridi's 'Against Digital Ontology'. Mind. Mach. **24**(1), 89–99 (2014)
15. J.R. Searle, Minds, brains, and programs. Behav. Brain Sci. **3**(3), 417–457 (1980)
16. J.R. Searle, Is the brain a digital computer? Proc. Am. Philos. Assoc. **64**, 21–37 (1990)
17. J.R. Searle, *The Rediscovery of the Mind* (MIT, Cambridge, MA, 1992)
18. S. Torrance, Thin phenomenality and machine consciousness, in *Proceeding of the AISB Symposium on Next Generation Approaches to Machine Consciousness*, ed. by R. Chrisley, R. Clowes, S. Torrance. University of Hertfordshire, Hertfordshire (2005)
19. A.M. Turing, Computing machinery and intelligence. Mind **49**, 433–460 (1950)
20. J.A. Wheeler, Information, physics, quantum: the search for links, in *Complexity, Entropy, and the Physics of Information*, ed. by W.H. Zureck (Westview Press, Boulder, CO, 1990)
21. K. Zuse, Rechnender Raum (Braunschweig: Vieweg). Eng. tr. with the title Calculating Space, MIT Technical Translation AZT-70-164-GEMIT, Massachusetts Institute of Technology (Project MAC), Cambridge, MA (1969)

Part IV
The Nature of Information: Complexity and Randomness

Complexity Barriers as Independence

Antonina Kolokolova

Abstract After many years of effort, the main questions of complexity theory remain unresolved, even though the concepts involved are simple. Understanding the main idea behind the statement of the "P vs. NP" problem does not require much background ("is it easier to check answers than to produce them?"). Yet, we are as far from resolving it as ever. Much work has been done to unravel the intricate structure in the complexity world, the "complexity zoo", contains hosts of inhabitants. But the main questions are still elusive.

So a natural question comes to mind: is there any intrinsic reason why this is still unknown? Is there any rationale why the proofs are out of our reach? Maybe we are not using the right techniques—or maybe we are not pushing our techniques far enough? After trying to prove a statement and failing, we try to prove its negation; after failing at that as well, we resort to looking for an explanation that might give us a hint about why our attempts are failing. Indeed, in the world of computational complexity there have been several results of this nature: results that state that current techniques are, in a precise mathematical sense, insufficient to resolve the main open problems. We call these results "barriers".

A pessimistic view of the barrier results would be that the questions are hard, intrinsically hard. But there is a more optimistic way of interpreting them. The fact that certain classes of proof techniques, ones that have specific properties, are eliminated gives us a direction in which to search for new techniques. It gives us a method for discovering ways of approaching questions in places where we might not have been looking, if not for the barrier results.

In this paper, we will focus on three major complexity barriers: Relativization (Baker et al., SIAM J Comput 4(4):431–442, 1975), Algebrization (Aaronson and Wigderson, ACM Trans Comput Theory 1(1), 2009), and Natural Proofs (Razborov and Rudich, J Comput Syst Sci 55:24–35, 1997). Interestingly enough, all three of those can be recast in the framework of independence of a theory of logic. That is, theories can be constructed which formalize (almost) all known techniques, yet for which the main open questions of complexity theory are independent.

A. Kolokolova (✉)
Department of Computer Science, Memorial University of Newfoundland St.John's, NL, Canada
e-mail: Kol@mun.ca

S.B. Cooper, M.I. Soskova (eds.), *The Incomputable*, Theory and Applications of Computability, DOI 10.1007/978-3-319-43669-2_10

1 Introduction

Complexity theory evolved from computability theory by restricting the notions of computable to computable *efficiently*. The main objects of complexity theory can be viewed as "scaled down" versions of basic objects of computability theory; here, "scaling down" amounts to limiting computational resources, such as bounding quantifiers to make quantified objects "small" and requiring all computation to be "efficient".

There are problems which are known to be decidable (that is, there is an algorithm producing a definite answer for every input), but for which the time to compute the answers is immense, exceeding the number of atoms in the universe even for relatively small inputs. Here we assume the customary way of measuring the computational complexity of a problem as a function of the input size. To make the notion of "computable" more realistic, it is natural to put a bound on the computational resources allotted to compute the answer: time for how long the computation is allowed to run (treated here as the length of the computation, that is, the number of steps of the algorithm), or amount of memory the algorithm is allowed to use. We bound the running time (memory) on all possible inputs of a given length, which is the worst-case complexity measure.

Intuitively, checking whether an input string is a palindrome is a simpler problem than deciding whether the input is a true statement of Presburger arithmetic, though the latter also has an algorithm producing an answer for every input. And in general, a scaled-down version of a problem of finding a solution out of infinitely many possibilities (i.e., finding an accepting computation of a Turing machine) can be formulated as finding a small solution (comparable to the size of the input) significantly faster than by brute-force search over all small answers.

1.1 The Early History of Efficient Time-Bounded Computation

What would be a reasonable bound on the length of computation which would make it "efficient"? In his 1956 letter to von Neumann,[1] Gödel asks whether there is an algorithm that, given an n-bit long encoding of a formula, can find a short proof in $k \cdot n$ or $k \cdot n^2$ steps, where k is a number (providing the $k \cdot n$ as a lower bound on such a computation); thus, here, "efficient" is kn or kn^2. There is a problem though with defining "efficient" as linear (or quadratic) in the length of the input: it is too model-dependent. For example, any single-tape Turing machine needs roughly n^2 time to decide if an n-bit string is a palindrome, whereas there is a two-tape Turing machine which accomplishes this task in about $2n$ time.

[1]This letter, though it outlines many of the central concepts of complexity theory, was only discovered in 1988, well after these concepts were defined independently.

A few more papers in the early 1960s dealt with the notions of efficient computation and computational difficulty of problems. Hartmanis and Stearns [31, 32] introduced time and space complexity classes and proved, using computability-theoretic techniques, that giving a Turing machine more time or space allows it to solve more problems (when time and space bounds are computable functions). Edmonds [22] discussed the notion of efficient algorithms in the paper where he provided a better than brute-force search algorithm for finding a maximum matching in a general graph. And in his talk at the International Congress on Logic Methodology and Philosophy of Science in 1964, Cobham [16] formulated the notion of efficient computation that became the standard in complexity theory: the number of steps of a computation should be bounded by some fixed polynomial function of the length of an input. There, Cobham also presented a recursion-theoretic characterization of polynomial time which we will discuss at greater length later in this paper.[2]

As the class of recursive (decidable) languages scales down to the class of polynomial-time computable languages (which we will denote by P), the recursively enumerable (semi-decidable) languages become NP, non-deterministic polynomial-time computable languages. That is, this is a class of languages such that for every string in the language there is a certificate (of size at most polynomial in the length of the input) which can be verified by a (polynomial-time) computable predicate. Rather than asking for the existence of a finite computation that can be computably verified, we ask for the existence of a polynomial-length witness which can be checked in time polynomial in the length n of the input string. Note that there are only exponentially many strings of length polynomial in n, so finding such a witness is always computable in exponential time (EXP): just try all possible small witnesses and see if one of them works.

Is trying all possible witnesses necessary, or is it always possible to do significantly better? This question is sometimes referred to as the problem of avoiding exhaustive search, also known as a "perebor problem" (проблема перебора) in Russian-language literature. This question was formalized as the "P vs. NP" question in the seminal paper of Stephen Cook [17] in which the notion of NP-completeness was first presented; independently, very similar notions have been formulated by Leonid Levin [44]. Unlikely as it seems, this question is wide open: though we know from the Hartmanis-Stearns paper [32] that $\mathsf{P} \subsetneq \mathsf{EXP}$, that is, there are languages computable in exponential time but not in polynomial time, and that $\mathsf{P} \subseteq \mathsf{NP} \subseteq \mathsf{EXP}$, we are still not able to rule out either of these two inclusions being equalities.

[2]This paragraph is by no means intended to give a full history of development of these concepts. There are numerous books and surveys that address this much better. My main goal is to give a quick introduction to how complexity classes came from the corresponding classes in computability theory, and motivate why we focus on polynomial-time computation.

1.2 Bounding Other Resources

There are many more beasts of the complexity zoo roaming in this range between P and EXP. Probably the most notable one is PSPACE, the class of languages computable by algorithms using only the amount of memory polynomial in the size of the input. The canonical complete problem for this class is True Quantified Boolean Formulae (TQBF) with arbitrary nesting depth of quantifiers; a number of other PSPACE-complete problems ask if there is a winning strategy in a two-player game: there, the alternating quantifiers code alternating moves by the two players.

Naturally, P $\subseteq$ PSPACE, since even when visiting a different memory location at every time step, there are only polynomially many locations that can be touched. Similarly, NP $\subseteq$ PSPACE: notice that the algorithm that checks all possible witnesses can be implemented with only the polynomial amount of memory: each check takes polynomial time (and thus polynomial space), and the counter to keep track of the witnesses is bounded by the length of the largest witness. Also, PSPACE $\subseteq$ EXP, since for a given language L and its input, there are exponentially many possible configurations (say, of the tape of a Turing machine deciding this language with polynomial amount of space). Then the question of whether this Turing machine accepts becomes a reachability question: is there a path from the start configuration to some accepting configuration in the configuration graph of this computation, where there is an edge from one configuration to another if the latter can be obtained from the former by a single transition? Since reachability for polynomial-size graphs can be tested in polynomial time using numerous algorithms starting from depth first search or breadth first search, reachability for exponential-size graphs such as this configuration graph can be decided in exponential time. Note that this also works for non-deterministic settings, as when there are multiple possible transitions out of a given configuration, so even non-deterministic polynomial-space computable languages can be computed in EXP. It turns out that, for polynomial space, non-determinism does not help: a classic result of Savich [53] shows that non-deterministic polynomial-space computation can be simulated by a deterministic computation with only polynomial overhead. Thus, P $\subseteq$ NP $\subseteq$ PSPACE $=$ NPSPACE $\subseteq$ EXP, where only the first and last classes are known to be distinct.

2 The First Barrier: Relativization

Hartmanis and Stearns [31, 32] were the first to define a notion of a complexity class (as a class of recursive sequences computable by a multi-tape Turing machine within a specified time bound given as a function of the input size). With this definition, they showed that such complexity classes form a strict hierarchy: given more time, Turing machines can compute more complex languages. The main tools they used were the familiar methods from computability theory: simulation and diagonalization. Essentially, they showed how, given a computable function $T(n)$, to construct a language computable in time $(T(n))^2$ which differs from any language computable in time $T(n)$. The quadratic time bound was later improved to $T(n) \log T(n)$ by Hennie and Stearns [34]; subsequently, hierarchy theorems were proven for non-deterministic computation [18] and space complexity [55].

The proofs of these hierarchy theorems, as is usually the case with proofs based on diagonalization, have a property which makes the results stronger: they are insensitive to the presence of oracles. More precisely, an "oracle", introduced by Turing [59], is just a language, and an oracle Turing machine for a given oracle can query whether a given string is in the language in constant (unit) time. This immediately allows oracle Turing machines to compute incomputable languages by taking the language in question as an oracle (e.g., oracle Turing machines can decide the Halting problem by a single query to the language of all positive instances of the Halting problem). Thus, any language can be decided by a Turing machine with an oracle to this language, trivially. However, there is no language powerful enough so that with this language as an, oracle Turing machines can decide everything. And the proof of this is the same as the standard diagonalization proof of existence of undecidable languages: the list of all Turing machines with an access to the same oracle is countable. Therefore, replacing the list of all Turing machines with the list of all Turing machines with an oracle A does not change the proof. We will call proofs which have this property, insensitivity to the presence of oracles, *relativizable*.

The Hartmanis-Stearns proof of the hierarchy theorem uses just the diagonalization and so is relativizable. Given an oracle, there is a hierarchy of time and space complexity classes with respect to this oracle. However, when types of resources are mixed, the situation gets more complicated, leading to the first barrier that we will consider, the *relativization barrier*.

2.1 Relativization: The Baker, Gill, and Solovay Theorem

If there is a purely diagonalization-based proof of $\mathsf{P} \neq \mathsf{NP}$, then this proof should be insensitive to the presence of oracles. So, P with an oracle A (denoted by P^A) should not be equal to NP^A for any oracle A. However, Baker, Gill and Solovay in their 1975 paper presented oracles A and B such that, with respect to A, $\mathsf{P} = \mathsf{NP}$, and with respect to B, $\mathsf{P} \neq \mathsf{NP}$.[3] Thus, in particular, neither the proof of $\mathsf{P} = \mathsf{NP}$ nor of $\mathsf{P} \neq \mathsf{NP}$ can be done using relativizing techniques such as diagonalization.

The first oracle, A, for which $\mathsf{P}^A = \mathsf{NP}^A$, is chosen to be a language powerful enough that it can be used to solve problems in both P and NP, and in a class with good closure properties. In particular, A can be any language that is complete for

[3]According to Baker, Gill and Solovay, an oracle for which $\mathsf{P} = \mathsf{NP}$ was independently constructed by Meyer and Fischer, and also by H. B. Hunt III; however, they did not use the oracle described here. For $\mathsf{P} \neq \mathsf{NP}$, they attributed some constructions to Ladner, and some again to Meyer/Fischer.

PSPACE. It is possible to decide languages in NP^A in NPSPACE by simulating all polynomially many queries in PSPACE for each non-deterministic branch of computation. Note that, since PSPACE = NPSPACE by Savitch's theorem, NP^A is already in PSPACE. Finally, PSPACE $\subseteq \mathsf{P}^A$, as every language in PSPACE can be decided by a reduction to A (in polynomial time) by definition of its completeness. Therefore, $\mathsf{P}^A = \mathsf{NP}^A$.

The oracle B for which P and NP are separate is constructed in [11] using diagonalization ideas. Consider, for a given set B, a language $L(B)$ which contains all strings of a given length if there is a string of that length in B, and no strings of that length otherwise (that is, $L(B) = \{x | \exists y \in B\ (|x| = |y|)\}$). This language is in NP^B, because it is possible, given x, to check if any y of the same length is in B, checking one such y on any computation path. The interesting part is to construct B such that $L(B) \notin \mathsf{P}^B$.

The set of strings which could potentially be in the oracle is divided into blocks by size, where each block is designed to fool the ith polynomial-time Turing machine M_i. Consider the behaviour of M_i on the string 0^n, for n exponentially increasing at each step and such that 2^n is larger than the limit on M_i's running time. If M_i accepts 0^n, then ith block will contain no strings (so for all x such that $|x| = n$, $x \notin L(B)$, yet M_i accepts 0^n). Otherwise, it will contain one string, which is the smallest string of length n not queried by M_i on 0^n; it exists because 2^n is larger than M_i's running time. Since M_i does not query this string, its presence or absence in B will not affect M_i's not accepting 0^n. But if M_i were deciding $L(B)$, it would have to accept all strings of length n, including 0^n. Finally, because each n is chosen to be exponentially larger than the n from the previous stage, but all M_j run in polynomial time, no previous M_j can query strings of length n, so adding strings of length n does not change their behavior.

In the same paper, Baker, Gill and Solovay construct a number of oracles corresponding to various scenarios of relationships between these classes. Is NP closed under complementation when it is not P? For both there is an oracle with respect to which this scenario is true. If not, then is P the subset of NP closed under intersection? Again, for both of them an oracle world making it true can be constructed. A number of subsequent results showed that for a vast majority of complexity questions there are contrary relativizations: existence of one-way functions, power of probabilistic algorithms (BPP vs. P), interactive games (IP vs. PSPACE), etc.

This surprising result can be immediately recast to show, for example, that with respect to some oracle C, P^C vs. NP^C is independent of the axioms of ZFC (or other axiomatizable consistent formal theories). More specifically, Hartmanis and Hopcroft [30] construct a recursive set C using the [11] oracles A and B as follows. Let C be the language $\mathcal{L}(M)$ of a Turing machine M which accepts x if there exists either a proof of $\mathsf{P}^{\mathcal{L}(M)} = \mathsf{NP}^{\mathcal{L}(M)}$ among the first x proofs in the theory and $x \in B$, or a proof of $\mathsf{P}^{\mathcal{L}(M)} \neq \mathsf{NP}^{\mathcal{L}(M)}$ among the first x proofs in the theory and $x \in A$. The existence of such an M comes from the recursion theorem. Let $C = \mathcal{L}(M)$. But now C is essentially A, except for the finite part, if there is a proof of $\mathsf{P}^C \neq \mathsf{NP}^C$, and essentially B if there is a proof of $\mathsf{P}^C = \mathsf{NP}^C$, contradicting the [11] theorem that the opposite is true for oracles A and B.

2.2 Polynomial Time as a Black Box: The Recursion-Theoretic Definition

Hartmanis and Hopcroft use results from [11] to define a complexity-theoretic problem independent of Zermelo-Fraenkel set theory. A different logic question that the relativization barrier inspires is whether it itself can be restated as an independence result. It does have a similar flavour: no technique with a certain property can be used to resolve questions. Thus, intuitively, a theory formalizing the reasoning having this property would only prove relativizing results, and all non-relativizing results would be independent from it. This intuition has been made precise in the unpublished manuscript[4] by Arora, Impagliazzo and Vazirani [6].

But which property would such a theory formalize? One way to summarize the techniques that give relativizing results is to say that they treat computation as a black box. Such techniques rely on the closure properties of the corresponding classes of functions, and on properties such as the existence of a universal function for a class (useful for simulation). However, they do not consider the inner workings of a model of computation such as a Turing machine: in most such results, a Turing machine can be readily replaced with a Random Access Machine or a recursion-theoretic characterization of the corresponding function class. Consider, for example, the difference between lambda calculus and the Turing machine models. In the former framework, the only information about the functions is their recursive definition. So they are, computationally, black boxes. There is no extra information. But in the Turing machine computation, there is a lot of additional information: for example, subsequent steps of computation only affect a small number of adjacent cells.

So, a theory formalizing only relativizing techniques can be based on reasoning that only works with some generic recursion-theoretic definition of a class of functions in a complexity class. And in [6], Arora, Impagliazzo and Vazirani explored that idea by building upon Cobham's [16] definition of polynomial-time computable functions.

Definition 2.1 Let FP' be a class of functions satisfying the following properties:

1. FP' contains basic functions[5]: constant, addition, subtraction, length $|x|$, $BIT(x,j)$, projection, multiplication and the "smash" function $2^{|x|^2}$.

---

[4]This paper has never been published, although it is mentioned in a number of published works. The standard graduate complexity theory textbook by Arora and Barak [4] devotes a page to it; Fortnow discusses it in his 1994 [23] paper. Most publications on the algebrization barrier, including [2], reference [6] and the follow-up paper [36], rely heavily on the results of [6] to formalize algebrization; it is mentioned in a number of other publications by Aaronson. Sometimes this manuscript is dated 1993 and/or cited with a title "On the role of the Cook-Levin theorem in complexity theory".

[5]In fact, successors s_0, s_1 and smash are sufficient.

2. FP' is closed under function composition $f \circ g$ and pairing $\langle f(x), g(x) \rangle$ of functions.
3. FP' is closed under the limited recursion on notation: for functions $g, h_0, h_1 \in FP'$ and constants c, d,

$$f(x, 0) = g(x) \tag{1}$$

$$f(x, 2k) = h_0(x, k, f(x, k)) \qquad f(x, 2k+1) = h_1(x, k, f(x, k)) \tag{2}$$

$$|f(x, k)| \leq c|x|^d \tag{3}$$

Here, $c|x|^d$ provides a polynomial bound on the length of the output of $f(x, k)$ to avoid using repeated squaring to define exponentiation. In this case, we use a true polynomial bound rather than a function from FP'.

Cobham's celebrated result states that the class FP of polynomial-time computable functions is the minimal class FP' satisfying Definition 2.1. But what can be said about classes FP' satisfying these axioms which are not minimal? Such classes might contain, in addition to polynomial-time functions, spurious functions of arbitrary complexity and their closure under polynomial-time operations. This already becomes reminiscent of the idea of an oracle: indeed, adding a characteristic function of an oracle to FP and closing under the operations (composition, pairing, limited recursion on notation) gives an FP' satisfying the definition.

Suppose now that FP' is a class satisfying the definition. Can it be viewed as FP^O for some oracle O? First of all, if the spurious functions produce a huge (non-polynomial length) output, then they would not fit into the oracle framework: an oracle polynomial-time Turing machine gets one bit of information from each query to the oracle, and has only a polynomial in input length amount of time to write its answer on the tape. But what if every function in FP' produces an output of polynomial length, in the same manner as the output of the function in the limited recursion on notation is polynomially bounded? Then, an oracle can be constructed with the use of one more device: a universal function for FP'. This universal function, $U(i, t, x)$, is defined to compute $f_i(x)$ within the time bound t, for every $f_i \in FP'$. Note that without the time bound there is no universal function for FP that would be in FP, even for the actual class of polynomial-time computable functions: there is no specific polynomial that can bound the running time of every polynomial-time function by the time hierarchy theorem. However, for every polynomial-time function, there is a bound $2^{|x|^c}$ for some constant c depending on the function which would allow the universal function to be computed within the time polynomial in its parameters. Now, provided $U(i, t, x)$ is in FP' and is a universal function for FP', an oracle O that on a query (i, t, x, j) outputs the jth bit of $U(i, t, x)$ is sufficient to compute all functions in FP' in FP^O.

2.3 *Relativization as Independence: The Arora, Impagliazzo and Vazirani [6] Framework*

Let T be a theory (for example, the Peano Arithmetic). Rather than reasoning about the actual polynomial-time functions in T, we would like to reason about functions in FP' given by the recursive definition above: this will be essentially the only information about these functions given to the theory. The question now becomes: what kind of results can be proven in T given only this limited information about FP'? It does not matter too much what is the reasoning power of T itself: we can take T to be as powerful as needed to formalize the combinatorial arguments a proof might require. The main restriction is the black-box view of FP' functions.

More precisely, define *RCT* (for "relativized complexity theory") to be T (Peano Arithmetic) augmented with function symbols $f_1, f_2, \ldots$ satisfying the axioms from Definition 2.1, together with two more axioms discussed above, one bounding the length of FP' functions and another giving the existence of a universal function for FP'.

$$\forall f \in FP' \exists c, d \forall x \quad |f(x)| \le c|x|^d \qquad \text{(Axiom Length)}$$

$$\exists U \in FP' \forall f \in FP' \exists i, c \forall x \quad f(x) = U(i, 2^{|x|^c}, x) \qquad \text{(Axiom } \mathcal{U}\text{)}$$

The resulting theory can be viewed as a two-sorted theory, with a number sort and a function sort. Functions here are defined over integers; multiple arguments can be represented using pairing. The notation $f(x)$ is a shortcut for $Apply(f, x)$; the latter notation allows for the theory to remain first-order with the introduction of the function symbols.

Recall that a standard model of a theory of arithmetic interprets number variables as natural numbers, and function and relation symbols as functions and relations over natural numbers, with symbols such as $+$, $\times$, 0 getting their usual meaning. A standard model of *RCT* can be fully described by interpretations of all symbols in T together with interpretations of all $f_i \in FP'$. Thus, interpreting Peano Arithmetic symbols in the usual manner, a standard model of *RCT* can be viewed as a set of interpretations of functions in FP'. This, together with the discussion above, gives the following correspondence: any standard model with FP' a set of additional functions satisfies *RCT* if and only if there is a set $O \subset \mathbb{N}$ such that $FP' = FP^O$.

For the proof, it is enough to consider the same arguments as above for encoding "spurious functions" as an oracle by the $O(i, t, x, j) = j$th bit of $U(i, t, x)$ for one direction, and adding a characteristic function of O as a "spurious function" to FP to obtain FP' as the closure for the other.

In this framework, the results from [11] become statements about independence: there exist two models of *RCT*, corresponding to FP^A and FP^B, which give contrary relativizations of the P vs. NP question. On the other hand, by assumption about the power of the underlying theory, proofs of theorems such as the Hartmanis/Stearns hierarchy theorem [32] are formalizable in *RCT*.

3 Non-relativizing Results and the Next Barrier

The result of Baker, Gill, and Solovay [11] shows that we cannot rely on the simulation and diagonalization techniques alone to resolve major complexity questions. But what other options are there? What other techniques can be used that avoid this barrier? Intuitively, these have to be techniques that look at the computation more closely than in the black-box way, techniques that analyse intrinsic properties of computation of specific computational models. Thus, the choice of the computational model will matter for approaching these problems.

Already, Stockmeyer [56] has talked about the significance of a representation of a machine model for complexity-theoretic results. The convenience of various representations for complexity results, even as differences between different ways to write a configuration of a Turing machine, has been studied on its own, in particular in [60].

After [11] appeared, a targeted search for non-relativizing results ensued. But even before, there were results in complexity theory that avoided the relativization barrier, the prime example being the Cook-Levin theorem. A number of important non-relativizing results came in the early 1990s, mainly in the setting of computation recast as "proof checking", including the $\mathsf{IP} = \mathsf{PSPACE}$ proof of [45, 54] (especially surprising since even with respect to a random oracle, $\mathsf{IP} \neq \mathsf{PSPACE}$ with probability 1 [15]) and the PCP theorem [5, 7].

3.1 *Interactive Proof Systems*

Both the $\mathsf{IP} = \mathsf{PSPACE}$ and the PCP theorem view, though in a somewhat different manner, solving a decision problem as a dialogue between a resource-bounded verifier and a powerful prover. In that setting, NP can be viewed as a class of problems where it is possible to verify in polynomial time that a given proof (i.e., a solution) is correct; for an input not in the language, no alleged proof will pass the scrutiny of the verifier. Interactive proofs generalize this in two ways. First, they allow for multiple rounds of interaction, with the verifier asking the prover for more and more answers to its questions. Alone, this generalization does not give extra power: the resulting class of problems is still NP, as the whole protocol of the interaction, being of polynomial length, can serve as a proof. More importantly, the verifier is allowed to use randomness; but then it may make mistakes. However, as long as the probability of the verifier making a mistake is less than $1/3$, we consider the interaction to be successful.

Definition 3.1 A language L is in the class IP if there exists a verifier such that, on every input x,

- if $x \in L$, then there is a prover (interaction strategy) that can convince the verifier with probability $> 2/3$ (over the verifier's randomness),

- if $x \notin L$, no prover can convince the verifier that it is in the language with probability $> 1/3$.

Here, the constant $1/3$ is not essential: any $1/2 - \epsilon$ would do, for ϵ within an inverse polynomial of the length of the input, the reason being that repeating the protocol with a different choice of randomness decreases the error probability.

A classic example of a problem with a natural interactive proof protocol is the graph (non-)isomorphism: checking whether two graphs are the same up to permutations of the vertices. In the graph non-isomorphism problem, the input consists of encodings of two graphs, G_1, G_2, and the algorithm should accept if G_1 and G_2 are not isomorphic. This problem is not known to be in P, but neither is it believed to be NP-complete.[6] To see that its complement, graph isomorphism, is in NP, note that giving a permutation of vertices of G_1 that makes it G_2 is sufficient for the certificate, with the verifier checking that all the edges now match. Now, suppose the verifier randomly picks one of G_1, G_2, permutes its vertices to obtain a new graph G_3, and then sends G_3 to the prover, asking to identify whether G_3 is a permuted version of G_1 or of G_2. The protocol, due to [27], is described below; there, r denotes the verifier's random bits.

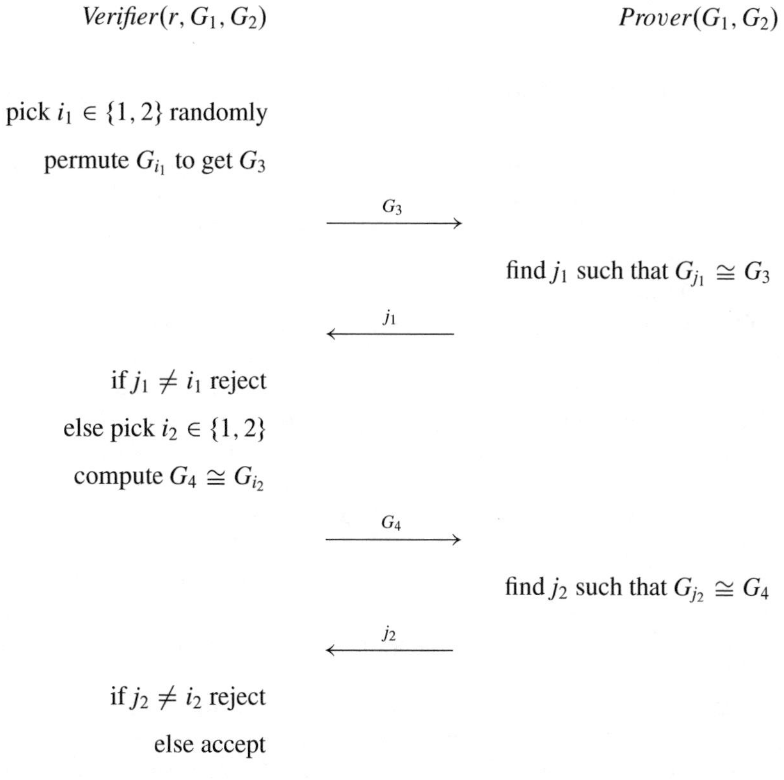

$Verifier(r, G_1, G_2)$		$Prover(G_1, G_2)$
pick $i_1 \in \{1, 2\}$ randomly		
permute G_{i_1} to get G_3		
	$\xrightarrow{G_3}$	
		find j_1 such that $G_{j_1} \cong G_3$
	$\xleftarrow{j_1}$	
if $j_1 \neq i_1$ reject		
else pick $i_2 \in \{1, 2\}$		
compute $G_4 \cong G_{i_2}$		
	$\xrightarrow{G_4}$	
		find j_2 such that $G_{j_2} \cong G_4$
	$\xleftarrow{j_2}$	
if $j_2 \neq i_2$ reject		
else accept		

[6]The recent breakthrough result by Babai [9] puts Graph Isomorphism in quasi-polynomial time (i.e., time $n^{\mathrm{poly}\log(n)}$), which is tantalizingly close to polynomial time.

If the two graphs are isomorphic, then G_3 could have come from either of them, so there is 1/2 chance that the prover answers correctly. However, if G_1 and G_2 are different, then, being computationally powerful, the prover can say with certainty which of G_1, G_2 the verifier picked. Thus, if the graphs are non-isomorphic, the prover answers correctly with probability 1, and if they are isomorphic, with probability 1/2. Now, as 1/2 is not good enough for our definition of IP, suppose the verifier repeats this procedure by picking a graph and permuting the vertices twice with independent randomness. This creates two graphs, G_3 and G_4, and the verifier then asks the prover for the origin of each. If G_1, G_2 are isomorphic, then the probability that the prover is correct in both cases is only 1/4. Note that here it is not necessary to use multiple rounds of interaction: even though the verifier can ask about G_3, receive the answer, and then ask about G_4, it is just as easy to send G_3 and G_4 to the prover simultaneously.

Interactive proofs can be simulated in PSPACE: since each interaction protocol is of polynomial length, it is possible to go over all such protocols for a given input x and a given verifier, and find the answers given by the best prover. With those, it is possible to count in PSPACE the fraction of random strings for which the verifier accepts or rejects.

3.2 *Power of* IP *in Oracle Worlds*

The main non-relativizing result that we will discuss is IP $=$ PSPACE. However, there are oracles (and, in fact, nearly all oracles are such) for which the inclusion of IP in PSPACE is strict. The construction of the first such oracle goes back to Fortnow and Sipser [25]; they present an oracle C with respect to which there is a language in coNP (the class of languages with complements in NP), but not in IP. Guided by this result, they conjecture that IP does not contain coNP, and that proving such a result would require non-relativizing techniques—as with respect to the oracle A, for which $\mathsf{P}^A = \mathsf{NP}^A$, IP contains coNP since it contains P. The IP $=$ PSPACE result shows that this intuition is misleading, as PSPACE is believed to be so much more powerful than coNP, containing all of the polynomial-time hierarchy, and more.

The idea behind Fortnow/Sipser's oracle construction is similar to that behind the construction of the oracle B from [11], for which $\mathsf{P}^B \neq \mathsf{NP}^B$. As for B, the oracle C is constructed using diagonalization ideas. A language $L(C)$ will contain all strings 1^n for which all strings of length n are in C, that is, $L(C) = \{1^n \mid \forall x, |x| = n, x \in C\}$. It is easy to see that this language is in coNP^C, as it is enough to ask the oracle for all strings of length n, and accept if all answers are "yes". Now, as for the construction of B from [11], the oracle C is constructed in blocks, where the ith block is constructed to fool the ith potential verifier V_i. Assume that the running time of V_i is within n^i for all i; then V_i cannot pose queries longer than n^i. Let N_i be the length on which V_i will be forced to accept or reject incorrectly with probability higher than $2/3$ (there, N_i should be large enough so that $N_{i-1}^{i-1} < N_i$, and $2^{N_i} > 3N_i^i$).

First, put into C all strings that V_i asks the oracle about. If there are no provers that make V_i accept 1^{N_i} with probability at least $2/3$, then put all the remaining strings of length N_i into C, in which case $1^{N_i} \in L(C)$ and V_i cannot be convinced of that. Also, put into C all strings V_i would ever ask about for any other provers other than ones considered in previous blocks.

Now, suppose that there is a prover P that will make V_i accept 1^{N_i} with probability at least $2/3$. One cannot guarantee that there is a string of length N_i that has not been queried by V_i for any choice of V_i's random bits; however, one can argue that a string x that is queried the least appears in less than $1/3$ possible computation paths, as N_i was chosen to be such that $2^{N_i} > 3N_i^i$. Remove x from C; this will be the only string queried by V_i over all possible interactions with P that are not in C. Now, even if querying this string makes V_i reject, it will affect less than $1/3$ of possible computation paths of V_i on 1^{N_i}. So V_i will still accept 1^{N_i} with probability greater than $1/3$, completing the proof.

Even though there is an oracle with respect to which IP is quite weak, one could say that this is an outlier, and maybe in general the power of IP with respect to an arbitrarily picked oracle will be more in sync with its power in the real world. This intuition, that behaviour with respect to a random oracle was indicative of the situation in the real world, became known as the random oracle hypothesis, following the result from Bennett and Gill [13] that $\mathsf{P}^A \neq \mathsf{NP}^A$ with probability 1 over oracles A. There, a random oracle is constructed by putting each string x in the oracle with probability $1/2$. However, a 1994 paper by Chang, Chor, Goldreich, Hartmanis, Håstad, Ranjan and Rohatgi, titled "Random oracle hypothesis is false" [15], shows that with respect to a randomly chosen oracle A, $\mathsf{coNP}^A \subsetneq \mathsf{IP}^A$, and so the behavior of IP^C for the oracle C of [25] is typical rather than exceptional.

3.3 *Arithmetization*

Many of these new non-relativizing results rely on a technique called *arithmetization*. In arithmetization, a Boolean formula is interpreted as a polynomial, which for 0/1 values of variables (corresponding to false/true in the Boolean case) evaluate to 0 or 1. However, one can evaluate this polynomial on any integers or elements of a given finite field, and obtain a host of values giving more information about the structure of the formula. More precisely, to convert a Boolean formula such as $(\bar{x} \vee y) \wedge z$ to a polynomial, the conjunction $\wedge$ is interpreted as multiplication, and negation $\bar{x}$ as $(1-x)$. Then, $(x \vee y)$ is converted, as $\overline{(\bar{x} \wedge \bar{y})}$, to $1-(1-x)(1-y) = x + y - x \cdot y$. So the formula $(\bar{x} \vee y) \wedge z$ becomes a three-variate polynomial $(1-x\cdot(1-y))\cdot z$. It can be checked that a formula resulting from this transformation indeed evaluates to 0/1 values when inputs are 0s and 1s, and it evaluates to 1 exactly when the original formula is true.

Arithmetization is a powerful algorithmic technique: for example, if the resulting polynomial is multilinear (that is, no variable occurs with degree more than 1), then counting the fraction of assignments of this formula that are satisfying can be done by evaluating this polynomial with every variable set to $1/2$ [41]. In converting a formula to a polynomial, though, both $\wedge$ and $\vee$ translations involve multiplication; thus, resulting polynomials, though of polynomial degree and described by polynomial-size circuits, are quite unlikely to be multilinear. One can always create a multilinear polynomial of a formula by taking a sum over terms corresponding to each satisfying assignment or taking a Fourier representation, but such a sum is likely to have more than a polynomial number of coefficients.

3.4 IP *Protocol for Counting Satisfying Assignments*

In general, counting the number of satisfying assignments is believed to be a harder problem than determining whether a formula is satisfiable. In fact, this problem (and, equivalently, computing the Permanent of a Boolean matrix) is at least as hard as any other problem in the polynomial-time hierarchy. The latter is the celebrated theorem by Toda [58]; see [24] for a short proof. The problem of counting the number of satisfying assignments, sometimes referred to as *#SAT*, is in PSPACE, and the technique for proving that it can be done using interactive proofs is very similar, except for a few details, to that for proving IP = PSPACE. Historically, this was one of the results leading the proof that IP = PSPACE; a quite entertaining account of the history of this result is presented in Babai's "E-mail and the unexpected power of interaction" [8].

Theorem 3.2 ([45]) *#SAT* $\in$ IP.

Proof To simplify the notation, let us use a decision version $\#SAT_D$ of *#SAT*, where, given a formula ϕ and a number K, the goal is to check whether K is the number of satisfying assignments of ϕ. To show that $\#SAT_D$ is in the class IP of problems that have interactive proofs, we will explicitly exhibit an interactive proof protocol for it. That is, we will describe which protocol the verifier should follow to check if an alleged proof presented by the prover is correct; if presented with a correct proof in the format it expects, the verifier will accept with probability 1, and for any incorrect proof, it will reject with probability significantly higher than $2/3$.

The number of satisfying assignments of a formula $\phi(x_1, \ldots, x_n)$ is equal to the sum, over all possible assignments, of the polynomial $p(x_1, \ldots, x_n)$ obtained by arithmetizing ϕ. Thus computing the number of satisfying assignments amounts to computing

$$\sum_{x_1 \in \{0,1\}} \sum_{x_2 \in \{0,1\}} \cdots \sum_{x_n \in \{0,1\}} p(x_1, \ldots, x_n).$$

For example, if the original formula is $\phi = (x_1 \vee \neg x_2)$, then the resulting polynomial is $1 - (1 - x_1)x_2 = 1 - x_2 + x_1x_2$, and the number of satisfying assignments of ϕ is

$$\sum_{x_1 \in \{0,1\}} \sum_{x_2 \in \{0,1\}} 1 - x_2 + x_1x_2 = 3.$$

The first key idea of the protocol is checking the value of the sum one variable at a time: setting

$$p_1(x_1) = \sum_{x_2 \in \{0,1\}} \cdots \sum_{x_n \in \{0,1\}} p(x_1, \dots, x_n),$$

if p_1 is correct then it is sufficient to compute $p_1(0) + p_1(1)$ to obtain the answer. The second idea is recursing on evaluating the sum with a random number inserted in place of x_1. More specifically, the protocol proceeds as follows.

1. *Verifier* asks for the coefficients of the univariate polynomial

$$p_1(x_1) = \sum_{x_2 \in \{0,1\}} \cdots \sum_{x_n \in \{0,1\}} p(x_1, \dots, x_n).$$

2. *Prover* sends the (alleged) coefficients of $c_{1,0}, \dots c_{1,m}$ of p_1 to Verifier.
3. *Verifier* checks that $K = p_1(0) + p_1(1)$. If not, it rejects as it knows that Prover must have been lying.
4. *Verifier* picks a random number $0 \leq a_1 \leq 2^n$ and sends it to Prover.
5. Now, repeat the following from $i = 2$ to $i = n$, or until Verifier rejects.

   (a) *Prover* sends the alleged coefficients of

$$p_i(x_i) = \sum_{x_{i+1} \in \{0,1\}} \cdots \sum_{x_n \in \{0,1\}} p(a_1, \dots, a_{i-1}, x_i, \dots, x_n).$$

   Note that p_i is a univariate polynomial in x_i.

   (b) *Verifier* checks that $p_{i-1}(a_{i-1}) = p_i(0) + p_i(1)$. If not, reject.

   (c) *Verifier* picks a random $0 \leq a_i \leq 2^n$ and sends it to Prover.

6. *Prover* sends the alleged number $c = p(a_1, \dots, a_n)$
7. *Verifier* checks that $p(a_1, \dots, a_n) = c$. If not, reject. If so, then Verifier concludes that K is indeed the correct number of assignments of ϕ.

As an example, suppose that $\phi = (x_1 \vee \neg x_2)$, with $K = 3$ assignments, and $p(x_1, x_2) = 1 - x_2 + x_1 x_2$. Here, $p_1(x_1) = 1 + x_1$, and so the correct prover sends back $c_0 = 1, c_1 = 1$. Now, the verifier checks that $p_1(0) + p_1(1) = 1 + 2 = 3$; this check passes. Then the verifier picks a number a_1; for example, $a_1 = 6$, and sends that to the prover. The prover now has to compute the coefficients of $p_2(x_2) = 1 + 5x_2$ and send them to the verifier. The verifier checks that $p_1(6) = p_2(0) + p_2(1) = 7$. It then picks another number, say $a_2 = 4$, and sends it to the prover. The prover then computes $p(a_1, a_2) = 21$, and sends 21 to the verifier. Now, the verifier checks that $p(a_1, a_2) = 21$, and concludes that 3 is indeed the number of the satisfying assignments to ϕ.

If the prover always produces the correct coefficients and K is the correct number of assignments, then it is clear that the verifier will accept. To show that the probability of accepting an incorrect value is small, note that two distinct univariate polynomials of degree $d+1$ may agree on at most d values of the variable, and the probability of randomly picking one of these d values is at most $d/2^n$. This is the probability that any check $p_{i-1}(a_{i-1}) = p_i(0) + p_i(1)$ passes incorrectly. Thus, if K is wrong, then the probability that the verifier rejects is at least $(1 - d/2^n)^n$, which is much larger than $2/3$. □

3.5 IP = PSPACE

A very similar protocol can be used to check whether a quantified Boolean formula $\Phi = \forall x_1 \exists x_2 \forall x_3 \ldots Q x_n \phi(\bar{x})$ is true: here, Q is either $\forall$ or $\exists$, depending on n. There, existential quantifiers correspond to a sum, and universal quantifiers to a product of the values of the arithmetized formula under them. For example, suppose that $\Phi = \forall x_1 \exists x_2 \forall x_3 (x_1 \vee \neg x_2) \wedge (\neg x_1 \vee \neg x_3)$. Then the polynomial p_ϕ for the formula $(x_1 \vee \neg x_2) \wedge (\neg x_1 \vee \neg x_3)$ is $(1 - x_2 + -x_1 x_2)(1 - x_1 x_3)$. Now, given Boolean values for x_1 and x_2, the formula $\forall x_3 (x_1 \vee \neg x_2) \wedge (\neg x_1 \vee \neg x_3)$ is true when polynomial $p_\phi(x_1, x_2, 0) \cdot p_\phi(x_1, x_2, 1) = (1 - x_2 + x_1 x_2)(1 - x_2 + x_1 x_2) \cdot (1 - x_1)$ is 1; false corresponds to 0. Proceeding in this manner, we obtain an arithmetic expression which is equal to 1 iff Φ is true, and to 0 iff Φ is false.

Note that the universal quantifier created a product of polynomials, thus doubling the degree. If there are n universal quantifiers, then the resulting degree can be exponential in the number of variables; thus the protocol above would need exponentially many coefficients to describe such a polynomial. To get around this problem, intermediate polynomials can be converted into a multilinear form by noticing that

$$\begin{aligned} p(x_1, \ldots, x_{i-1}, x_i, x_{i+1}, \ldots, x_n) =& x_i \cdot p(x_1, \ldots, x_{i-1}, 1, x_{i+1}, \ldots, x_n) + \\ &+ (1 - x_i) \cdot p(x_1, \ldots, x_{i-1}, 0, x_{i+1}, \ldots, x_n) \end{aligned}$$

over Boolean values of x_i.

With this modification, essentially the same protocol can be used to check if Φ is true, except the verifier would check if the sum $p_{i-1}(a_{i-1}) = p_i(0) + p_i(1)$ is right when x_i is existentially quantified, and if the product $p_{i-1}(a_{i-1}) = p_i(0)p_i(1)$ is correct for a universally quantified x_i.

3.6 Algebrization Barrier

Although there are results on the complexity of oracles with respect to which IP = PSPACE holds soon after the results themselves have been proven (for example, [23]), the barrier itself was defined in 2008 by Aaronson and Wigderson [2]. There, the main idea was to allow access not just to an oracle A as a language, but also its algebraic extension $\tilde{A}$: here, $\tilde{A}$ is a low-degree polynomial which agrees with A on Boolean inputs. We view an oracle access for a formula or a polynomial as having operations of the form $A(b_1, \ldots, b_k)$, where $b_1, \ldots, b_k$ form a query.

Aaronson and Wigderson define an inclusion of complexity classes $C_1 \subseteq C_2$ (for example, $\mathsf{PSPACE}^A \subseteq \mathsf{IP}^{\tilde{A}}$) to be algebrizing when $C_1^A \subseteq C_2^{\tilde{A}}$ no matter how $\tilde{A}$ is chosen, as long as it is a low-degree polynomial extension of A. As for separations, $C_1 \not\subset C_2$ is algebrizing whenever, again for any polynomial extension $\tilde{A}$, $C_1^{\tilde{A}} \not\subset C_2^A$. That is, an inclusion in their setting algebrizes if C_2 can simulate C_1^A with a little more powerful access to the oracle A; and a separation algebrizes if a more powerful access to A for C_1 makes it impossible to simulate C_1 in C_2 with the conventional access.

With these definitions, they show that algebrization indeed provides a pretty precise boundary of the current techniques: most known non-relativizing results algebrize, and most open questions do not. In particular, the proof of $\mathsf{PSPACE}^A \subseteq \mathsf{IP}^{\tilde{A}}$ uses the same protocol as the original proof, with the verifier relying on the oracle access to $\tilde{A}$ to evaluate $p_\phi(a_1, \ldots, a_n)$. However, questions such as P vs. NP do not algebrize: an oracle A and its algebraic extension $\tilde{A}$ can be constructed for which $\mathsf{NP}^{\tilde{A}} \subseteq \mathsf{P}^A$, whereas for a different B with its extension $\tilde{B}$, $\mathsf{NP}^B \not\subset \mathsf{P}^{\tilde{B}}$. A number of other open questions are similarly non-algebrizing. The main tool used for proving such independence results is communication complexity.

But these results need a different kind of oracle on the two sides of an inclusion or a separation. For what kinds of oracles would it be possible to use the same language as an oracle on both sides? In 1994, Fortnow [23] showed how to construct a language that encodes its algebraic extension, and proved that with respect to these kinds of languages, $\mathsf{IP} = \mathsf{PSPACE}$. He constructs such a "self-algebrizing" language inductively. (A multilinear extension of $TQBF^L$ gives another example of a language that encodes its own algebraic structure.) Let L be a language, and let $\tilde{L}$ be the unique multilinear extension of L. Start by setting $A(0, x_1, \ldots, x_n) = L(x_1, \ldots, x_n)$. Now, if $\tilde{L}(x_1, \ldots, x_n) > 0$, put $(1, x_1, \ldots, x_n)$ in A. Finally, $\forall i \geq 0$, set $A(i+2, x_1, \ldots, x_n)$ to be the ith bit of $\tilde{L}(x_1, \ldots, x_n)$. Thus, a value of $\tilde{L}(x_1, \ldots, x_n)$ could be obtained by using A as an oracle.

Though $\mathsf{IP} = \mathsf{PSPACE}$ with respect to any oracle of this form, it is not clear whether there are such oracles giving differing outcomes for major open questions such as P vs. NP.

3.7 *Axiomatizing Algebrization*

To create a theory capturing algebrizing techniques, we start with the theory *RCT* for relativizing techniques, and use additional axioms to keep only the standard models with enough algebraic structure [36]. To achieve this goal, we add an axiom stating that NP is a class of languages that have solutions verifiable by polynomial-time computable low-degree polynomials in polynomially many variables. There, the "polynomial-time computable" is in the setting of *RCT*, that is, functions definable by Cobham axioms without minimality. We call the resulting theory *ACT*, for "algebraic complexity theory".

Now, the models of *ACT* can still contain spurious functions interpreted as polynomial-time, as long as they are closed under polynomial-time operations. However, now such functions used as polynomial-time verifiers have to be representable by low-degree polynomials. Alternatively, we can add an axiom redefining polynomial time, by stating that every polynomial-time function has a unique polynomial-time computable witness.

For example, consider the classic NP-complete problem Independent Set: given a graph and a number k, check if there are k vertices in the graph such that there are no edges between any of these k vertices; this set of vertices is an independent set in the graph. Let input variables $x_{i,j}$ be 1 if there is an edge between vertices i and j, and be 0 otherwise. Let witness variables y_i be 1 iff vertex i is in the independent set, and 0 otherwise. Now, the polynomial

$$f(x, y) = \Pi_{i,j}(x_{i,j} + y_i + y_j - 3) \cdot \Pi_{t<k}(\Sigma_{i=1}^{n} y_i - t)$$

will be non-zero if and only if there is an independent set of size at least k in the graph. There, the first product will be 0 if there is an edge between two vertices i, j for which $y_i = y_j = 1$, and the second product is 0 if there are only $t < k$ variables y_i that are 1.

This view of computation is less black-box than the original Cobham axioms: we do require polynomial-time functions to have some structure. However, there are still abitrarily powerful functions that could show up in models of *ACT*. In particular, a characteristic function for any of the oracles built using Fortnow's construction described above, the "self-algebrizing oracles", would be a function with respect to which this axiom is satisfied, that is, NP has witnesses definable by low-degree polynomial-time computable polynomial families.

Now, this theory is clearly more powerful than *RCT*. It is possible to show that nearly all results that algebrize in the sense of [2] are provable in *ACT*, and open questions that require non-algebrizing techniques are independent of *ACT*. In particular, IP = PSPACE is provable in *ACT*, and P vs. NP is independent of it.

One notable exception is MIP = NEXP by Babai, Fortnow and Lund [10]: when a randomized polynomial-time verifier is allowed to interact with several provers that do not talk to each other, it becomes possible to solve not just PSPACE problems, but ones as hard as non-deterministic exponential time. This equality is also independent of *ACT*, even though Aaronson and Wigderson show in [2] that a version of this statement algebrizes. However, they only allow a NEXP Turing machine to asking oracle queries of polynomial length, to make it "more fair" for MIP, where the verifier cannot possibly ask longer queries. But such a restriction makes NEXP^A much too weak [35]. Consider an oracle $A = \{\langle M, x, 1^t \rangle \mid M$ is a non-deterministic Turing machine and M^A accepts x on some path with all oracle queries shorter than $t\}$. With respect to this oracle, $\mathsf{NEXP}^A = \mathsf{P}^A$ when

NEXP^A is restricted to ask only polynomially long queries. This goes against the time hierarchy theorem, which relativizes.

This axiomatic approach is one example where considering a complexity barrier from the logic viewpoint, as an independence of a logic theory, allows for a more convenient setting. Moreover, with the closure under logic operations, it becomes possible to argue about composite statements such as "$\mathsf{BPP} = \mathsf{P}$ and $\mathsf{P} \neq \mathsf{NP}$": this statement is independent of *ACT*, as well.

4 Efficient Reasoning, Circuits and Natural Proofs

4.1 Bounded Arithmetic

Before, we talked about creating specific theories of arithmetic to formalize polynomial-time computation and its relativizations. However, there is an area of mathematical logic that is specifically developed to study the reasoning corresponding to efficient computation, such as reasoning with polynomial-time definable concepts. Starting with Parikh's [46] fragment of Peano Arithmetic, where induction is limited to bounded formulas, followed by Cook's equational theory PV [19] and then Buss' theories [14], bounded arithmetic (term coined by Buss) became one of the standard ways to work with complexity theory concepts in the logic framework. By contrast with *RCT* and *ACT* described above, the polynomial-time computable objects are now indeed polynomial-time, and, moreover, the reasoning power of bounded arithmetic theories is severely restricted by allowing only reasoning with such "efficient" concepts.

This is accomplished by restricting, for example, the induction axiom of Peano Arithmetic to formulas where all quantifiers are bounded by terms in the language. The resulting theory, $I\Delta_0$, captures the linear-time hierarchy, but a number of theories with richer language (including $x\#y = 2^{|x|\cdot|y|}$) capture exactly polynomial-time computable functions. Another way is to construct two-sorted theories operating on strings as well as numbers, and lengths of strings, as well as values of numbers, bounded by the term. In this case, there is no need to introduce # in the language. There is a direct correspondence between these theories and Buss' hierarchies via "RSUV isomorphism" [48, 57]. One advantage of such a characterization is that "reasoning with concepts from a class C" is easy to state using finite model theory characterizations of the corresponding class [21, 42]: for example, induction over Grädel's second-order Horn formulas [28] gives a theory of polynomial-time reasoning [20].

We will skip the formal definitions; see Krajicek [43], Buss [14], Pudlák and Hajek [29], and Cook and Nguyen [21] for more information.

These theories are too weak to prove the totality of exponentiation, that is, that for every x there exists $y = 2^x$. And overall, independence results for them mean "not provable with computationally easy reasoning". For example, in Buss' theory

S_2^1, which operates with NP-definable predicates and has induction on the length of a number, any bounded existential statement ϕ for which S_2^1 proves that $\phi \in \mathsf{NP} \cap \mathsf{coNP}$ can be witnessed in polynomial time. This result, which is known as Buss' witnessing theorem, immediately implies that S_2^1 cannot prove that $\mathsf{P} \neq \mathsf{NP} \cap \mathsf{coNP}$.

So what kinds of results, in particular complexity results, can be proven using that kind of limited reasoning? It is possible that these theories, or just slight extensions of them, are sufficient to formalize the known complexity results: see, for example, [49], Krajicek [43], Cook and Nguyen [21] or Pudlák [47]. Some of the more recent results about provability and unprovability of complexity-theoretic statements in bounded arithmetic are due to Jerabek (formalizing probabilistic reasoning and pseudorandomness in an extension of S_2^1 with a dual weak pigeonhole principle) [37–40]; many other related results are in the works of Krajicek, Cook, Buss, Razborov, and others.

One may ask whether independence of P vs. NP from a very strong classical theory, such as ZFC or Peano arithmetic, is a possibility that should be considered instead of focusing on these very restrictive theories. A possible reason why this is quite unlikely is given by Ben-David and Halevi [12]. They show that if P vs. NP is independent of Peano arithmetic augmented with all Π_1 sentences true in the standard model, then NP is "essentially polynomial-time", as there exists a deterministic algorithm for SAT that makes few mistakes, with complexity $n^{f(n)}$, where $f(n)$ is a very slow growing function. An expository paper by Aaronson [1] contains an excellent discussion of this subject.

4.2 Circuit Lower Bounds

The 1990s saw a flurry of beautiful results in another area of computational complexity theory: circuit complexity. There, the requirement that there be a single Turing machine solving a problem for all input lengths in bounded time is relaxed to consider a family of computational devices solving the problem, one for each input size, with the sizes of the devices growing only slowly (polynomially) with the number of input bits. The setting is often referred to as *non-uniform*. Non-uniformity does make for a much more powerful model: for example, the unary Halting Problem, asking if the nth Turing machine halts on blank input, with n encoded in unary or as $|x|$, is trivially solvable in this model using only constant size devices.

In particular, devices that we will consider here are Boolean circuits: acyclic graphs with inputs as sources, a single output as a sink, and every node (gate) computing a Boolean function over values of gates with edges coming into it. In the most basic case, the Boolean functions at gates are *AND*, *OR* and *NOT*. For example, this circuit computes the majority of three bits.

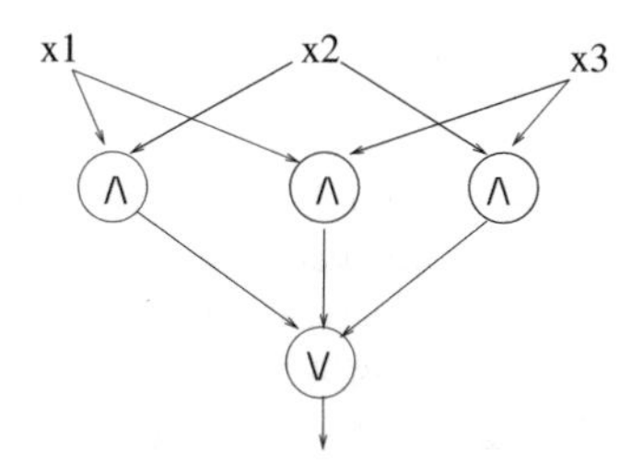

As time and space are in the uniform setting, in the circuit setting the main resources are size (the number of gates) and depth (the length of the longest path from the inputs to the output). A class of languages computed by a family of polynomial-size circuits is denoted by $\mathsf{P/poly}$; this is a non-uniform analogue of P. Trivially, $\mathsf{P} \subset \mathsf{P/poly}$, but whether $\mathsf{NP} \subset \mathsf{P/poly}$ is a major open question: a negative answer would be a stronger result than $\mathsf{P} \neq \mathsf{NP}$.

However, restricting the depth of the circuits to be constant (here, we are assuming that gates can have arbitrary fan-in) gives a complexity class AC^0 for which the lower bounds are known. In particular, the function $PARITY(x)$, outputting 1 iff the number of 1s in binary string x is odd, is not AC^0-computable [3, 26, 33]. Even allowing parity or modulo a prime gates to appear in these circuits results in a class with strong lower bounds. But how complex is the reasoning needed to prove those statements? And could these techniques be extended to argue about NP vs. $\mathsf{P/poly}$?

4.3 Natural Proofs

These questions fascinated researchers in proof complexity for many years, with a number of interesting results proven. But focusing on complexity barriers, let us turn to a series of papers by Razborov [49–51] that have tried to address these questions, using extensively the framework of bounded arithmetic. By far the most well known of them is a pure complexity paper, though: "Natural proofs" by Razborov and Rudich [51]. Introducing the notion of "natural" proofs, they come to a somewhat discouraging conclusion that the circuit lower bound proofs known at the time are "natural", and such proofs cannot resolve NP vs. $\mathsf{P/poly}$, albeit under a believable cryptographic conjecture.

Intuitively, if a proof of a lower bound for a given complexity class is "natural enough" it would contain an algorithm to distinguish easy problems that are in this class from hard problems that are not, based on whether or not a problem possesses a given "natural" property. Note that, in particular, algorithms from the class for which there is a natural proof would not be able to generate distributions of strings computationally indistinguishable from random (uniform).

Definition 4.1 A property (set) F of Boolean functions is natural if it (or its subset) satisfies three conditions:

1. *Usefulness*: functions in F are infinitely often not in a complexity class C (that is, F can be used to prove lower bounds on C).
2. *Largeness*: A large fraction of all functions are in F.
3. *Constructivity*: Given a truth table of a function f, it is computationally easy to check whether $f \in F$. If it can be checked in a class C', we say that a proof is C'-natural.

For example, the proof that $PARITY(x)$ is not in AC^0 [26] works by showing that any function in AC^0 becomes constant if enough of its input variables are set (with high probability over the choice of the subset of variables). The $PARITY(x)$ function, however, does not become constant even if one bit is left unset, and thus $PARITY(x) \notin \mathsf{AC}^0$. So the natural property F is that a function does not become constant under any restriction of a large enough fraction of its input variables. This property is useful against AC^0, as functions in AC^0 do not have it. It has largeness by the counting argument (most functions on $n-k$ variables are not constant). And it has constructivity, even AC^0-constructivity: given a truth table of a function on n variables, which is of length 2^n, a depth-3 circuit of size $2^{O(n)}$ can check whether this truth table satisfies the property: just consider all possible restrictions of $n-k$ variables (roughly $2^{O(n)}$ of them), and check that not all input bits corresponding to this restriction are the same.

Razborov and Rudich proceed to show that a number of circuit lower bound proofs are indeed natural. The main result of [51] states, though, that there is no $\mathsf{P/poly}$-natural proof useful against $\mathsf{P/poly}$, provided that there is an exponentially hard pseudo-random generator computable in $\mathsf{P/poly}$. Pseudo-random generators are functions $PRG\colon \{0,1\}^k \to \{0,1\}^{2k}$; such PRGs are s-hard if s is the minimal size of a circuit that can distinguish a random $2k$ bit string from the output of a k-bit generator with probability $\geq 1/s$. The main idea of the proof is to use the $\mathsf{P/poly}$-computable check from the constructivity property to distinguish between the PRG output and the random string for any given $\mathsf{P/poly}$-computable PRG.

It is believed that PRGs based on factoring or the discrete logarithm problem (solving $b^k = g$ over a finite group) are exponentially hard. With that, Razborov and Rudich also show unconditionally that there is no $\mathsf{P/poly}$-natural proof that the discrete logarithm problem requires exponential size circuits.

4.4 Natural Proof as Independence in Bounded Arithmetic

The formalization of the notion of natural proofs as independence in bounded arithmetic was presented by Razborov in [49]. The theories he considers are extensions of Buss' S^1_2; there are technical details in allowing these theories to talk about functions. The connection between natural proofs and these theories is not as tight as for the RCT/ACT, though.

Recall that the language of Buss' S_2^1 is the language of arithmetic plus $x\#y = 2^{|x|\cdot|y|}$, and it is axiomatized by the basic axioms describing the operators together with induction on the length of the number. The formulas allowed in the induction are of the form $\exists\bar{x} \leq t(n)\phi(\bar{x}, n)$, where all quantifiers in ϕ are bounded by a term in the *lengths* of the input variables n ("strictly bounded" quantifiers) in length. Generally, theories S_2^i allow up to i alternations of bounded (rather than strictly bounded) quantifiers in the induction.

In Razborov's setting, there is a free relational variable γ added to the theories. This variable is interpreted as encoding a Boolean circuit. There are bounded existential formulas defining various properties of the circuit encoded by γ: its type, its size, the function it computes, etc. With an extension of these definitions, he proves that $S_2^2(\gamma)$ cannot disprove that γ encodes a circuit of superpolynomial size, under the same assumption as for natural proofs of existence of strong PRGs. For weaker systems, he proves similar statements under weaker assumptions, for some even unconditionally. The proofs rely on a communication complexity characterization of the circuit size. However, this is not quite the same setting as proving lower bounds in S_2^1 itself, rather than arguing about a given circuit γ.

5 Conclusion: Avoiding Barriers

One can look at complexity barriers from either the pessimistic or the optimistic viewpoint. A pessimist would say that the results beyond the barriers, non-relativizing and non-algebrizing results, are intrinsically hard to prove. Surely, it shows just how formidable a problem is when one can prove that "nearly all current techniques" are inadequate for resolving it. And even though there is some truth to this viewpoint, and barrier results historically came from trying to understand the failed attempts to resolve these open problems, there is a bright side to the barriers.

The optimistic view of the barrier results is finding precise properties of techniques that would make it possible to resolve the open problems. For example, relativization tells us that it is fruitless to treat computation as a black box, and representation matters. Algebrization tells us that representing polynomial-time functions by low-degree polynomials, powerful as it is, is not enough to lead us all the way to resolving P vs. NP. However, it does tell us where to look: the less "black-box" our view of computation, the more we can show about it. Besides, the barriers do not tell us to throw out relativizing and algebrizing techniques altogether; they just point out that any meaningful approach to resolving open problems should use at least some non-relativizing and non-algebrizing component.

And indeed, there are already such components known to us. The $\mathsf{MIP} = \mathsf{NEXP}$ result mentioned above relies on verifying computations by 3*CNF* formulas in a different way than by treating these formulas as polynomials. The *PCP* theorem, stating that any language in NP has proofs that can be verified using $O(\log n)$ randomness and examining just a constant number of bits of the proof, uses an especially fine-grained view of a computational process, and is neither algebrizing in the sense of [2] nor provable in *ACT*.

Besides, proving a relativizing result can be more useful, in the sense that it automatically generalizes for any oracle. In particular, sometimes it is convenient to consider circuits with *SAT* or *TQBF* oracle gates, and any relativizing result about the respective family of circuits generalizes to the circuits with such gates.

In the case of natural proofs, we know the techniques providing proofs that are not natural, most notably diagonalization and counting arguments. Moreover, there are results proved using techniques that avoid both barriers, such as Santhanam's proof that PromiseMA does not have circuits of size n^k for a fixed k [52]. And a recent result by Williams [61] that there are problems in NEXP not solvable by bounded-depth circuits with arbitrary $\mod d$ gates uses both a non-relativizing element and a non-naturalizing element (diagonalization) to avoid both barriers simultaneously.

Viewing complexity barriers as independent of logic theories allows us to make precise what exactly the barriers capture. It is a very natural setting, and a convenient way to specify what exactly is meant by classes of techniques such as relativizing techniques. Besides, we can always hope that the formidable machinery of logic will come to our service if only we phrase the right questions in the right framework.

Acknowledgements I am very grateful to a number of people for suggestions and comments for this survey. I am especially thankful to Valentine Kabanets and Russell Impagliazzo, who introduced me to the whole topic of axiomatic approach to barriers, answered many of my questions and gave lots of insightful suggestions. I am also grateful to Avi Wigderson, Rahul Santhanam, Toni Pitassi, Sam Buss and Peter van Emde Boas for discussions that greatly influenced this paper.

References

1. S. Aaronson, Is P versus NP formally independent? Bull. Eur. Assoc. Theor. Comput. Sci. **81**, 109–136 (2003)
2. S. Aaronson, A. Wigderson, Algebrization: a new barrier in complexity theory. ACM Trans. Comput. Theory **1**(1) (2009). Preliminary version in STOC'08
3. M. Ajtai, σ_1^1-formulae on finite structures. Ann. Pure Appl. Log. **24**(1), 1–48 (1983)
4. S. Arora, B. Barak, *Computational Complexity: A Modern Approach*, 1st edn. (Cambridge University Press, New York, 2009)
5. S. Arora, S. Safra, Probabilistic checking of proofs: a new characterization of NP. J. Assoc. Comput. Mach. **45**(1), 70–122 (1998). Preliminary version in FOCS'92
6. S. Arora, R. Impagliazzo, U. Vazirani, Relativizing versus nonrelativizing techniques: the role of local checkability. Manuscript (1992)
7. S. Arora, C. Lund, R. Motwani, M. Sudan, M. Szegedy, Proof verification and the hardness of approximation problems. J. Assoc. Comput. Mach. **45**(3), 501–555 (1998). Preliminary version in FOCS'92
8. L. Babai, E-mail and the unexpected power of interaction, in *Proceedings, Fifth Annual Structure in Complexity Theory Conference, 1990* (IEEE, Barcelona, 1990), pp. 30–44
9. L. Babai, Graph isomorphism in quasipolynomial time, in *Proceedings of the Forty-Eighth Annual ACM Symposium on Theory of Computing* (2016)
10. L. Babai, L. Fortnow, C. Lund, Non-deterministic exponential time has two-prover interactive protocols. Comput. Complex. **1**(1), 3–40 (1991)

11. T.P. Baker, J. Gill, R. Solovay, Relativizations of the P=?NP question. SIAM J. Comput. **4**(4), 431–442 (1975)
12. S. Ben-David, S. Halevi, On the Independence of P versus NP, Tech. Report 714, Technion (1992)
13. C.H. Bennett, J. Gill, Relative to a random Oracle A, $P^A \neq NP^A \neq coNP^A$ with probability 1. SIAM J. Comput. **10**(1), 96–113 (1981)
14. S.R. Buss, *Bounded Arithmetic* (Bibliopolis, Naples, 1986)
15. R. Chang, B. Chor, O. Goldreich, J. Hartmanis, J. Håstad, D. Ranjan, P. Rohatgi, The random oracle hypothesis is false. J. Comput. Syst. Sci. **49**(1), 24–39 (1994)
16. A. Cobham, The intrinsic computational difficulty of functions, in *Proceedings of the 1964 International Congress for Logic, Methodology, and Philosophy of Science*, ed. by Y. Bar-Hillel (North-Holland, Amsterdam, 1964), pp. 24–30
17. S.A. Cook, The complexity of theorem-proving procedures, in *Proceedings of the Third Annual ACM Symposium on Theory of Computing* (1971), pp. 151–158
18. S.A. Cook, A hierarchy for nondeterministic time complexity, in *Proceedings of the Fourth Annual ACM Symposium on Theory of Computing, STOC '72* (1972), pp. 187–192
19. S.A. Cook, Feasibly constructive proofs and the propositional calculus, in *Proceedings of the Seventh Annual ACM Symposium on Theory of Computing* (1975), pp. 83–97
20. S.A. Cook, A. Kolokolova, A second-order system for polytime reasoning based on Grädel's theorem. Ann. Pure Appl. Log. **124**, 193–231 (2003)
21. S.A. Cook, P. Nguyen, Logical foundations of proof complexity, in *Perspectives in Logic of the Association for Symbolic Logic* (Cambridge University Press, Cambridge, 2008)
22. J. Edmonds, Paths, trees, and flowers. Can. J. Math. **17**, 449–467 (1965)
23. L. Fortnow, The role of relativization in complexity theory. Bull. Eur. Assoc. Theor. Comput. Sci. **52**, 229–244 (1994). Columns: Structural Complexity
24. L. Fortnow, A simple proof of Toda's theorem. Theory Comput. **5**(7), 135–140 (2009)
25. L. Fortnow, M. Sipser, Are there interactive protocols for co-NP Languages? Inf. Process. Lett. **28**, 249–251 (1988)
26. M. Furst, J.B. Saxe, M. Sipser, Parity, circuits, and the polynomial-time hierarchy. Math. Syst. Theory **17**(1), 13–27 (1984)
27. O. Goldreich, S. Micali, A. Wigderson, Proofs that yield nothing but their validity or all languages in NP have zero-knowledge proof systems. J. Assoc. Comput. Mach. **38**, 691–729 (1991)
28. E. Grädel, *The Expressive Power of Second Order Horn Logic*. Lecture Notes in Computer Science, vol. 480 (Springer, Heidelberg, 1991), pp. 466–477
29. P. Hájek, P. Pudlák, Metamathematics of first-order arithmetic, in *Perspectives in Mathematical Logic* (Springer, Berlin/Heidelberg, 1998)
30. J. Hartmanis, J.E. Hopcroft, Independence results in computer science. ACM SIGACT News **8**(4), 13–24 (1976)
31. J. Hartmanis, R.E. Stearns, Computational complexity of recursive sequences, in *Proceedings of the Fifth Annual IEEE Symposium on Foundations of Computer Science* (1964), pp. 82–90
32. J. Hartmanis, R.E. Stearns, On the computational complexity of algorithms. Trans. Am. Math. Soc. **117**, 285–306 (1965)
33. J. Håstad, Almost optimal lower bounds for small depth circuits, in *Randomness and Computation*, ed. by S. Micali. Advances in Computing Research, vol. 5 (JAI Press, Greenwich, 1989), pp. 143–170
34. F.C. Hennie, R.E. Stearns, Two-tape simulation of multitape Turing machines. J. Assoc. Comput. Mach. **13**(4), 533–546 (1966)
35. R. Impagliazzo, V. Kabanets, A. Kolokolova, An axiomatic approach to algebrization (in preparation)
36. R. Impagliazzo, V. Kabanets, A. Kolokolova, An axiomatic approach to algebrization, in *Proceedings of the Forty-First Annual ACM Symposium on Theory of Computing* (2009), pp. 695–704

37. E. Jeřábek, Dual weak pigeonhole principle, Boolean complexity, and derandomization. Ann. Pure Appl. Log. **129**, 1–37 (2004)
38. E. Jeřábek, Approximate counting in bounded arithmetic. J. Symb. Log. **72**(3), 959–993 (2007)
39. E. Jeřábek, On independence of variants of the weak pigeonhole principle. J. Log. Comput. **17**(3), 587–604 (2007)
40. E. Jeřábek, Approximate counting by hashing in bounded arithmetic. J. Symb. Log. **74**(3), 829–860 (2009)
41. A. Juma, V. Kabanets, C. Rackoff, A. Shpilka, The black-box query complexity of polynomial summation. Comput. Complex. **18**(1), 59–79 (2009)
42. A. Kolokolova, Expressing versus proving: relating forms of complexity in logic. J. Log. Comput. **22**(2), 267–280 (2012)
43. J. Krajíček, Bounded arithmetic, propositional logic and complexity theory, in *Encyclopedia of Mathematics and Its Applications* (Cambridge University Press, Cambridge, 1995)
44. L. Levin, Universal sorting problems. Problems. Inf. Transm. **9**, 265–266 (1973)
45. C. Lund, L. Fortnow, H. Karloff, N. Nisan, Algebraic methods for interactive proof systems. J. Assoc. Comput. Mach. **39**(4), 859–868 (1992)
46. R. Parikh, Existence and feasibility of arithmetic. J. Symb. Log. **36**, 494–508 (1971)
47. P. Pudlák, *Logical Foundations of Mathematics and Computational Complexity: A Gentle Introduction*. Springer Monographs in Mathematics (Springer, New York, 2013)
48. A.A. Razborov, An equivalence between second-order bounded domain bounded arithmetic and first-order bounded arithmetic, in *Arithmetic, Proof Theory and Computational Complexity*, ed. by P. Clote, J. Krajíček (Clarendon Press, Oxford, 1993), pp. 247–277
49. A.A. Razborov, Bounded arithmetic and lower bounds in boolean complexity, in *Feasible Mathematics II* (Springer, New York, 1995), pp. 344–386
50. A.A. Razborov, Unprovability of lower bounds on circuit size in certain fragments of bounded arithmetic. Izv. Math. **59**(1), 205–227 (1995)
51. A.A. Razborov, S. Rudich, Natural proofs. J. Comput. Syst. Sci. **55**, 24–35 (1997)
52. R. Santhanam, Circuit lower bounds for Merlin-Arthur classes, in *Proceedings of the Thirty-Ninth Annual ACM Symposium on Theory of Computing* (2007), pp. 275–283
53. W.J. Savitch, Relationships between nondeterministic and deterministic space complexities. J. Comput. Syst. Sci. **4**(2), 177–192 (1970)
54. A. Shamir, IP=PSPACE. J. Assoc. Comput. Mach. **39**(4), 869–877 (1992)
55. R.E. Stearns, J. Hartmanis, P.M. Lewis II, Hierarchies of memory limited computations, in *SWCT (FOCS)* (1965), pp. 179–190
56. L.J. Stockmeyer, The complexity of decision problems in automata theory and logic, Ph.D. thesis, Massachusetts Institute of Technology, 1974
57. G. Takeuti, RSUV isomorphism, in *Arithmetic, Proof Theory and Computational Complexity*, ed. by P. Clote, J. Krajíček (Clarendon Press, Oxford, 1993), pp. 364–386
58. S. Toda, PP is as hard as the polynomial-time hierarchy. SIAM J. Comput. **20**(5), 865–877 (1991)
59. A.M. Turing, Systems of logic based on ordinals, in *Proceedings of the London Mathematical Society. Second Series*, vol. 45 (1939), pp. 161–228
60. P. van Emde Boas, Turing machines for dummies—why representations do matter, in *SOFSEM* (2012), pp. 14–30
61. R. Williams, Nonuniform ACC circuit lower bounds. J. ACM **61**(1), 2 (2014)

Quantum Randomness: From Practice to Theory and Back

Cristian S. Calude

> *Phenomena that we cannot predict must be judged random.*
>
> P. Suppes

Abstract There is a huge demand for "random" bits. Random number generators use software or physical processes to produce "random" bits. While it is known that programs cannot produce high-quality randomness—their bits are pseudo-random—other methods claim to produce "true" or "perfect" random sequences. Such claims are made for quantum random generators, but, if true, can they be proved? This paper discusses this problem—which is important not only philosophically, but also from a practical point of view.

1 "Babylon Is Nothing but an Infinite Game of Chance"

A mythical Babylon in which everything is dictated by a *universal lottery* is sketched in *The Lottery in Babylon* [8], a short story published in 1941 (first English translation in 1962) by Jorge Luis Borges. A normally operated lottery—with tickets, winners, losers, and money rewards—starts adding punishments to rewards and finally evolves into an all-encompassing "Company" whose decisions are mandatory for all but a small elite. The Company acts at random and in secrecy. Most Babylonians have two only options: to accept the all-knowing, all-powerful, but mysterious Company, or to deny its very existence (no such Company). While various possible philosophical interpretations of the story have been discussed, there is a large consensus that the Company symbolises the power and pervasiveness of *randomness*. Indeed, randomness is the very stuff of life, impinging on everything, fortunes and misfortunes, from the beginning to the end. It causes fear and anxiety, but also makes for fun; and, most interestingly, it provides efficient tools used since ancient Athens.

C.S. Calude (✉)
Department of Computer Science, University of Auckland, Private Bag 92019, Auckland, New Zealand
e-mail: cristian@cs.auckland.ac.nz; www.cs.auckland.ac.nz/~cristian

S.B. Cooper, M.I. Soskova (eds.), *The Incomputable*, Theory and Applications of Computability, DOI 10.1007/978-3-319-43669-2_11

2 A Case Study: Security

It is difficult to deny that security is one of the key issue of our time. Here is an example related to the NSA scandal (June 2013)[1] presented in [18]. The CNN report, significantly sub-titled "Tapping the strange science of quantum mechanics, physicists are creating new data security protocols that even the NSA can't crack", starts with ... Snowden.

> The news out of Moscow of late has been dominated by Edward Snowden, the American leaker of secret state documents who is currently seeking temporary asylum in Russia. Meanwhile, across town and to much less fanfare, Dr. Nicolas Gisin found himself explaining last week the solution to the very problems of data security and privacy intrusion Snowden brought to light in exposing the vast reach of the National Security Agency's data collection tools: *data encryption that is unbreakable now and will remain unbreakable in the future.*[2]

According to Wikipedia, "Cryptography is the practice and study of techniques for secure communication in the presence of third parties (called adversaries)". A cryptosystem is a suite of algorithms used to implement a particular form of encryption and decryption. Modern cryptography is dominated by three main approaches: (a) the information-theoretic approach, in which the adversary should have not enough information to break a cryptosystem, (b) the complexity-theoretic approach, in which the adversary should have not enough computational power to break a cryptosystem and (c) the quantum physics approach, in which the adversary would need to break some physical laws to break a cryptosystem. The third approach is called quantum cryptography; by contrast, the first two approaches are referred to as classical cryptography.

Cryptographic algorithms require a method of generating a secret key from "random" bits. The encryption algorithm uses the key to encrypt and decrypt messages, which are sent over unsecured communication channels. The strength of the system ultimately depends on the strength of the key used, i.e. on the difficulty for an eavesdropper to guess or calculate it. Vulnerabilities of classical cryptography are well documented, but quantum cryptography was (and, as we will see below, continues to be) believed to be unbreakable: Heisenberg's Uncertainty Principle guarantees that an adversary cannot look into the series of photons which transmit the key without either changing or destroying them. The difference between classical and quantum cryptography rests on keys: classical keys are vulnerable, but keys formed with quantum random bits have been claimed to be unbreakable because *quantum randomness is true randomness* (see [31]). In the words

[1]Borges names "Qaphqa", an obvious code for Kafka, the "sacred latrine" allowing access to the Company.

[2]My Italics.

from [18]:

> "It sounds like there's some quantum magic in this new technology, but of course it's not magic, it's just very modern science", Gisin says. But next to classical communication and encryption methods, it might as well be magic. Classical cryptography generally relies on algorithms to randomly generate encryption and decryption keys enabling the sender to essentially scramble a message and a receiver to unscramble it at the other end. If a third-party ... obtains a copy of the key, that person can make a copy of the transmission and decipher it, or—with enough time and computing power—use powerful algorithms to break the decryption key. (This is what the NSA and other agencies around the world are allegedly up to.) But Gisin's quantum magic taps some of the stranger known phenomena of the quantum world to transmit encryption keys that cannot be copied, stolen, or broken without rendering the key useless.
>
> The primary quantum tool at work in ID Quantique's quantum communication scheme is known as "entanglement", a phenomenon in which two particles—in this case individual photons—are placed in a correlated state. Under the rules of quantum mechanics, these two entangled photons are inextricably linked; a change to the state of one photon will affect the state of the other, regardless of whether they are right next to each other, in different rooms, or on opposite sides of the planet. One of these entangled photons is sent from sender to receiver, so each possesses a photon. These photons are not encoded with any useful information—that information is encoded using normal classical encryption methods—but with a decryption key created by a random number generator. (*True random number*[3] generators represent another technology enabled by quantum physics—more on that in a moment.)

The above quote contains errors and misleading statements, but we reproduce it here as it is illustrative of the way quantum cryptography is presented to the public. Really, how good is this technology?

Gruska [25] offers a cautious answer from a theoretical point of view:

> Goals of quantum cryptography have been very ambitious. Indeed, some protocols of quantum cryptography provably achieve so-called unconditional secrecy, a synonym for absolute secrecy, also in the presence of eavesdroppers endowed with unlimited computational power and limited only by the laws of nature, or even only by foreseeable laws of nature not contradicting the non-signaling principle of relativity.

An answer from a practical point of view appears in [18]:

> "Security experts didn't learn anything from this Snowden story, it was already obvious that it is so easy to monitor all the information passing through the Internet", Gisin says. "No security expert can pretend to be surprised by his revelation. And I'm not a national security expert, but I don't think the Americans are the only ones who are doing this—the Russians are doing it, the Chinese are doing it, everybody is spying on the others and that's always been the case and it always will be. One way to be a step ahead of the others is to use quantum cryptography, because *for sure the programs that the Americans and others are using will not be able to crack it.*[4]

---

[3]My Italics.

[4]My Italics.

3 True Randomness

The "magic" of the quantum technology capable of producing unbreakable security depends on the possibility of producing *true random* bits. What does "true randomness"[5] mean? The concept is not formally defined, but a common meaning is the *lack of any possible correlations*. Is this indeed theoretically possible? The answer is *negative*: *there is no true randomness, irrespective of the method used to produce it*. The British mathematician and logician Frank P. Ramsey was the first to demonstrate it in his study of *conditions under which order must appear* (see [24, 32]); other proofs have been given in the framework of algorithmic information theory [9].

We will illustrate Ramsey's theory later in this section. For now, let's ask a more pragmatic question: Are these mathematical results relevant for the theory or practice of quantum cryptography? Poor quality randomness is, among other issues, the cause of various failures of quantum cryptographic systems. After a natural euphoria period when quantum cryptography was genuinely considered to be "unbreakable", scientists started to exercise one of the most important attitudes in science: skepticism. And, indeed, weaknesses of quantum cryptography have been discovered; they are not new and they are not a few. Issues were found as early as 2008 [13], even earlier. In 2010, V. Makarov and his colleagues published the details of a traceless attack against a class of quantum cryptographic systems [29] which includes the products commercialised by ID Quantique[6] (Geneva) (www.idquantique.com) and MagiQ Technologies (Boston) (http://www.magiqtech.com): both companies claim to produce *true randomness*. Recent critical weaknesses of a new class of quantum cryptographic schemes called "device-independent" protocols—that rely on public communication between secure laboratories—are described in [7].

Geneva is only 280 km from Zurich, but the views on quantum cryptography of ID Quantique and physicist R. Renner, from the Institute of Theoretical Physics in Zurich, are quite different. Recognising the weaknesses of quantum cryptography, R. Renner has embarked on a program to evaluate the failure rate of different quantum cryptography systems. He was quoted in (http://www.sciencedaily.com/releases/2013/05/130528122435.htm):

> The security of Quantum Key Distribution systems is never absolute.

Renner's work was presented at the 2013 Conference on Lasers and Electro-Optics (San Jose, California, USA, [14]). Not surprisingly, even before presenting his guest lecture on June 11, Renner's main findings made the news: [17, 30] are two examples. Commenting on the "timeslicing" BB84 protocol, K. Svozil, cited in [26], said: "The newly proposed [quantum] protocol is 'breakable' by middlemen attacks" in the same way as BB84: "complete secrecy" is an illusion. See also [28].

---

[5] Also called perfect randomness.

[6] Featured in Sect. 2.

Why would some physicists claim that quantum randomness is *true randomness*? According to ID Quantique website (www.idquantique.com).

> Existing randomness sources can be grouped in two classes: software solutions, which can only generate pseudo-random bit streams, and physical sources. In the latter, most random generators rely on classical physics to produce what looks like a random stream of bits. In reality, determinism is hidden behind complexity. Contrary to classical physics, quantum physics is fundamentally random. It is the only theory within the fabric of modern physics that integrates randomness.

Certainly, this statement is not a proper scientific justification. Randomness in quantum mechanics comes from measurement, which is part of the interpretation of quantum mechanics. To start with we need to assume that measurement yields a physically meaningful and unique result. This may seem rather self-evident, but it is not true of interpretations of quantum mechanics such as the many-worlds, where measurement is just a process by which the apparatus or experimenter becomes entangled with the state being "measured"; in such an interpretation it does not make sense to talk about the unique "result" of a measurement.

If the only basis for claiming that quantum randomness is better than pseudo-randomness is the fact that the first is true randomness, then the claim is very weak. After all, experimentally, both types of randomness are far from being perfect [12]; we need much more understanding of randomness to be able to say something non-trivial about quantum randomness.

Interestingly, Ramsey theory provides arguments for the impossibility of true randomness resting on the sole fact that any model of randomness has to satisfy the common intuition that "randomness means no correlations, no patterns". The question becomes:

> Are there binary (finite) strings or (infinite) sequences with no patterns/correlations?

Ramsey theory answers the above question in the negative; measure-theoretical arguments have been also found in algorithmic information theory [9]. Here is an illustration of the Ramsey-type argument.

Let $s_1 \cdots s_n$ be a binary string. A monochromatic arithmetic progression of length k is a substring $s_i s_{i+t} s_{i+2t} \cdots s_{i+(k-1)t}$, $1 \leq i$ and $i + (k-1)t \leq n$, with all characters equal (0 or 1) for some $t > 0$. The theorem below states that *all* binary strings with sufficient length have monochromatic arithmetic progressions of any given length. The importance of the theorem lies in the fact that *all* strings display one of the simplest types of correlation, without being constant.

Van der Waerden finite theorem. *For every natural k there is a natural $n > k$ such that every string of length n contains a monochromatic arithmetic progression of length k.*

The Van der Waerden number, $W(k)$, is the smallest n such that every string of length n contains a monochromatic arithmetic progression of length k. For example, $W(3) = 9$. The string 01100110 contains no arithmetic progression of length 3 because the positions 1, 4, 5, 8 (for 0) and 2, 3, 6, 7 (for 1) do not contain an arithmetic progression of length 3; hence $W(3) > 8$. However, both strings 011001100 and 011001101 do: 1, 5, 9 for 0 and 3, 6, 9 for 1. In fact, one can easily test that every string of length 9 contains three terms of a monochromatic arithmetic progression, so $W(3)=9$.

How long should a string be to display a monochromatic arithmetic progression, i.e. how big is $W(k)$? In [22] it was proved that $W(k) < 2^{2^{2^{2^{2^{k+9}}}}}$, but it was conjectured to be much smaller in [23]: $W(k) < 2^{k^2}$.

The Van der Waerden result is true for infinite binary sequences as well:

Van der Waerden infinite theorem. *Every infinite binary sequence contains arbitrarily long monochromatic arithmetic progressions.*

This is one of the many results in Ramsey theory [32]. Graham and Spencer, well-known experts in this field, subtitled their *Scientific American* presentation of Ramsey Theory [24] with a sentence similar in spirit to Renner's (quoted above):

> Complete disorder is an impossibility. Every large set of numbers, points or objects necessarily contains a highly regular pattern.

The adjective "large" applies to both finite and infinite sets.[7] The simplest finite example is *the pigeonhole principle*: A set of N objects is partitioned into n classes. Here "large" means $N > n$. Conclusion: a class contains at least two objects. Example: "Of three ordinary people, two must have the same sex" (D. J. Kleitmen). *The infinite pigeonhole principle*: A set of objects is partitioned into finitely many classes. Here "large" means that the set is infinite while the number of classes which is finite. Conclusion: a class is infinite.

Randomness comes from different sources and means different things in different fields. Algorithmic information theory [9, 19] is a mathematical theory in which, in contrast to probability theory, the randomness of individual objects is studied. Given the impossibility of true randomness, the effort is directed towards studying degrees of randomness. The main point of algorithmic information theory (a point emphasised from a philosophical point of view in [20]) is:

> Randomness means unpredictability with respect to some fixed theory.

The quality of a particular type of randomness depends on the power of the theory to detect correlations, which determines how difficult predictability is (see more in [4, 6]). For example, finite automata detect less correlations than Turing machines. Consequently, finite automata, based unpredictability is weaker than Turing machine, based unpredictability: there are (many) sequences computable by Turing machines (hence, predictable, not random) that are unpredictable, random, for finite automata.

In analogy with the notion of incomputability (see [15]), one can prove that there is a never-ending hierarchy of stronger (better quality) and stronger forms of randomness.

---

[7] We identify a binary finite string and an infinite sequence with sets of positive integers.

4 Is Quantum Randomness "Better" Than Pseudo-Randomness

The intuition confirmed by experimental results reported in [12] suggests that the quality of quantum randomness is *better* than that of pseudo-randomness. Is there any solid basis to compare quantum randomness and pseudo-randomness?

Although in practice only finitely many bits are necessary, to be able to evaluate and compare the quality of randomness we need to consider infinite sequences of bits. In [1–5, 5, 6, 10–12] the first steps in this direction have been made.

Pseudo-random sequences are obviously Turing computable (i.e. they are produced by an algorithm); they are easily predictable once we know the seed and the algorithm generating the sequence, so, not surprisingly, their quality of randomness is low. Is quantum randomness Turing computable?

How can one prove such a result? As we have already observed in the previous section, we need to make some physical assumptions to base our mathematical reasoning on. To present these assumptions we need a few notions specific to quantum mechanics; we will adopt them in the form presented in [1].

In what follows we only consider pure quantum states. Projection operators—projecting on to the linear subspace spanned by a non-zero vector $|\psi\rangle$—will be denoted by $P_\psi = \frac{|\psi\rangle\langle\psi|}{\langle\psi|\psi\rangle}$.

We fix a positive integer n. Let $\mathcal{O} \subseteq \{P_\psi \mid |\psi\rangle \in \mathbb{C}^n\}$ be a non-empty set of *projection observables* in the Hilbert space $\mathbb{C}^n$, and $\mathcal{C} \subseteq \{\{P_1, P_2, \ldots P_n\} \mid P_i \in \mathcal{O} \text{ and } \langle i|j\rangle = 0 \text{ for } i \neq j\}$ be a set of measurement contexts over $\mathcal{O}$. A *context* $C \in \mathcal{C}$ is thus a maximal set of compatible (i.e. they can be simultaneously measured) projection observables. Let $v : \{(o, C) \mid o \in \mathcal{O}, C \in \mathcal{C} \text{ and } o \in C\} \stackrel{o}{\to} B$ be a partial function (i.e., it may be undefined for some values in its domain) called *assignment function*. For some $o, o' \in \mathcal{O}$ and $C, C' \in \mathcal{C}$ we say $v(o, C) = v(o', C')$ if $v(o, C), v(o', C')$ are both defined and have equal values.

Value definiteness corresponds to the notion of predictability in classical determinism: an observable is value definite if v assigns it a definite value—i.e. is able to predict in advance, independently of measurement, the value obtained via measurement. Here is the formal definition: an observable $o \in C$ is *value definite* in the context C under v if $v(o, C)$ is defined; otherwise o is *value indefinite* in C. If o is value definite in all contexts $C \in \mathcal{C}$ for which $o \in C$ then we simply say that o is value definite under v. The set $\mathcal{O}$ is *value definite* under v if every observable $o \in \mathcal{O}$ is value definite under v.

Non-contextuality corresponds to the classical notion that the value obtained via measurement is independent of other compatible observables measured alongside it. Formally, an observable $o \in \mathcal{O}$ is *non-contextual* under v if for all contexts $C, C' \in \mathcal{C}$ with $o \in C, C'$ we have $v(o, C) = v(o, C')$; otherwise, v is *contextual*. The set of observables $\mathcal{O}$ is *non-contextual* under v if every observable $o \in \mathcal{O}$ which is not value indefinite (i.e. value definite in *some* context) is non-contextual under v; otherwise, the set of observables $\mathcal{O}$ is *contextual*.

To be in agreement with quantum mechanics we restrict the assignment functions to admissible ones: v is *admissible* if the following hold for all $C \in \mathcal{C}$: (a) if there exists an $o \in C$ with $v(o, C) = 1$, then $v(o', C) = 0$ for all $o' \in C \setminus \{o\}$; (b) if there exists an $o \in C$ such that $v(o', C) = 0$ for all $o' \in C \setminus \{o\}$, then $v(o, C) = 1$.

We are now ready to list the physical assumptions. A *value indefinite quantum experiment* is an experiment in which a particular value indefinite observable in a standard (von Neumann type) quantum mechanics is measured, subject to the following assumptions **(A1)–(A5)** (for a detailed motivation we refer you to [1]).

We exclude interpretations of quantum mechanics, such as the many-worlds interpretation, where there is no unique "result" of a measurement.

(A1) Measurement assumption. *Measurement yields a physically meaningful and unique result.*

We restrict the set of assignments to those which agree with quantum mechanics.

(A2) Assignment assumption. *The assignment function v is a* faithful *representation of a realisation r_ψ of a state $|\psi\rangle$, that is, the measurement of observable o in the context C on the physical state r_ψ yields the result $v(o, C)$ whenever o has a definite value under v.*

We assume a classical-like behaviour of measurement: the values of variables are intrinsic and independent of the device used to measure the m.

(A3) Non-contextuality assumption. *The set of observables $\mathcal{O}$ is non-contextual.*

The following assumption reflects another agreement with quantum mechanics.

(A4) Eigenstate assumption. *For every (normalised) quantum state $|\psi\rangle$ and faithful assignment function v, we have $v(P_\psi, C) = 1$ and $v(P_\phi, C) = 0$, for any context $C \in \mathcal{C}$, with $P_\psi, P_\phi \in C$.*

The motivation for the next assumption comes from the notion of "element of physical reality" described by Einstein, Podolsky and Rosen in [21, p. 777]:

> If, without in any way disturbing a system, we can predict with certainty (i.e., with probability equal to unity) the value of a physical quantity, then there exists an element of physical reality [8] [(e.p.r.)] corresponding to this physical quantity.

The last assumption is a weak form of e.p.r. in which *prediction is certain* (not only with probability 1) and *given by some function which can be proved to be computable*.

(A5) Elements of physical reality (e.p.r.) assumption. *If there exists a computable function $f : \mathbf{N} \times \mathcal{O} \times \mathcal{C} \to B$ such that for infinitely many $i \geq 1, f(i, o_i, C_i) = x_i$, then there is a definite value associated with o_i at each step* [i.e., $v_i(o_i, C_i) = f(i, o_i, C_i)$].

To use the e.p.r. assumption we need *to prove* the existence of a computable function f such that for infinitely many $i \geq 1, f(i, o_i, C_i) = x_i$.

Can projection observables be value definite and non-contextual? The following theorem answers this question in the negative.

Kochen-Specker theorem. *In a Hilbert space of dimension $n > 2$ there exists a set of projection observables $\mathcal{O}$ on $\mathbb{C}^n$ and a set of contexts over $\mathcal{O}$ such that there is no admissible assignment function v under which $\mathcal{O}$ is both non-contextual and value definite.*

[8] An element of physical reality corresponds to the notion of a definite value, possibly contextual.

The Kochen-Specker theorem [27]—proved almost 50 years ago—is a famous result showing a contradiction between two basic assumptions of a hypothetical hidden variable theory intended to reproduce the results of quantum mechanics: (a) all hidden variables corresponding to quantum mechanical observables have definite values at any given time, and (b) the values of those variables are intrinsic and independent of the device used to measure them. The result is important in the debate on the (in)completeness of quantum mechanics created by the EPR paradox [21].

Interestingly, the theorem, which is considered a topic in the foundations of quantum mechanics, with more philosophical flavour and little presence in mainstream quantum mechanical textbooks, has actually an *operational* importance. Indeed, using the assumption **(A3)**, the Kochen-Specker theorem states that some projection observables have to be value indefinite.

Why should we care about a value indefinite observable? Because a way "to see" the randomness in quantum mechanics is by measuring such an observable. Of course, we need to be able to *certify* that a given observable is value indefinite. Unfortunately, the theorem gives no indication which observables are value indefinite. We know that not all projection observables are value indefinite [1], but can we be sure that a specific observable is value indefinite observable? The following result from [1, 5] answers this question in the affirmative:

Localised Kochen-Specker theorem. *Assume* **(A1)–(A4)**. *Let $n \geq 3$ and $|\psi\rangle, |\phi\rangle \in \mathbb{C}^n$ be unit vectors such that $0 < |\langle\psi|\phi\rangle| < 1$. We can effectively find a finite set of one-dimensional projection observables $\mathcal{O}$ containing P_ψ and P_ϕ for which there is no admissible value assignment function on $\mathcal{O}$ such that $v(P_\psi) = 1$ and P_ϕ is value definite.*

An operational form of the localised Kochen-Specker theorem capable of identifying a value indefinite observable is given by:

Operational Kochen-Specker theorem. *Assume* **(A1)–(A4)**. *Consider a quantum system prepared in the state $|\psi\rangle$ in $\mathbb{C}^3$, and let $|\phi\rangle$ be any state such that $0 < |\langle\psi|\phi\rangle| < 1$. Then the projection observable $P_\phi = |\phi\rangle\langle\phi|$ is value indefinite.*

The operational Kochen-Specker theorem allows us to identify and then measure a value indefinite observable, a crucial point in what follows. Consider a system in which a value indefinite quantum experiment is prepared, measured, rinsed and repeated ad infinitum. The infinite sequence $\mathbf{x} = x_1x_2\ldots$ obtained by concatenating the outputs of these measurements is called *value indefinite quantum random sequence*, shortly, *quantum random sequence*. We are now able to give a mathematical argument showing that quantum randomness produced by a specific type of experiment is better than pseudo-randomness [1, 5, 11]:

Incomputability theorem. *Assume* **(A1)–(A5)** *for $\mathbb{C}^3$. Then, every quantum random sequence is Turing incomputable.*

In fact, a stronger result is true [1, 5, 11]:

Strong incomputability theorem. *Assume* **(A1)–(A5)** *for $\mathbb{C}^3$. Then, every quantum random sequence is bi-immune, that is, every Turing machine cannot compute exactly more than finitely many bits of the sequence.*

Bi-immunity ensures that any adversary can be sure of no more than finitely many exact values—guessed or computed—of any given quantum random sequence. This is indeed a good certificate of quality for this type of quantum randomness.

Finally, we ask the following natural question: is quantum indeterminacy a pervasive phenomenon or just an accidental one? The answer was provided in [3]:

Value indefiniteness theorem. *Assume* **(A1)–(A4)** *for* $\mathbb{C}^n$ *for* $n \geq 3$. *The set of value indefinite observables in* $\mathbb{C}^n$ *has constructive Lebesgue measure 1.*

The above theorem provides the strongest conceivable form of quantum indeterminacy: once a single arbitrary observable is fixed to be value definite, almost (i.e. with Lebesgue measure 1) all remaining observables are value indefinite.

5 A Quantum Random Number Generator

The theoretical results discussed in the previous section have practical value only if one can design a quantum random number generator in which a value indefinite observable is measured, a guarantee for its strong incomputability. In particular, a quantum random number generator has to act in $\mathbb{C}^n$ with $n \geq 3$ (see [1, 2]).

A quantum random number generator [1] designed in terms of generalised beam splitters satisfies these requirements; its blueprint is presented in Fig. 1. The

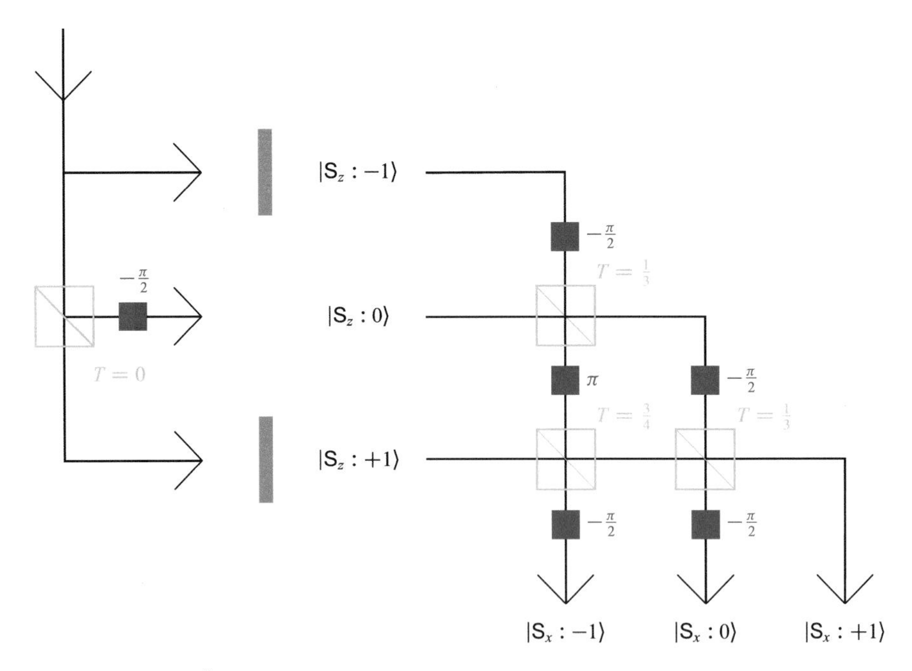

Fig. 1 QRNG in $\mathbb{C}^3$

configuration indicates the preparation and the measurement stages, including filters blocking $|\mathsf{S}_z : -1\rangle$ and $|\mathsf{S}_z : +1\rangle$. (For ideal beam splitters, these filters would not be required.) The measurement stage (right array) realises a unitary quantum gate U_x, corresponding to the projectors onto the S_x state observables for spin state measurements along the x-axis, in terms of generalised beam splitters. More details about its implementation and practical issues are presented in [1].

6 Conclusion and Open Questions

The practice of generating and commercialising quantum random bits raises many questions about the quality of randomness it produces. Based on certain natural physical hypotheses, we have described a procedure to generate quantum random bits that *provably* are not reproducible by any Turing machine, an example of incomputability in nature (see [16]). In particular, this proves that this type quantum randomness is superior in quality to pseudo-randomness. A quantum random generator which produces bi-immune sequences has been described.

This is just the start of a program for better understanding and producing quantum randomness. Many problems remain open, and here are some of them. Does a variant of the (strong) incomputability theorem, possibly with additional physical assumptions, hold true in $\mathbb{C}^2$? Does Quantis, the quantum random generator produced by ID Quantique which operates in $\mathbb{C}^2$, produce bi-immune sequences? Are other physical assumptions sufficient for proving an analogue of the operational Kochen-Specker theorem? Can other physical assumptions lead to different types of quantum random generators producing bi-immune sequences? How random is a sequence produced by an experiment certified by the operational Kochen-Specker theorem? Is quantum randomness unique or of different forms and qualities?

Acknowledgements The author has been supported in part by the Marie Curie FP7-PEOPLE-2010-IRSES Grant RANPHYS and has benefitted from discussions and collaboration with A. Abbott, J. Conder, B. Cooper, M. Dinneen, M. Dumitrescu, J. Gruska, G. Longo, K. Svozil and K. Tadaki. I thank them all.

References

1. A.A. Abbott, C.S. Calude, J. Conder, K. Svozil, Strong Kochen-Specker theorem and incomputability of quantum randomness. Phys. Rev. A **86**, 6 (2012). doi:10.1103/PhysRevA.00.002100
2. A. Abbott, C.S. Calude, K. Svozil, A quantum random number generator certified by value indefiniteness. Math. Struct. Comput. Sci. **24**(3), e240303 (2014). doi:10.1017/S0960129512000692
3. A. Abbott, C.S. Calude, K. Svozil, Value-indefinite observables are almost everywhere. Phys. Rev. A **89**(3), 032109–032116 (2014). doi:10.1103/PhysRevA.89.032109

4. A.A. Abbott, C.S. Calude, K. Svozil, On the unpredictability of individual quantum measurement outcomes, in *Fields of Logic and Computation II*, ed. by L.D. Beklemishev, A. Blass, N. Dershowitz, B. Finkbeiner, W. Schulte. Lecture Notes in Computer Science, vol. 9300 (Springer, Heidelberg, 2015), pp. 69–86
5. A. Abbott, C.S. Calude, K. Svozil, A variant of the Kochen-Specker theorem localising value indefiniteness. J. Math. Phys. **56**, 102201 (2015). http://dx.doi.org/10.1063/1.4931658
6. A. Abbott, C.S. Calude, K. Svozil, A non-probabilistic model of relativised predictability in physics. Information (2015). http://dx.doi.org/10.1016/j.ic.2015.11.003
7. J. Barrett, R. Colbeck, A. Kent, Memory Attacks on Device-Independent Quantum Cryptography, arXiv:1201.4407v6 [quant-ph], 6 Aug 2013
8. J.L. Borges, The lottery in Babylon, in *Everything and Nothing*, ed. by D.A. Yates, J.E. Irby, J.M. Fein, E. Weinberger (New Directions, New York, 1999), pp. 31–38
9. C.S. Calude, *Information and Randomness: An Algorithmic Perspective*, 2nd edn. (Springer, Berlin, 2002)
10. C.S. Calude, Algorithmic randomness, quantum physics, and incompleteness, in *Proceedings of Conference on "Machines, Computations and Universality" (MCU'2004)*, ed. by M. Margenstern. Lectures Notes in Computer Science, vol. 3354 (Springer, Berlin, 2005), pp. 1–17
11. C.S. Calude, K. Svozil, Quantum randomness and value indefiniteness. Adv. Sci. Lett. **1**, 165–168 (2008)
12. C.S. Calude, M.J. Dinneen, M. Dumitrescu, K. Svozil, Experimental evidence of quantum randomness incomputability. Phys. Rev. A **82**, 022102, 1–8 (2010)
13. J. Cederlof, J.-A. Larsson, Security aspects of the authentication used in quantum cryptography. IEEE Trans. Inf. Theory **54**(4), 1735–1741 (2008)
14. Conference on Lasers and Electro-Optic 2013, http://www.cleoconference.org/home/news-and-press/cleo-press-releases/cleo-2013-the-premier-international-laser-and-elec/ (2013)
15. S.B. Cooper, *Computability Theory* (Chapman & Hall/CRC, London, 2004)
16. S.B. Cooper, P. Odifreddi, Incomputability in nature, in *Computability and Models: Perspectives East and West*, ed. by S.B. Cooper, S.S. Goncharov (Plenum Publishers, New York, 2003), pp. 137–160
17. B. Day, Just how secure is quantum cryptography? *Optical Society*, http://www.osa.org/en-us/about_osa/newsroom/newsreleases/2013/just_how_secure_is_quantum_cryptography/, 28 May 2013
18. C. Dillow, Zeroing in on unbreakable computer security, http://tech.fortune.cnn.com/2013/07/29/from-russia-unbreakable-computer-code/, 29 July 2013
19. R. Downey, D. Hirschfeldt, *Algorithmic Randomness and Complexity* (Springer, Heidelberg, 2010)
20. A. Eagle, Randomness is unpredictability. Br. J. Philos. Sci. **56**, 749–90 (2005)
21. A. Einstein, B. Podolsky, N. Rosen, Can quantum-mechanical description of physical reality be considered complete? Phys. Rev. **47**, 777–780 (1935)
22. T. Gowers, A new proof of Szemerédi's theorem. Geom. Funct. Anal. **11**(3), 465–588 (2001)
23. R. Graham, Some of my favorite problems in Ramsey theory. INTEGERS Electron. J. Comb. Number Theory **7**(2), #A2 (2007)
24. R. Graham, J.H. Spencer, Ramsey theory. Sci. Am. **262**(7), 112–117 (1990)
25. J. Gruska, From classical cryptography to quantum physics through quantum cryptography. J. Indian Inst. Sci. **89**(3), 271–282 (2009)
26. M. Hogenboom, 'Uncrackable' codes set for step up. BBC News, 4 September 2013, http://www.bbc.co.uk/news/science-environment-23946488
27. S. Kochen, E.P. Specker, The problem of hidden variables in quantum mechanics. J. Math. Mech. **17**, 59–87 (1967)
28. H.-W. Li, Z.-Q. Yin, S. Wang, Y.-J. Qian, W. Chen, G.-C. Guo, Z.-F. Han, Randomness determines practical security of BB84 quantum key distribution. Sci. Rep. **5**, 16200 (2015). doi:10.1038/srep16200

29. L. Lydersen, C. Wiechers, C. Wittmann, D. Elser, J. Skaar, V. Makarov, Hacking commercial quantum cryptography systems by tailored bright illumination. Nat. Photon. **4**(10), 686–689 (2010). Supplementary information, http://www.nature.com/nphoton/journal/v4/n10/abs/nphoton.2010.214.html
30. A. Mann, The laws of physics say quantum cryptography is unhackable. It's not! Wired Science, http://www.wired.com/wiredscience/2013/06/quantum-cryptography-hack/, 21 March 2013
31. Nature, True randomness demonstrated, http://www.nature.com/nature/journal/v464/n7291/edsumm/e100415-06.html, 15 April 2010
32. A. Soifer, Ramsey theory before Ramsey, prehistory and early history: an essay, in *Ramsey Theory: Yesterday, Today, and Tomorrow*, ed. by A. Soifer. Progress in Mathematics, vol. 285 (Springer, Berlin, 2011), pp. 1–26

Calculus of Cost Functions

André Nies

Abstract Cost functions provide a framework for constructions of sets Turing below the halting problem that are close to computable. We carry out a systematic study of cost functions. We relate their algebraic properties to their expressive strength. We show that the class of additive cost functions describes the K-trivial sets. We prove a cost function basis theorem, and give a general construction for building computably enumerable sets that are close to being Turing complete.

1991 *Mathematics Subject Classification.* Primary: 03F60; Secondary: 03D30

1 Introduction

In the time period from 1986 to 2003, several constructions of computably enumerable (c.e.) sets appeared. They turned out to be closely related.

(a) Given a Martin-Löf random (ML-random for short) Δ_2^0 set Y, Kučera [16] built a c.e. incomputable set $A \leq_T Y$. His construction is interesting because in the case $Y <_T \emptyset'$, it provides a c.e. set A such that $\emptyset <_T A <_T \emptyset'$ without using injury to requirements as in the traditional proofs. ($\emptyset'$ denotes the halting problem.)
(b) Kučera and Terwijn [19] built a c.e. incomputable set A that is low for ML-randomness: every ML-random set is already ML-random relative to A.
(c) A is called K-trivial if $K(A \upharpoonright_n) \leq K(n) + O(1)$, where K denotes prefix-free descriptive string complexity. This means that the initial segment complexity of A grows as slowly as that of a computable set. Downey et al. [8] gave a very short construction (almost a "definition") of a c.e., but incomputable, K-trivial set.

A. Nies (✉)
Department of Computer Science, University of Auckland, Auckland, New Zealand
e-mail: andre@cs.auckland.ac.nz

S.B. Cooper, M.I. Soskova (eds.), *The Incomputable*, Theory and Applications of Computability, DOI 10.1007/978-3-319-43669-2_12

The sets in (a) and (b) enjoy a so-called lowness property, which says that the set is very close to computable. Such properties can be classified according to various paradigms introduced in [13, 24]. The set in (a) obeys the *Turing-below-many* paradigm, which says that A is close to being computable because it is easy for an oracle set to compute it. A frequent alternative is the *weak-as-an-oracle* paradigm: A is weak in a specific sense when used as an oracle set in a Turing machine computation. An example is the oracle set in (b), which is so weak that it is useless as an extra computational device when testing for ML-randomness. On the other hand, K-triviality in (c) is a property stating that the set is far from random: by the Schnorr-Levin Theorem, for a random set Z the initial segment complexity grows fast in that $K(Z \upharpoonright_n) \geq n - O(1)$. For background on the properties in (a)–(c), see [7] and [23, Chap. 5].[1]

A central point for starting our investigations is the fact that the constructions in (a)–(c) look very similar. In hindsight this is not surprising: the classes of sets implicit in (a)–(c) coincide! Let us discuss why.

(b) coincides with (c): Nies [22], with some assistance from Hirschfeldt, showed that lowness for ML-randomness is the same as K-triviality. For this he introduced a method now known as the "golden run".
(a) coincides with (b): The construction in (a) is only interesting if $Y \not\geq_T \emptyset'$. Hirschfeldt, Nies and Stephan [14] proved that if A is a c.e. set such that $A \leq_T Y$ for some ML-random set $Y \not\geq_T \emptyset'$, then A is K-trivial, confirming the intuition that sets of the type built by Kučera are close to computable. They asked whether, conversely, for every K-trivial set A there is an ML-random set $Y \geq_T A$ with $Y \not\geq_T \emptyset'$. This question became known as the ML-covering problem. Recently, the question was answered in the affirmative by combining the work of seven authors in two separate papers. In fact, there is a single ML-random Δ^0_2 set $Y \not\geq_T \emptyset'$ that is Turing above all the K-trivials. A summary is given in [2].

The common objective for these constructions is to ensure lowness of A dynamically by restricting the overall manner in which numbers can be enumerated into A. This third lowness paradigm has been called *inertness* in [24]: a set A is close to computable because it is computably approximable with a small number of changes.

The idea is implemented as follows. The enumeration of a number x into A at stage s bears a cost $\mathbf{c}(x, s)$, a non-negative rational that can be computed from x and s. We have to enumerate A in such a way that the sum of all costs is finite. A construction of this type will be called a *cost function construction*. If we enumerate at a stage more than one number into A, only the cost for enumerating the least number is charged.

[1] We note that the result (c) has a complicated history. Solovay [26] built a Δ^0_2 incomputable set A that is K-trivial. Constructing a c.e. example of such a set was attempted in various sources such as [4], and in unpublished work of Kummer.

1.1 Background on Cost Functions

The general theory of cost functions began in [23, Sect. 5.3]. It was further developed in [6, 10, 13]. We use the language of [23, Sect. 5.3] which already allows for the constructions of Δ^0_2 sets. The language is enriched by some notation from [6]. We will see that most examples of cost functions are based on randomness-related concepts.

Definition 1.1 A *cost function* is a computable function

$$\mathbf{c} : \mathbb{N} \times \mathbb{N} \to \{x \in \mathbb{Q}\colon\ x \geq 0\}.$$

Recall that a *computable approximation* is a computable sequence of finite sets $\langle A_s \rangle_{s\in\mathbb{N}}$ such that $\lim_s A_s(x)$ exists for each x.

Definition 1.2

(i) Given a computable approximation $\langle A_s \rangle_{s\in\mathbb{N}}$ and a cost function $\mathbf{c}$, for $s > 0$ we let

$$\mathbf{c}_s(A_s) = \mathbf{c}(x, s) \text{ where } x < s\ \wedge\ x \text{ is least s.t. } A_{s-1}(x) \neq A_s(x);$$

if there is no such x, we let $\mathbf{c}_s(A_s) = 0$. This is the cost of changing A_{s-1} to A_s. We let

$$\mathbf{c}\langle A_s \rangle_{s\in\mathbb{N}} = \sum_{s>0} \mathbf{c}_s(A_s)$$

be the total cost of all the A-changes. We will often write $\mathbf{c}\langle A_s \rangle$ as a shorthand for $\mathbf{c}\langle A_s \rangle_{s\in\mathbb{N}}$.

(ii) We say that $\langle A_s \rangle_{s\in\mathbb{N}}$ *obeys* $\mathbf{c}$ if $\mathbf{c}\langle A_s \rangle$ is finite. We denote this by

$$\langle A_s \rangle \models \mathbf{c}.$$

(iii) We say that a Δ^0_2 set A *obeys* $\mathbf{c}$, and write $A \models \mathbf{c}$, if some computable approximation of A obeys $\mathbf{c}$.

A cost function $\mathbf{c}$ acts like a global restraint, which is successful if the condition $\mathbf{c}\langle A_s \rangle < \infty$ holds. Kučera's construction mentioned in (a) above needs to be recast in order to be viewed as a cost function construction [10, 23]. In contrast, (b) and (c) can be directly seen as cost function constructions. In each of (a)–(c) above, one defines a cost function $\mathbf{c}$ such that any set A obeying $\mathbf{c}$ has the lowness property in question. For, if $A \models \mathbf{c}$, then one can enumerate an auxiliary object that has in some sense a bounded weight.

In (a), this object is a Solovay test that accumulates the errors in an attempted computation of A with oracle Y. Since Y passes this test, Y computes A.

In (b), one is given a $\Sigma^0_1(A)$ class $\mathcal{V} \subseteq 2^\omega$ such that the uniform measure $\boldsymbol{\lambda}\mathcal{V}$ is less than 1, and the complement of $\mathcal{V}$ consists only of ML-randoms. Using the fact that A obeys $\mathbf{c}$, one builds a Σ^0_1 class $\mathcal{S} \subseteq 2^\omega$ containing $\mathcal{V}$ such that still $\boldsymbol{\lambda}\mathcal{S} < 1$. This implies that A is low for ML-randomness.

In (c), one builds a bounded request set (i.e., Kraft-Chaitin set) which shows that A is K-trivial.

The cost function in (b) is adaptive in the sense that $\mathbf{c}(x, s)$ depends on A_{s-1}. In contrast, the cost functions in (a) and (c) can be defined in advance, independently of the computable approximation of the set A that is built.

The main existence theorem, which we recall as Theorem 2.7 below, states that for any cost function $\mathbf{c}$ with the limit condition $\lim_x \liminf_s \mathbf{c}(x, s) = 0$, there is an incomputable c.e. set A obeying $\mathbf{c}$. The cost functions in (a)–(c) all have the limit condition. Thus, by the existence theorem, there is an incomputable c.e. set A with the required lowness property.

Besides providing a unifying picture of these constructions, cost functions have many other applications. We discuss some of them.

Weak 2-randomness is a notion stronger than ML-randomness: a set Z is weakly 2-random if Z is in no Π^0_2 null class. In 2006, Hirschfeldt and Miller gave a characterization of this notion: an ML-random is weakly 2-random if and only if it forms a minimal pair with $\emptyset'$. The implication from left to right is straightforward. The converse direction relies on a cost function related to the one for Kučera's result (a) above. (For details, see, e.g., [23, Theorem 5.3.6].) Their result can be seen as an instance of the randomness enhancement principle [24]: the ML-random sets get more random as they lose computational complexity.

The author [22] proved that the single cost function $\mathbf{c}_{\mathcal{K}}$ introduced in [8] (see Sect. 2.3 below) characterises the K-trivials. As a corollary, he showed that every K-trivial set A is truth-table below a c.e. K-trivial D. The proof of this corollary uses the general framework of change sets spelled out in Proposition 2.14 below. While this is still the only known proof yielding $A \leq_{\mathrm{tt}} D$, Bienvenu et al. [3] have recently given an alternative proof using Solovay functions in order to obtain the weaker reduction $A \leq_T D$.

In model theory, one asks whether a class of structures can be described by a first-order theory. Analogously, we ask whether an ideal of the Turing degrees below $\mathbf{0}'$ is given by obedience to all cost functions of an appropriate type. For instance, the K-trivials are axiomatized by $\mathbf{c}_{\mathcal{K}}$.

Call a cost function $\mathbf{c}$ *benign* if from n one can compute a bound on the number of disjoint intervals $[x, s)$ such that $\mathbf{c}(x, s) \geq 2^{-n}$. Figueira et al. [9] introduced the property of being strongly jump traceable (s.j.t.), which is an extreme lowness property of an oracle A, even stronger than being low for K. Roughly speaking, A is s.j.t. if the jump $J^A(x)$ is in T_x whenever it is defined, where $\langle T_x \rangle$ is a uniformly c.e. sequence of sets such that any given order function bounds the size of almost all the T_x. Greenberg and Nies [10] showed that the class of benign cost functions axiomatizes the c.e. strongly jump traceable sets.

Greenberg et al. [13] used cost functions to show that each strongly jump traceable c.e. set is Turing below each ω-c.e. ML-random set. As a main result, they also obtained the converse. In fact, they showed that any set that is below each superlow ML-random set is s.j.t.

The question remained whether a general s.j.t. set is Turing below each ω-c.e. ML-random set. Diamondstone et al. [6] showed that each s.j.t. set A is Turing below a c.e. s.j.t. set D. To do so, as a main technical result they provided a benign cost function $\mathbf{c}$ such that each set A obeying $\mathbf{c}$ is Turing below a c.e. set D which obeys every cost function that A obeys. In particular, if A is s.j.t., then $A \models \mathbf{c}$, so the c.e. cover D exists and is also s.j.t. by the above-mentioned result of Greenberg and Nies [10]. This gives an affirmative answer to the question. Note that this answer is analogous to the result [2] that every K-trivial is below an incomplete random.

1.2 Overview of Our Results

The main purpose of the paper is a systematic study of cost functions and the sets obeying them. We are guided by the above-mentioned analogy from first-order model theory: cost functions are like sentences, sets are like models, and obedience is like satisfaction. So far this analogy has been developed only for cost functions that are monotonic (that is, non-increasing in the first component, non-decreasing in the stage component). In Sect. 3 we show that the conjunction of two monotonic cost functions is given by their sum, and implication $\mathbf{c} \to \mathbf{d}$ is equivalent to $\underline{\mathbf{d}} = O(\underline{\mathbf{c}})$ where $\underline{\mathbf{c}}(x) = \sup_s \mathbf{c}(x, s)$ is the limit function.

In Sect. 4 we show that a natural class of cost functions introduced in Nies [24] characterizes the K-trivial sets: a cost function $\mathbf{c}$ is additive if $\mathbf{c}(x, y) + \mathbf{c}(y, z) = \mathbf{c}(x, z)$ for all $x < y < z$. We show that such a cost function is given by an enumeration of a left-c.e. real, and that implication corresponds to Solovay reducibility on left-c.e. reals. Additive cost functions have been used prominently in the solution of the ML-covering problem [2]. The fact that a given K-trivial A obeys every additive cost function is used to show that $A \leq_T Y$ for the Turing incomplete ML-random set constructed by Day and Miller [5].

Section 5 contains some more applications of cost functions to the study of computational lowness and K-triviality. For instance, strengthening the result in [13] mentioned above, we show that each c.e. s.j.t. set is below any complex ω-c.e. set Y, namely, a set Y such that there is an order function g with $g(n) \leq^+ K(Y \upharpoonright_n)$ for each n. In addition, the use of the reduction is bounded by the identity. Thus, the full ML-randomness assumed in [13] was too strong a hypothesis. We also discuss the relationship of cost functions and a weakening of K-triviality.

In the remaining part of the paper we obtain two existence theorems. Section 6 shows that given an arbitrary monotonic cost function $\mathbf{c}$, any nonempty Π^0_1 class contains a Δ^0_2 set Y that is so low that each c.e. set $A \leq_T Y$ obeys $\mathbf{c}$. In Sect. 7 we relativize a cost function $\mathbf{c}$ to an oracle set Z, and show that there is a c.e. set D such that $\emptyset'$ obeys $\mathbf{c}^D$ relative to D. This much harder "dual" cost function construction can be used to build incomplete c.e. sets that are very close to computing $\emptyset'$. For instance, if $\mathbf{c}$ is the cost function $\mathbf{c}_{\mathcal{K}}$ for K-triviality, then D is LR-complete.

2 Basics

We provide a formal background, basic facts and examples relating to the discussion above. We introduce classes of cost functions: monotonic and proper cost functions. We formally define the limit condition, and give a proof of the existence theorem.

2.1 Some Easy Facts on Cost Functions

Definition 2.1 We say that a cost function $\mathbf{c}$ is *nonincreasing in the main argument* if

$$\forall x, s\, [\mathbf{c}(x+1, s) \leq \mathbf{c}(x, s)].$$

We say that $\mathbf{c}$ is *nondecreasing in the stage* if $\mathbf{c}(x, s) = 0$ for $x > s$ and

$$\forall x, s\, [\mathbf{c}(x, s) \leq \mathbf{c}(x, s+1)].$$

If $\mathbf{c}$ has both properties we say that $\mathbf{c}$ is *monotonic*. This means that the cost $\mathbf{c}(x, s)$ does not decrease when we enlarge the interval $[x, s]$.

Fact 2.2 *Suppose $A \models \mathbf{c}$. Then for each $\epsilon > 0$ there is a computable approximation $\langle A_s \rangle_{s \in \mathbb{N}}$ of A such that $\mathbf{c}\langle A_s \rangle_{s \in \mathbb{N}} < \epsilon$.* □

Proof Suppose $\langle \widehat{A}_s \rangle_{s \in \mathbb{N}} \models \mathbf{c}$. Given x_0, consider the modified computable approximation $\langle A_s \rangle_{s \in \mathbb{N}}$ of A that always outputs the final value $A(x)$ for each $x \leq x_0$. That is, $A_s(x) = A(x)$ for $x \leq x_0$, and $A_s(x) = \widehat{A}_s(x)$ for $x > x_0$. Choosing x_0 sufficiently large, we can ensure $\mathbf{c}\langle A \rangle_s < \epsilon$. □

Definition 2.3 Suppose that a cost function $\mathbf{c}(x, t)$ is non-increasing in the main argument x. We say that $\mathbf{c}$ is *proper* if $\forall x\, \exists t\, \mathbf{c}(x, t) > 0$.

If a cost function that is non-increasing in the main argument is not proper, then every Δ^0_2 set obeys $\mathbf{c}$. Usually we will henceforth assume that a cost function $\mathbf{c}$ is proper. Here is an example of how being proper helps.

Fact 2.4 *Suppose that $\mathbf{c}$ is a proper cost function and $S = \mathbf{c}\langle A_s \rangle < \infty$ is a computable real. Then A is computable.*

Proof Given an input x, compute a stage t such that $\delta = \mathbf{c}(x, t) > 0$ and $S - \mathbf{c}\langle A_s \rangle_{s \leq t} < \delta$. Then $A(x) = A_t(x)$. □

A *computable enumeration* is a computable approximation $\langle B_s \rangle_{s \in \mathbb{N}}$ such that $B_s \subseteq B_{s+1}$ for each s.

Fact 2.5 *Suppose $\mathbf{c}$ is a monotonic cost function and $A \models \mathbf{c}$ for a c.e. set A. Then there is a computable* enumeration *$\langle \widetilde{A}_s \rangle$ that obeys $\mathbf{c}$.*

Proof Suppose $\langle A_s \rangle \models \mathbf{c}$ for a computable approximation $\langle A_s \rangle$ of A. Let $\langle B_t \rangle$ be a computable enumeration of A. Define $\langle \widetilde{A}_s \rangle$ as follows. Let $\widetilde{A}_0(x) = 0$; for $s > 0$ let $\widetilde{A}_s(x) = \widetilde{A}_{s-1}(x)$ if $\widetilde{A}_{s-1}(x) = 1$; otherwise let $\widetilde{A}_s(x) = A_t(x)$, where $t \geq s$ is least such that $A_t(x) = B_t(x)$.

Clearly, $\langle \widetilde{A}_s \rangle$ is a computable enumeration of A. If $\widetilde{A}_s(x) \neq \widetilde{A}_{s-1}(x)$ then $A_{s-1}(x) = 0$ and $A_s(x) = 1$. Therefore $\mathbf{c}\langle \widetilde{A}_s \rangle \leq \mathbf{c}\langle A_s \rangle < \infty$. □

2.2 *The Limit Condition and the Existence Theorem*

For a cost function $\mathbf{c}$, let

$$\underline{\mathbf{c}}(x) = \liminf_s \mathbf{c}(x, s). \tag{1}$$

Definition 2.6 We say that a cost function $\mathbf{c}$ satisfies the limit condition if $\lim_x \underline{\mathbf{c}}(x) = 0$; that is, for each e, for almost every x we have

$$\exists^\infty s\, [\mathbf{c}(x, s) \leq 2^{-e}].$$

In previous works such as [23], the limit condition was defined in terms of $\sup_s \mathbf{c}(x, s)$, rather than $\liminf_s \mathbf{c}(x, s)$. The cost functions previously considered were usually nondecreasing in the stage component, in which case $\sup_s \mathbf{c}(x, s) = \liminf_s \mathbf{c}(x, s)$, and hence the two versions of the limit condition are equivalent. Note that the limit condition is a Π^0_3 condition on cost functions that are nondecreasing in the stage, and Π^0_4 in general.

The basic existence theorem says that a cost function with the limit condition has a c.e., incomputable model. This was proved by various authors for particular cost functions. The following version of the proof appeared in [8] for the particular cost function $\mathbf{c}_{\mathcal{K}}$ defined in Sect. 2.3 below, and then in full generality in [23, Theorem 5.3.10].

Theorem 2.7 *Let* $\mathbf{c}$ *be a cost function with the limit condition.*

(i) *There is a simple set* A *such that* $A \models \mathbf{c}$. *Moreover,* A *can be obtained uniformly in (a computable index for)* $\mathbf{c}$.
(ii) *If* $\mathbf{c}$ *is nondecreasing in the stage component, then we can make* A *promptly simple.*

Proof

(i) We meet the usual simplicity requirements:

$$S_e : \#W_e = \infty \Rightarrow W_e \cap A \neq \emptyset.$$

To do so, we define a computable enumeration $\langle A_s\rangle_{s\in\mathbb{N}}$ as follows. Let $A_0 = \emptyset$. At stage $s > 0$, for each $e < s$, if S_e has not been met so far and there is an $x \geq 2e$ such that $x \in W_{e,s}$ and $\mathbf{c}(x,s) \leq 2^{-e}$, put x into A_s. Declare S_e met.

To see that $\langle A_s\rangle_{s\in\mathbb{N}}$ obeys $\mathbf{c}$, note that at most one number is put into A for the sake of each requirement. Thus $\mathbf{c}\langle A_s\rangle \leq \sum_e 2^{-e} = 2$.

If W_e is infinite, then there is an $x \geq 2e$ and an $s > x$ such that $x \in W_{e,s}$ and $\mathbf{c}(x,s) \leq 2^{-e}$, because $\mathbf{c}$ satisfies the limit condition. So we meet S_e. Clearly, the construction of A is uniform in an index for the computable function $\mathbf{c}$.

(ii) Now we meet the prompt simplicity requirements;

$$PS_e : \#W_e = \infty \Rightarrow \exists s\, \exists x\, [x \in W_{e,s} - W_{e,s-1} \wedge x \in A_s].$$

Let $A_0 = \emptyset$. At stage $s > 0$, for each $e < s$, if PS_e has not been met so far and there is an $x \geq 2e$ such that $x \in W_{e,s} - W_{e,s-1}$ and $\mathbf{c}(x,s) \leq 2^{-e}$, put x into A_s. Declare PS_e met.

If W_e is infinite, there is an $x \geq 2e$ in W_e such that $\mathbf{c}(x,s) \leq 2^{-e}$ for all $s > x$, because $\mathbf{c}$ satisfies the limit condition and is nondecreasing in the stage component. We enumerate such an x into A at the stage $s > x$ when x appears in W_e if PS_e has not been met yet by stage s. Thus A is promptly simple. □

Theorem 2.7(i) was strengthened in [23, Theorem 5.3.22]. As before, let $\mathbf{c}$ be a cost function with the limit condition. Then for each low c.e. set B, there is a c.e. set A obeying $\mathbf{c}$ such that $A \not\leq_T B$. The proof of [23, Theorem 5.3.22] is for the case of the stronger version of the limit condition $\lim_x \sup_s \mathbf{c}(x,s) = 0$, but in fact works for the version given above.

The assumption that B is c.e. is necessary: there is a low set Turing above all the K-trivial sets by [18], and the K-trivial sets can be characterized as the sets obeying the cost function $\mathbf{c}_{\mathcal{K}}$ of Sect. 2.3 below.

The following fact implies the converse of Theorem 2.7 in the monotonic case.

Fact 2.8 *Let* $\mathbf{c}$ *be a monotonic cost function. If a computable approximation* $\langle A_s\rangle_{s\in\mathbb{N}}$ *of an incomputable set* A *obeys* $\mathbf{c}$, *then* $\mathbf{c}$ *satisfies the limit condition.*

Proof Suppose the limit condition fails for e. There is an s_0 such that

$$\sum_{s\geq s_0}\sum_{x<s} \mathbf{c}_s(A_s) \leq 2^{-e}.$$

To compute A, on input n compute $s > \max(s_0, n)$ such that $\mathbf{c}(n,s) > 2^{-e}$. Then $A_s(n) = A(n)$. □

Convention 2.9 For a *monotonic* cost function $\mathbf{c}$, we may forthwith assume that $\underline{\mathbf{c}}(x) < \infty$ for each x. For, firstly, if $\forall x\, [\underline{\mathbf{c}}(x) = \infty]$, then $A \models \mathbf{c}$ implies that A is computable. Thus, we may assume there is an x_0 such that $\underline{\mathbf{c}}(x)$ is finite for all $x \geq x_0$ since $\underline{\mathbf{c}}(x)$ is nonincreasing. Secondly, changing values $\mathbf{c}(x,s)$ for the finitely many $x < x_0$ does not alter the class of sets A obeying $\mathbf{c}$. So fix some rational $q > \mathbf{c}(x_0)$ and, for $x < x_0$, redefine $\mathbf{c}(x,s) = q$ for all s.

2.3 The Cost Function for K-Triviality

Let $K_s(x) = \min\{|\sigma| \colon \mathbb{U}_s(\sigma) = x\}$ be the value of the prefix-free descriptive string complexity of x at stage s. We use the conventions $K_s(x) = \infty$ for $x \geq s$ and $2^{-\infty} = 0$. Let

$$\mathbf{c}_{\mathcal{K}}(x, s) = \sum_{w=x+1}^{s} 2^{-K_s(w)}. \tag{2}$$

Sometimes $\mathbf{c}_{\mathcal{K}}$ is called the *standard cost function*, mainly because it was the first example of a cost function that received attention. Clearly, $\mathbf{c}_{\mathcal{K}}$ is monotonic. Note that $\underline{\mathbf{c}}_{\mathcal{K}}(x) = \sum_{w>x} 2^{-K(w)}$. Hence $\mathbf{c}_{\mathcal{K}}$ satisfies the limit condition: given $e \in \mathbb{N}$, since $\sum_w 2^{-K(w)} \leq 1$, there is an x_0 such that

$$\sum_{w \geq x_0} 2^{-K(w)} \leq 2^{-e}.$$

Therefore $\underline{\mathbf{c}}_{\mathcal{K}}(x) \leq 2^{-e}$ for all $x \geq x_0$.

The following example illustrates that in Definition 1.2, obeying $\mathbf{c}_{\mathcal{K}}$, say, strongly depends on the chosen enumeration. Clearly, if we enumerate $A = \mathbb{N}$ by putting in x at stage x, then the total cost of changes is zero.

Proposition 2.10 *There is a computable enumeration $\langle A_s \rangle_{s \in \mathbb{N}}$ of $\mathbb{N}$ in the order $0, 1, 2, \ldots$ (i.e., each A_s is an initial segment of $\mathbb{N}$) such that $\langle A_s \rangle_{s \in \mathbb{N}}$ does not obey $\mathbf{c}_{\mathcal{K}}$.*

Proof Since $K(2^j) \leq^+ 2 \log j$, there is an increasing computable function f and a number j_0 such that $\forall j \geq j_0 \; K_{f(j)}(2^j) \leq j - 1$. Enumerate the set $A = \mathbb{N}$ in order, but so slowly that for each $j \geq j_0$ the elements of $(2^{j-1}, 2^j]$ are enumerated only after stage $f(j)$, one by one. Each such enumeration costs at least $2^{-(j-1)}$, so the cost for each interval $(2^{j-1}, 2^j]$ is 1. □

Intuitively speaking, an infinite c.e. set A can obey the cost function $\mathbf{c}_{\mathcal{K}}$ only because during an enumeration of x at stage s one merely pays the current cost $\mathbf{c}_{\mathcal{K}}(x, s)$, not the limit cost $\underline{\mathbf{c}}_{\mathcal{K}}(x)$.

Fact 2.11 *If a c.e. set A is infinite, then $\sum_{x \in A} \underline{\mathbf{c}}_{\mathcal{K}}(x) = \infty$.*

Proof Let f be a 1–1 computable function with range A. Let L be the bounded request set $\{\langle r, \max_{i \leq 2^{r+1}} f(i) \rangle \colon r \in \mathbb{N}\}$. Let M be a machine for L according to the Machine Existence Theorem, also known as the Kraft-Chaitin Theorem. See, e.g., [23, Chap. 2] for background. □

In [22] (also see [23, Chap. 5]) it is shown that A is K-trivial iff $A \models \mathbf{c}_{\mathcal{K}}$. So far, the class of K-trivial sets has been the only known natural class that is characterized by a single cost function. However, recent work with Greenberg and Miller [12] show that for a c.e. set A, being below both halves Z_0, Z_1 of some Martin-Löf-random $Z = Z_0 \oplus Z_1$ is equivalent to obeying the cost function $\mathbf{c}(x, s) = \sqrt{\Omega_s - \Omega_x}$.

2.4 Basic Properties of the Class of Sets Obeying a Cost Function

In this subsection, unless otherwise stated, cost functions will be monotonic. Recall from Definition 2.3 that a cost function $\mathbf{c}$ is called *proper* if $\forall x \exists t\, \mathbf{c}(x,t) > 0$. We investigate the class of models of a proper cost function $\mathbf{c}$. We also assume Convention 2.9 that $\underline{\mathbf{c}}(x) < \infty$ for each x.

The first two results together show that $A \models \mathbf{c}$ implies that A is weak truth-table below a c.e. set C such that $C \models \mathbf{c}$. Recall that a Δ^0_2 set A is called ω-c.e. if there is a computable approximation $\langle A_s \rangle$ such that the number of changes $\#\{s\colon A_s(x) \neq A_{s-1}(x)\}$ is computably bounded in x; equivalently, $A \leq_{\mathrm{wtt}} \emptyset'$ (see [23, 1.4.3]).

Fact 2.12 *Suppose that* $\mathbf{c}$ *is a proper monotonic cost function. Let* $A \models \mathbf{c}$. *Then* A *is* ω-*c.e.*

Proof Suppose $\langle A_s \rangle \models \mathbf{c}$. Let g be the computable function given by $g(x) = \mu t.\, \mathbf{c}(x,t) > 0$. Let $\hat{A}_s(x) = A_{g(x)}(x)$ for $s < g(x)$, and $\hat{A}_s(x) = A_s(x)$ otherwise. Then the number of times $\hat{A}_s(x)$ can change is bounded by $\mathbf{c}\langle A_s \rangle / \mathbf{c}(x, g(x))$. □

Let V_e denote the eth ω-c.e. set (see [23, p. 20]).

Fact 2.13 *For each cost function* $\mathbf{c}$, *the index set* $\{e\colon\ V_e \models \mathbf{c}\}$ *is* Σ^0_3.

Proof Let D_n denote the nth finite set of numbers. We may view the ith partial computable function Φ_i as a (possibly partial) computable approximation $\langle A_t \rangle$ by letting $A_t \simeq D_{\Phi_i(t)}$ (the symbol $\simeq$ indicates that 'undefined' is a possible value). Saying that Φ_i is total and a computable approximation of V_e is a Π^0_2 condition of i and e. Given that Φ_i is total, the condition that $\langle A_t \rangle \models \mathbf{c}$ is Σ^0_2. □

The *change set* (see [23, 1.4.2]) for a computable approximation $\langle A_s \rangle_{s \in \mathbb{N}}$ of a Δ^0_2 set A is a c.e. set $C \geq_T A$ defined as follows: if $s > 0$ and $A_{s-1}(x) \neq A_s(x)$ we put $\langle x, i \rangle$ into C_s, where i is least such that $\langle x, i \rangle \notin C_{s-1}$. If A is ω-c.e. via this approximation, then $C \geq_{tt} A$. The change set can be used to prove the implication of the Shoenfield Limit Lemma that $A \in \Delta^0_2$ implies $A \leq_T \emptyset'$; moreover, if A is ω-c.e., then $A \leq_{\mathrm{wtt}} \emptyset'$.

Proposition 2.14 ([23], Sect. 5.3) *Let the cost function* $\mathbf{c}$ *be non-increasing in the first component. If a computable approximation* $\langle A_s \rangle_{s \in \mathbb{N}}$ *of a set* A *obeys* $\mathbf{c}$, *then its change set* C *obeys* $\mathbf{c}$ *as well.*

Proof Since $x < \langle x, i \rangle$ for each x, i, we have

$$C_{s-1}(x) \neq C_s(x) \rightarrow A_{s-1} \upharpoonright_x \neq A_s \upharpoonright_x$$

for each x, s. Then, since $\mathbf{c}(x, s)$ is non-increasing in x, we have $\mathbf{c}\langle C_s \rangle \leq \mathbf{c}\langle A_s \rangle < \infty$. □

This yields a limitation on the expressiveness of cost functions. Recall that A is superlow if $A' \leq_{tt} \emptyset'$.

Corollary 2.15 *There is no cost function* $\mathbf{c}$ *monotonic in the first component such that* $A \models \mathbf{c}$ *iff* A *is superlow.*

Proof Otherwise, for each superlow set A there is a c.e. superlow set $C \geq_T A$. This is clearly not the case: for instance, A could be ML-random, and hence of diagonally non-computable degree, so that any c.e. set $C \geq_T A$ is Turing complete. □

For $X \subseteq \mathbb{N}$ let $2X$ denote $\{2x \colon x \in X\}$. Recall that $A \oplus B = 2A \cup (2B + 1)$. We now show that the class of sets obeying $\mathbf{c}$ is closed under $\oplus$ and closed downward under a restricted form of weak truth-table reducibility.

Clearly, $E \models \mathbf{c} \wedge F \models \mathbf{c}$ implies $E \cup F \models \mathbf{c}$.

Proposition 2.16 *Let the cost function* $\mathbf{c}$ *be monotonic in the first component. Then* $A \models \mathbf{c} \wedge B \models \mathbf{c}$ *implies* $A \oplus B \models \mathbf{c}$.

Proof Let $\langle A_s \rangle$ be a computable approximation of A. By the monotonicity of $\mathbf{c}$ we have $\mathbf{c}\langle A_s \rangle \geq \mathbf{c}(2A_s)$. Hence $2A \models \mathbf{c}$. Similarly, $2B + 1 \models \mathbf{c}$. Thus $A \oplus B \models \mathbf{c}$. □

Recall that there are superlow c.e. sets A_0, A_1 such that $A_0 \oplus A_1$ is Turing complete (see [23, 6.1.4]). Thus the foregoing result yields a stronger form of Corollary 2.15: no cost function characterizes superlowness within the c.e. sets.

3 Look-Ahead Arguments

This core section of the paper introduces an important type of argument. Suppose we want to construct a computable approximation of a set A that obeys a given monotonic cost function. If we can anticipate that $A(x)$ needs to be changed in the future, we try to change it as early as possible, because earlier changes are cheaper. Such an argument will be called a *look-ahead argument*. (Also see the remark before Fact 2.11.) The main application of this method is to characterize logical properties of cost functions algebraically.

3.1 Downward Closure Under $\leq_{ibT}$

Recall that $A \leq_{ibT} B$ if $A \leq_{wtt} B$, with the use function bounded by the identity. We now show that the class of models of $\mathbf{c}$ is downward closed under $\leq_{ibT}$.

Proposition 3.1 *Let* $\mathbf{c}$ *be a monotonic cost function. Suppose that* $B \models \mathbf{c}$ *and* $A = \Gamma^B$ *via a Turing reduction* Γ *such that each oracle query on an input* x *is at most* x. *Then* $A \models \mathbf{c}$.

Proof Suppose $B \models \mathbf{c}$ via a computable approximation $\langle B_s \rangle_{s \in \mathbb{N}}$. We define a computable increasing sequence of stages $\langle s(i) \rangle_{i \in \mathbb{N}}$ by $s(0) = 0$ and

$$s(i+1) = \mu s > s(i) \, [\Gamma^B \upharpoonright_{s(i)} [s] \downarrow].$$

In other words, $s(i+1)$ is the least stage s greater than $s(i)$ such that at stage s, $\Gamma^B(n)$ is defined for each $n < s(i)$. We will define $A_{s(k)}(x)$ for each $k \in \mathbb{N}$. Thereafter we let $A_s(x) = A_{s(k)}(x)$, where k is maximal such that $s(k) \leq s$.

Suppose $s(i) \leq x < s(i+1)$. For $k < i$, let $A_{s(k)}(x) = v$, where $v = \Gamma^B(x)[s(i+2)]$. For $k \geq i$, let $A_{s(k)}(x) = \Gamma^B(x)[s(k+2)]$. (Note that these values are defined. Taking the $\Gamma^B(x)$ value at the large stage $s(k+2)$ represents the look-ahead.)

Clearly, $\lim_s A_s(x) = A(x)$. We show that $\mathbf{c}\langle A_s \rangle \leq \mathbf{c}\langle B_t \rangle$. Suppose that x is least such that $A_{s(k)}(x) \neq A_{s(k)-1}(x)$. By the use bound on the reduction procedure Γ, there is a $y \leq x$ such that $B_t(y) \neq B_{t-1}(y)$ for some t, $s(k+1) < t \leq s(k+2)$. Then $\mathbf{c}(x, s(k)) \leq \mathbf{c}(y, t)$ by monotonicity of $\mathbf{c}$. Therefore $\langle A_s \rangle \models \mathbf{c}$. □

3.2 *Conjunction of Cost Functions*

In the remainder of this section we characterize conjunction and implication of monotonic cost functions algebraically. Firstly, we show that a set A is a model of $\mathbf{c}$ and $\mathbf{d}$ if and only if A is a model of $\mathbf{c} + \mathbf{d}$. Then we show that $\mathbf{c}$ implies $\mathbf{d}$ if and only if $\underline{\mathbf{d}} = O(\underline{\mathbf{c}})$.

Theorem 3.2 *Let* $\mathbf{c}, \mathbf{d}$ *be monotonic cost functions. Then*

$$A \models \mathbf{c} \, \wedge \, A \models \mathbf{d} \, \Leftrightarrow \, A \models \mathbf{c} + \mathbf{d}.$$

Proof $\Leftarrow$: This implication is trivial.

$\Rightarrow$: We carry out a look-ahead argument of the type introduced in the proof of Proposition 3.1. Suppose that $\langle E_s \rangle_{s \in \mathbb{N}}$ and $\langle F_s \rangle_{s \in \mathbb{N}}$ are computable approximations of a set A such that $\langle E_s \rangle \models \mathbf{c}$ and $\langle F_s \rangle \models \mathbf{d}$. We may assume that $E_s(x) = F_s(x) = 0$ for $s < x$ because changing $E(x)$, say, to 1 at stage x will not increase the cost as $\mathbf{c}(x, s) = 0$ for $x > s$. We define a computable increasing sequence of stages $\langle s(i) \rangle_{i \in \mathbb{N}}$ by letting $s(0) = 0$ and

$$s(i+1) = \mu s > s(i) \, [E_s \upharpoonright_{s(i)} = F_s \upharpoonright_{s(i)}].$$

We define $A_{s(k)}(x)$ for each $k \in \mathbb{N}$. Thereafter we let $A_s(x) = A_{s(k)}(x)$, where k is maximal such that $s(k) \leq s$.

Suppose $s(i) \leq x < s(i+1)$. Let $A_{s(k)}(x) = 0$ for $k < i$. To define $A_{s(k)}(x)$ for $k \geq i$, let $j(x)$ be the least $j \geq i$ such that $v = E_{s(j+1)}(x) = F_{s(j+1)}(x)$.

$$A_{s(k)}(x) = \begin{cases} v & \text{if } i \leq k \leq j(x) \\ E_{s(k+1)}(x) = F_{s(k+1)}(x) & \text{if } k > j(x). \end{cases}$$

Clearly, $\lim_s A_s(x) = A(x)$. To show $(\mathbf{c}+\mathbf{d})\langle A_s \rangle < \infty$, suppose that $A_{s(k)}(x) \neq A_{s(k)-1}(x)$. The only possible cost if $i \leq k \leq j(x)$ is at stage $s(i)$ when $v = 1$. Such a cost is bounded by 2^{-x}. Now consider a cost if $k > j(x)$. There is a least y such that $E_t(y) \neq E_{t-1}(y)$ for some t, $s(k) < t \leq s(k+1)$. Then $y \leq x$, whence $\mathbf{c}(x, s(k)) \leq \mathbf{c}(y, t)$ by the monotonicity of $\mathbf{c}$. Similarly, using $\langle F_s \rangle$ one can bound the cost of changes due to $\mathbf{d}$. Therefore $(\mathbf{c}+\mathbf{d})\langle A_s \rangle \leq 4 + \mathbf{c}\langle E_s \rangle + \mathbf{d}\langle F_s \rangle < \infty$. □

3.3 Implication Between Cost Functions

Definition 3.3 For cost functions $\mathbf{c}$ and $\mathbf{d}$, we write $\mathbf{c} \to \mathbf{d}$ if $A \models \mathbf{c}$ implies $A \models \mathbf{d}$ for each (Δ^0_2) set A.

If a cost function c is monotonic in the stage component, then $\underline{\mathbf{c}}(x) = \sup_s \mathbf{c}(x, s)$. By Remark 2.9 we may assume $\underline{\mathbf{c}}(x)$ is finite for each x. We will show $\mathbf{c} \to \mathbf{d}$ is equivalent to $\underline{\mathbf{d}}(x) = O(\underline{\mathbf{c}}(x))$. In particular, whether or not $A \models \mathbf{c}$ only depends on the limit function $\underline{\mathbf{c}}$.

Theorem 3.4 *Let* $\mathbf{c}, \mathbf{d}$ *be cost functions that are monotonic in the stage component. Suppose* $\mathbf{c}$ *satisfies the limit condition in Definition 2.6. Then*

$$\mathbf{c} \to \mathbf{d} \;\Leftrightarrow\; \exists N\, \forall x \left[N\mathbf{c}(x) > \mathbf{d}(x)\right].$$

Proof $\Leftarrow$: We carry out yet another look-ahead argument. We define a computable increasing sequence of stages $s(0) < s(1) < \ldots$ by $s(0) = 0$ and

$$s(i+1) = \mu s > s(i). \forall x < s(i) \left[N\mathbf{c}(x, s) > \mathbf{d}(x, s)\right].$$

Suppose A is a Δ^0_2 set with a computable approximation $\langle A_s \rangle \models \mathbf{c}$. We show that $\langle \widetilde{A}_t \rangle \models \mathbf{d}$ for some computable approximation $\langle \widetilde{A}_t \rangle$ of A. As usual, we define $\widetilde{A}_{s(k)}(x)$ for each $k \in \mathbb{N}$. We then let $\widetilde{A}_s(x) = \widetilde{A}_{s(k)}(x)$, where k is maximal such that $s(k) \leq s$.

Suppose $s(i) \leq x < s(i+1)$. If $k < i+1$, let $\widetilde{A}_{s(k)}(x) = A_{s(i+2)}(x)$. If $k \geq i+1$, let $\widetilde{A}_{s(k)}(x) = A_{s(k+1)}(x)$.

Given k, suppose that x is least such that $\widetilde{A}_{s(k)}(x) \neq \widetilde{A}_{s(k)-1}(x)$. Let i be the number such that $s(i) \leq x < s(i+1)$. Then $k \geq i+1$. We have $A_t(x) \neq A_{t-1}(x)$ for some t such that $s(k) < t \leq s(k+1)$. Since $x < s(i+1) \leq s(k)$, by the monotonicity hypothesis this implies $N\mathbf{c}(x, t) \geq N\mathbf{c}(x, s(k)) > \mathbf{d}(x, s(k))$. So $\mathbf{d}\langle \widetilde{A}_s \rangle \leq N \cdot \mathbf{c}\langle A_s \rangle < \infty$. Hence $A \models \mathbf{d}$.

$\Rightarrow$: Recall from the proof of Fact 2.13 that we view the eth partial computable function Φ_e as a (possibly partial) computable approximation $\langle B_t \rangle$, where $B_t \simeq D_{\Phi_e(t)}$.

Suppose that $\exists N\, \forall x \left[N\underline{\mathbf{c}}(x) > \underline{\mathbf{d}}(x)\right]$ fails. We build a set $A \models \mathbf{c}$ such that for no computable approximation Φ_e of A do we have $\mathbf{d}\, \Phi_e \leq 1$. This suffices for the theorem by Fact 2.2. We meet the requirements

$$R_e \colon\; \Phi_e \text{ istotalandapproximates} A \;\Rightarrow\; \Phi_e \not\models \mathbf{d}.$$

The idea is to change $A(x)$ for some fixed x at sufficiently many stages s with $N\mathbf{c}(x,s) < \mathbf{d}(x,s)$, where N is an appropriate large constant. After each change we wait for recovery from the side of Φ_e. In this way our $\mathbf{c}$-cost of changes to A remains bounded, while the opponent's $\mathbf{d}$-cost of changes to Φ_e exceeds 1.

For a stage s, we let $\mathsf{init}_s(e) \leq s$ be the largest stage such that R_e has been initialized at that stage (or 0 if there is no such stage). Waiting for recovery is implemented as follows. We say that s is *e-expansionary* if $s = \mathsf{init}_s(e)$, or $s > \mathsf{init}_s(e)$ and, where u is the greatest e-expansionary stage less than s,

$$\exists t \in [u,s)\,[\Phi_{e,s}(t)\downarrow \wedge \Phi_{e,s}(t)\upharpoonright_u = A_u\upharpoonright_u].$$

The strategy for R_e can only change $A(x)$ at an e-expansionary stage u such that $x < u$. In this case it preserves $A_u\upharpoonright_u$ until the next e-expansionary stage. Then, Φ_e also has to change its mind on x: we have

$$x \in \Phi_e(u-1) \leftrightarrow x \notin \Phi_e(t) \text{ for some } t \in [u,s).$$

We measure the progress of R_e at stages s via a quantity $\alpha_s(e)$. When R_e is initialized at stage s, we set $\alpha_s(e)$ to 0. If R_e changes $A(x)$ at stage s, we increase $\alpha_s(e)$ by $\mathbf{c}(x,s)$. R_e is declared satisfied when $\alpha_s(e)$ exceeds 2^{-b-e}, where b is the number of times R_e has been initialized.

Construction of $\langle A_s\rangle$ *and* $\langle \alpha_s\rangle$ Let $A_0 = \emptyset$. Let $\alpha_0(e) = 0$ for each e.

Stage $s > 0$. Let e be least such that s is e-expansionary and $\alpha_{s-1}(e) \leq 2^{-b-e}$, where b is the number of times R_e has been initialized so far. If e exists, do the following.

Let x be least such that $\mathsf{init}_s(e) \leq x < s$, $\mathbf{c}(x,s) < 2^{-b-e}$ and

$$2^{b+e}\mathbf{c}(x,s) < \mathbf{d}(x,s).$$

If x exists, let $A_s(x) = 1 - A_{s-1}(x)$. Also, let $A_s(y) = 0$ for $x < y < s$. Let $\alpha_s(e) = \alpha_{s-1}(e) + \mathbf{c}(x,s)$. Initialize the requirements R_i for $i > e$ and let $\alpha_s(i) = 0$. (This preserves $A_s\upharpoonright_s$ unless R_e itself is later initialized.) We say that R_e *acts*.

Verification If s is a stage such that R_e has been initialized b times, then $\alpha_s(e) \leq 2^{-b-e+1}$. Hence the total cost of changes to A due to R_e is at most $\sum_b 2^{-b-e+1} = 2^{-e+2}$. Therefore $\langle A_s\rangle \models \mathbf{c}$.

We show that *each* R_e *only acts finitely often and is met.* Inductively, $\mathsf{init}_s(e)$ assumes a final value s_0. Let b be the number of times R_e has been initialized by stage s_0.

Since the condition $\exists N \forall x\,[N\underline{\mathbf{c}}(x) > \underline{\mathbf{d}}(x)]$ fails, there is $x \geq s_0$ such that for some $s_1 \geq x$ we have $\forall s \geq s_1\,[2^{b+e}\mathbf{c}(x,s) < \mathbf{d}(x,s)]$. Furthermore, since $\mathbf{c}$ satisfies the limit condition, we may suppose that $\underline{\mathbf{c}}(x) < 2^{-b-e}$. Choose x least.

If Φ_e is a computable approximation of A, there are infinitely many e-expansionary stages $s \geq s_1$. For each such s, we can choose this x at stage s in the construction. So we can add at least $\mathbf{c}(x,s_1)$ to $\alpha(e)$. Therefore $\alpha_t(e)$ exceeds the bound 2^{-b-e} for some stage $t \geq s_1$, whence R_e stops acting at t. Furthermore, since $\mathbf{d}$ is monotonic in the second component and by the initialization due to R_e, between stages s_0 and t we have caused $\mathbf{d}\,\Phi_e$ to increase by at least $2^{b+e}\alpha_t(e) > 1$. Hence R_e is met. □

The foregoing proof uses in an essential way the ability to change $A(x)$, for the same x, a multiple number of times. If we restrict implication to c.e. sets, the implication from left to right in Theorem 3.4 fails. For a trivial example, let $\mathbf{c}(x,s) = 4^{-x}$ and $\mathbf{d}(x,s) = 2^{-x}$. Then each c.e. set obeys $\mathbf{d}$, so $\mathbf{c} \to \mathbf{d}$ for c.e. sets. However, we do not have $\mathbf{d}(x) = O(\mathbf{c}(x))$.

We mention that Melnikov and Nies (unpublished, 2010) have obtained a sufficient algebraic condition for the non-implication of cost functions via a c.e. set. Informally speaking, the condition $\mathbf{d}(x) = O(\mathbf{c}(x))$ fails "badly".

Proposition 3.5 *Let* $\mathbf{c}$ *and* $\mathbf{d}$ *be monotonic cost functions satisfying the limit condition such that* $\sum_{x\in\mathbb{N}} \underline{\mathbf{d}}(x) = \infty$ *and, for each* $N > 0$,

$$\sum \underline{\mathbf{d}}(x)[\![N\underline{\mathbf{c}}(x) > \underline{\mathbf{d}}(x)]\!] < \infty.$$

Then there exists a c.e. set A *that obeys* $\mathbf{c}$*, but not* $\mathbf{d}$.

The hope is that some variant of this will yield an algebraic criterion for cost function implication restricted to the c.e. sets.

4 Additive Cost Functions

We discuss a class of very simple cost functions introduced in [24]. We show that a Δ^0_2 set obeys all of them if and only if it is K-trivial. There is a universal cost function of this kind, namely $\mathbf{c}(x,s) = \Omega_s - \Omega_x$. Recall Convention 2.9 that $\underline{\mathbf{c}}(x) < \infty$ for each cost function $\mathbf{c}$.

Definition 4.1 ([24]) We say that a cost function $\mathbf{c}$ is *additive* if $\mathbf{c}(x,s) = 0$ for $x > s$, and for each $x < y < z$ we have

$$\mathbf{c}(x,y) + \mathbf{c}(y,z) = \mathbf{c}(x,z).$$

Additive cost functions correspond to nondecreasing effective sequences $\langle\beta_s\rangle_{s\in\mathbb{N}}$ of non-negative rationals, that is, to effective approximations of left-c.e. reals β. Given such an approximation, $\langle\beta\rangle = \langle\beta_s\rangle_{s\in\mathbb{N}}$, let, for $x \leq s$,

$$\mathbf{c}_{\langle\beta\rangle}(x,s) = \beta_s - \beta_x.$$

Conversely, given an additive cost function $\mathbf{c}$, let $\beta_s = \mathbf{c}(0,s)$. Clearly, the two effective transformations are inverses of each other.

4.1 K-Triviality and the Cost Function $\mathbf{c}_{\langle\Omega\rangle}$

The standard cost function $\mathbf{c}_{\mathcal{K}}$ introduced in (2) is *not* additive. We certainly have $\mathbf{c}_{\mathcal{K}}(x, y) + \mathbf{c}_{\mathcal{K}}(y, z) \leq \mathbf{c}_{\mathcal{K}}(x, z)$, but by stage z there could be a shorter description of, say, $x + 1$ than at stage y, so that the inequality may be proper. On the other hand, let g be a computable function such that $\sum_w 2^{-g(w)} < \infty$; this implies that $K(x) \leq^+ g(x)$. The "analog" of $\mathbf{c}_{\mathcal{K}}$ when we write $g(x)$ instead of $K_s(x)$, namely $\mathbf{c}_g(x, s) = \sum_{w=x+1}^{s} 2^{-g(w)}$, is an additive cost function.

Also, $\mathbf{c}_{\mathcal{K}}$ is dominated by an additive cost function $\mathbf{c}_{\langle\Omega\rangle}$ we introduce next. Let $\mathbb{U}$ be the standard universal prefix-free machine (see, e.g., [23, Chap. 2]). Let $\langle\Omega\rangle$ denote the computable approximation of Ω given by $\Omega_s = \boldsymbol{\lambda}\,\mathrm{dom}(\mathbb{U}_s)$. (That is, Ω_s is the Lebesgue measure of the domain of the universal prefix-free machine at stage s.)

Fact 4.2 *For each $x \leq s$, we have $\mathbf{c}_{\mathcal{K}}(x, s) \leq \mathbf{c}_{\langle\Omega\rangle}(x, s) = \Omega_s - \Omega_x$.*

Proof Fix x. We prove the statement by induction on $s \geq x$. For $s = x$, we have $\mathbf{c}_{\mathcal{K}}(x, s) = 0$. Now,

$$\mathbf{c}_{\mathcal{K}}(x, s + 1) - \mathbf{c}_{\mathcal{K}}(x, s) = \sum_{w=x+1}^{s+1} 2^{-K_{s+1}(w)} - \sum_{w=x+1}^{s} 2^{-K_s(w)} \leq \Omega_{s+1} - \Omega_s,$$

because the difference is due to convergence at stage s of new $\mathbb{U}$-computations. □

Theorem 4.3 *Let A be Δ^0_2. Then the following are equivalent.*

(i) A is K-trivial.
(ii) A obeys each additive cost function.
(iii) A obeys $\mathbf{c}_{\langle\Omega\rangle}$, where $\Omega_s = \boldsymbol{\lambda}\,dom(\mathbb{U}_s)$.

Proof (ii) → (iii) is immediate, and (iii) → (i) follows from Fact 4.2. It remains to show that (i)→(ii).

Fix some computable approximation $\langle A_s\rangle_{s\in\mathbb{N}}$ of A. Let $\mathbf{c}$ be an additive cost function. We may suppose that $\underline{\mathbf{c}}(0) \leq 1$.

For $w > 0$, let $r_w \in \mathbb{N} \cup \infty$ be least such that $2^{-r_w} \leq c(w - 1, w)$ (where $2^{-\infty} = 0$). Then $\sum_w 2^{-r_w} \leq 1$. Hence by the Machine Existence Theorem we have $K(w) \leq^+ r_w$ for each w. This implies $2^{-r_w} = O(2^{-K(w)})$, so $\sum_{w>x} 2^{-r_w} = O(\underline{\mathbf{c}}_{\mathcal{K}}(x))$ and hence $\underline{\mathbf{c}}(x) = \sum_{w>x} c(w - 1, w) = O(\underline{\mathbf{c}}_{\mathcal{K}}(x))$. Thus $\mathbf{c}_{\mathcal{K}} \to \mathbf{c}$ by Theorem 3.4, whence the K-trivial set A obeys $\mathbf{c}$. (See [1] for a proof not relying on Theorem 3.4.) □

Because of Theorem 3.4, we have $\mathbf{c}_{\langle\Omega\rangle} \leftrightarrow \mathbf{c}_{\mathcal{K}}$. That is,

$$\Omega - \Omega_x \sim \sum_{w=x+1}^{\infty} 2^{-K(w)}.$$

This can easily be seen directly: for instance, $\mathbf{c}_{\mathcal{K}} \leq \mathbf{c}_{\langle\Omega\rangle}$ by Fact 4.2.

4.2 Solovay Reducibility

Let $\mathbb{Q}_2$ denote the dyadic rationals, and let the variable q range over $\mathbb{Q}_2$. Recall Solovay reducibility on left-c.e. reals: $\beta \le_S \alpha$ iff there is a partial computable $\phi\colon \mathbb{Q}_2 \cap [0,\alpha) \to \mathbb{Q}_2 \cap [0,\beta)$ and $N \in \mathbb{N}$ such that

$$\forall q < \alpha\big[\beta - \phi(q) < N(\alpha - q)\big].$$

Informally, it is easier to approximate β from the left, than α. See, e.g., [23, 3.2.8] for background.

We will show that reverse implication of additive cost functions corresponds to Solovay reducibility on the corresponding left-c.e. reals. Given a left-c.e. real γ, we let the variable $\langle\gamma\rangle$ range over the nondecreasing effective sequences of rationals converging to γ.

Proposition 4.4 *Let α, β be left-c.e. reals. The following are equivalent.*

(i) $\beta \le_S \alpha$
(ii) $\forall\langle\alpha\rangle\exists\langle\beta\rangle\,[c_{\langle\alpha\rangle} \to c_{\langle\beta\rangle}]$
(iii) $\exists\langle\alpha\rangle\exists\langle\beta\rangle\,[c_{\langle\alpha\rangle} \to c_{\langle\beta\rangle}]$.

Proof

(i) $\to$ (ii) Given an effective sequence $\langle\alpha\rangle$, by the definition of $\le_S$ there is an effective sequence $\langle\beta\rangle$ such that $\beta - \beta_x = O(\alpha - \alpha_x)$ for each x. Thus $\underline{\mathbf{c}}_{\langle\beta\rangle} = O(\underline{\mathbf{c}}_{\langle\rangle\alpha})$. Hence $\mathbf{c}_{\langle\alpha\rangle} \to \mathbf{c}_{\langle\beta\rangle}$ by Theorem 3.4.

(iii) $\to$ (i) Suppose we are given $\langle\alpha\rangle$ and $\langle\beta\rangle$ such that $\underline{\mathbf{c}}_{\langle\beta\rangle} = O(\underline{\mathbf{c}}_{\langle\alpha\rangle})$. Define a partial computable function ϕ by $\phi(q) = \beta_x$ if $\alpha_{x-1} \le q < \alpha_x$. Then $\beta \le_S \alpha$ via ϕ.

□

4.3 The Strength of an Additive Cost Function

Firstly, we make some remarks related to Proposition 4.4. For instance, it implies that an additive cost function can be weaker than $\mathbf{c}_{\langle\Omega\rangle}$ without being obeyed by all the Δ^0_2 sets.

Proposition 4.5 *There are additive cost functions* $\mathbf{c}, \mathbf{d}$ *such that* $\mathbf{c}_{\langle\Omega\rangle} \to \mathbf{c}$, $\mathbf{c}_{\langle\Omega\rangle} \to \mathbf{d}$ *and* $\mathbf{c}, \mathbf{d}$ *are incomparable under the implication of cost functions.*

Proof Let $\mathbf{c}, \mathbf{d}$ be cost functions corresponding to enumerations of Turing (and hence Solovay) incomparable left-c.e. reals. Now apply Proposition 4.4. □

Clearly, if β is a computable real then any c.e. set obeys $\mathbf{c}_{\langle\beta\rangle}$. The intuition we garner from Proposition 4.4 is that a more complex left-c.e. real β means that the sets $A \models \mathbf{c}_{\langle\beta\rangle}$ become less complex, and conversely. We give a little more evidence for this principle: if β is non-computable, we show that a set $A \models \mathbf{c}_{\langle\beta\rangle}$ cannot be weak truth-table complete. However, we also build a non-computable β and a c.e. Turing complete set that obeys $\mathbf{c}_{\langle\beta\rangle}$

Proposition 4.6 *Suppose β is a non-computable left-c.e. real and $A \models \mathbf{c}_{\langle\beta\rangle}$. Then A is not weak truth-table complete.*

Proof Assume for a contradiction that A is weak truth-table complete. We can fix a computable approximation $\langle A_s\rangle$ of A such that $\mathbf{c}_{\langle\beta\rangle}\langle A_s\rangle \le 1$. We build a c.e. set B. By the recursion theorem we can suppose we have a weak truth-table reduction Γ with computable use bound g such that $B = \Gamma^A$. We build B so that $\beta - \beta_{g(2^e+1)} \le 2^{-e}$, which implies that β is computable.

Let $I_e = [2^e, 2^{e+1})$. If ever a stage s appears such that $\beta_s - \beta_{g(2^e+1)} \le 2^{-e}$, then we start enumerating into $B \cap I_e$ sufficiently slowly so that $A \upharpoonright_{g(2^e+1)}$ must change 2^e times. To do so, each time we enumerate into B, we wait for a recovery of $B = \Gamma^A$ up to $2^{(e+1)}$. The A-changes we enforce yield a total cost > 1 for a contradiction.

□

Proposition 4.7 *There is a non-computable left-c.e. real β and a c.e. set $A \models \mathbf{c}_{\langle\beta\rangle}$ such that A is Turing complete.*

Proof We build a Turing reduction Γ such that $\emptyset' = \Gamma(A)$. Let $\gamma_{k,s} + 1$ be the use of the computation $\Gamma^{\emptyset'}(k)[s]$. We view γ_k as a movable marker as usual. The initial value is $\gamma_{k,0} = k$. Throughout the construction we maintain the invariant

$$\beta_s - \beta_{\gamma_{k,s}} \le 2^{-k}.$$

Let $\langle\phi_e\rangle$ be the usual effective list of partial computable functions. By convention, at each stage at most one computation $\phi_e(k)$ converges newly. To make β non-computable, it suffices to meet the requirements

$$R_k\colon\ \phi_k(k)\downarrow\ \Rightarrow\ \beta - \beta_{\phi_k(k)} \ge 2^{-k}.$$

Strategy for R_k. If $\phi_k(k)$ converges newly at stage s, do the following.

1. Enumerate $\gamma_{k,s}$ into A. (This incurs a cost of at most 2^{-k}.)
2. Let $\beta_s = \beta_{s-1} + 2^{-k}$.
3. Redefine γ_i $(i \ge k)$ to large values in an increasing fashion.

In the construction, we run the strategies for the R_k. If k enters $\emptyset'$ at stage s, we enumerate $\gamma_{k,s}$ into A.

Clearly each R_k acts at most once, and is met. Therefore β is non-computable. The markers γ_k reach a limit. Therefore $\emptyset' = \Gamma(A)$. Finally, we maintain the stage invariant, which implies that the total cost of enumerating A is at most 4. □

As pointed out by Turetsky, it can be verified that β is in fact Turing complete.

Next, we note that if we have two computable approximations from the left of the same real, we obtain additive cost functions with very similar classes of models.

Proposition 4.8 *Let $\langle\alpha\rangle, \langle\beta\rangle$ be left-c.e. approximations of the same real. Suppose that $A \models \mathbf{c}_{\langle\alpha\rangle}$. Then there is $B \equiv_m A$ such that $B \models \mathbf{c}_{\langle\beta\rangle}$. If A is c.e., then B can be chosen c.e. as well.*

Proof Firstly, suppose that A is c.e. By Fact 2.5, choose a computable enumeration $\langle A_s\rangle \models \mathbf{c}_{\langle\alpha\rangle}$.

By the hypothesis on the sequences $\langle\alpha\rangle$ and $\langle\beta\rangle$, there is a computable sequence of stages $s_0 < s_1 < \ldots$ such that $|\alpha_{s_i} - \beta_{s_i}| \leq 2^{-i}$. Let f be a strictly increasing computable function such that $\alpha_x \leq \beta_{f(x)}$ for each x.

To define B, if x enters A at stage s, let i be greatest such that $s_i \leq s$. If $f(x) \leq s_i$, put $f(x)$ into B at stage s_i.

Clearly,

$$\alpha_s - \alpha_x \geq \alpha_{s_i} - \alpha_x \geq \alpha_{s_i} - \beta_{f(x)} \geq \beta_{s_i} - \beta_{f(x)} - 2^{-i}.$$

So $\mathbf{c}_{\langle\beta\rangle}\langle B_s\rangle \leq \mathbf{c}_{\langle\alpha\rangle}\langle A_s\rangle + \sum_i 2^{-i}$.

Let R be the computable subset of A consisting of those x that are enumerated early, namely x enters A at a stage s and $f(x) > s_i$ where i is greatest such that $s_i \leq s$. Clearly, $B = f(A - R)$. Hence $B \equiv_m A$.

The argument can be adapted to the case that A is Δ^0_2. Given a computable approximation $\langle A_s\rangle$ obeying $\mathbf{c}_{\langle\alpha\rangle}$, let t be the least s_i such that $s_i \geq f(x)$. For $s \leq t$, let $B_s(f(x)) = A_t(x)$. For $s > t$, let $B_s(f(x)) = A_{s_i}(x)$, where $s_i \leq s < s_{i+1}$. □

5 Randomness, Lowness, and K-Triviality

Benign cost functions were briefly discussed in the introduction.

Definition 5.1 ([10]) A monotonic cost function $\mathbf{c}$ is called *benign* if there is a computable function g such that for all k,

$$x_0 < x_1 < \ldots < x_k \wedge \forall i < k\,[\mathbf{c}(x_i, x_{i+1}) \geq 2^{-n}] \text{ implies } k \leq g(n).$$

Clearly, such a cost function satisfies the limit condition. Indeed, $\mathbf{c}$ satisfies the limit condition if and only if the above holds for some $g \leq_T \emptyset'$. For example, the cost function $\mathbf{c}_{\mathcal{K}}$ is benign via $g(n) = 2^n$. Each additive cost function is benign where $g(n) = O(2^n)$. For more detail see [10] or [23, Sect. 8.5].

For definitions and background on the extreme lowness property called strong jump traceability, see [10, 13] or [23, Chap. 8]. We will use the main result in [10], already quoted in the introduction: a c.e. set A is strongly jump traceable iff A obeys each benign cost function.

5.1 A Cost Function Implying Strong Jump Traceability

The following type of cost functions first appeared in [10] and [23, Sect. 5.3]. Let $Z \in \Delta^0_2$ be ML-random. Fix a computable approximation $\langle Z_s \rangle$ of Z and let $\mathbf{c}_Z$ (or, more accurately, $\mathbf{c}_{\langle Z_s \rangle}$) be the cost function defined as follows. Let $\mathbf{c}_Z(x, s) = 2^{-x}$ for each $x \geq s$; if $x < s$, and $e < x$ is least such that $Z_{s-1}(e) \neq Z_s(e)$, we let

$$c_Z(x, s) = \max(c_Z(x, s-1), 2^{-e}). \qquad (3)$$

Then $A \models \mathbf{c}_Z$ implies $A \leq_T Z$ by the aforementioned result from [10], which is proved like its variant above.

A Demuth test is a sequence of c.e. open sets $(S_m)_{m \in \mathbb{N}}$ such that

- $\forall m \, \boldsymbol{\lambda} S_m \leq 2^{-m}$, and there is a function f such that S_m is the Σ^0_1 class $[W_{f(m)}]^{\prec}$;
- $f(m) = \lim_s g(m, s)$ for a computable function g such that the size of the set $\{s\colon\, g(m, s) \neq g(m, s-1)\}$ is bounded by a computable function $h(m)$.

A set Z passes the test if $Z \notin S_m$ for almost every m. We say that Z is Demuth random if Z passes each Demuth test. For background on Demuth randomness, see [23, p. 141].

Proposition 5.2 *Suppose Y is a Demuth random Δ^0_2 set and $A \models c_Y$. Then $A \leq_T Z$ for each ω-c.e. ML-random set Z.*

In particular, A is strongly jump traceable by [13].

Proof Let $G^s_e = [Y_t \upharpoonright_e]$, where $t \leq s$ is greatest such that $Z_t(e) \neq Z_{t-1}(e)$. Let $G_e = \lim_s G^s_e$. (Thus, we only update G_e when $Z(e)$ changes.) Then $(G_e)_{e \in \mathbb{N}}$ is a Demuth test. Since Y passes this test, there is e_0 such that

$$\forall e \geq e_0 \, \forall t \, [Z_t(e) \neq Z_{t-1}(e) \rightarrow \exists s > t \; Y_{s-1} \upharpoonright_e \neq Y_s \upharpoonright_e].$$

We use this fact to define a computable approximation $(\hat{Z}_u)$ of Z as follows: let $\hat{Z}_u(e) = Z(e)$ for $e \leq e_0$; for $e > e_0$, let $\hat{Z}_u(e) = Z_s(e)$, where $s \leq u$ is greatest such that $Y_{s-1} \upharpoonright_e \neq Y_s \upharpoonright_e$.

Note that $\mathbf{c}_{\hat{Z}}(x, s) \leq c_Y(x, s)$ for all x, s. Hence $A \models \mathbf{c}_{\hat{Z}}$ and therefore $A \leq_T Z$. □

Recall that some Demuth random set is Δ^0_2. Kučera and Nies [17] in their main result strengthened the foregoing proposition in the case of a c.e. set A: if $A \leq_T Y$ for some Demuth random set Y, then A is strongly jump traceable. Greenberg and Turetsky [11] obtained the converse of this result: every c.e. strongly jump traceable is below a Demuth random.

Remark 5.3 For each Δ^0_2 set Y, we have $\mathbf{c}_Y(x) = 2^{-F(x)}$, where F is the Δ^0_2 function such that

$$F(x) = \min\{e\colon\, \exists s > x \; Y_s(e) \neq Y_{s-1}(e)\}.$$

Thus F can be viewed as a modulus function in the sense of [25].

For a computable approximation Φ, define the cost function $\mathbf{c}_\Phi$ as in (3). The following (together with Rmk. 5.3) implies that any computable approximation Φ of an ML-random Turing incomplete set changes late at small numbers, because the convergence of Ω_s to Ω is slow.

Corollary 5.4 *Let $Y <_T \emptyset'$ be an ML-random set. Let Φ be any computable approximation of Y. Then $\mathbf{c}_\Phi \to \mathbf{c}_\mathcal{K}$ and therefore $O(c_\Phi(x)) = \mathbf{c}_{\langle\Omega\rangle}(x)$.*

Proof If $A \models \mathbf{c}_\Phi$ then $C \models \mathbf{c}_\Phi$, where $C \geq_T A$ is the change set of the given approximation of A as in Proposition 2.14. By [14] (also see [23, 5.1.23]), C and therefore A are K-trivial. Hence $A \models \mathbf{c}_{\langle\Omega\rangle}$. □

5.2 Strongly Jump Traceable Sets and d.n.c. Functions

Recall that we write $X \leq_{ibT} Y$ if $X \leq_T Y$ with use function bounded by the identity. When building prefix-free machines, we use the terminology of [23, Sect. 2.3], such as Machine Existence Theorem (also called the Kraft-Chaitin Theorem), bounded request set, etc.

Theorem 5.5 *Suppose an ω-c.e. set Y is diagonally noncomputable via a function that is weak truth-table below Y. Let A be a strongly jump traceable c.e. set. Then $A \leq_{ibT} Y$.*

Proof By [15] (also see [23, 4.1.10]), there is an order function h such that $2h(n) \leq^+ K(Y \upharpoonright_n)$ for each n. The argument of the present proof goes back to Kučera's injury-free solution to Post's problem (see [23, Sect. 4.2]). Our proof is phrased in the language of cost functions, extending the similar result in [10], where Y is ML-random (equivalently, the condition above holds with $h(n) = \lfloor n/2 \rfloor + 1$).

Let $\langle Y_s \rangle$ be a computable approximation via which Y is ω-c.e. To help with building a reduction procedure for $A \leq_{ibT} Y$, via the Machine Existence Theorem we give prefix-free descriptions of initial segments $Y_s \upharpoonright_e$. On input x, if at a stage $s > x$ e is least such that $Y(e)$ has changed between stages x and s, then we still hope that $Y_s \upharpoonright_e$ is the final version of $Y \upharpoonright_e$. So whenever $A(x)$ changes at such a stage s, we give a description of $Y_s \upharpoonright_e$ of length $h(e)$. By hypothesis, A is strongly jump traceable, and hence obeys each benign cost function. We define an appropriate benign cost function $\mathbf{c}$ so that a set A that obeys $\mathbf{c}$ changes little enough that we can provide all the descriptions needed.

To ensure that $A \leq_{ibT} Y$, we define a computation $\Gamma(Y \upharpoonright_x)$ with output $A(x)$ at the least stage $t \geq x$ such that $Y_t \upharpoonright_x$ has the final value. If Y satisfies the hypotheses of the theorem, $A(x)$ cannot change at any stage $s > t$ (for almost all x), for otherwise $Y \upharpoonright_e$ would receive a description of length $h(e) + O(1)$, where e is least such that $Y(e)$ has changed between x and s.

We give the details. Firstly we give a definition of a cost function $\mathbf{c}$ which generalizes the definition in (3). Let $\mathbf{c}(x,s) = 0$ for each $x \geq s$. If $x < s$, and $e < x$ is least such that $Y_{s-1}(e) \neq Y_s(e)$, let

$$\mathbf{c}(x,s) = \max(\mathbf{c}(x,s-1), 2^{-h(e)}). \tag{4}$$

Since Y is ω-c.e., $\mathbf{c}$ is benign. Thus each strongly jump traceable c.e. set obeys $\mathbf{c}$ by the main result in [10]. So it suffices to show that $A \models \mathbf{c}$ implies $A \leq_{ibT} Y$ for any set A. Suppose that $\mathbf{c}\langle A_s \rangle \leq 2^u$. Enumerate a bounded request set L as follows. When $A_{s-1}(x) \neq A_s(x)$ and e is least such that $e = x$ or $Y_{t-1}(e) \neq Y_t(e)$ for some $t \in [x, s)$, put the request $\langle u + h(e), Y_s \upharpoonright_e \rangle$ into L. Then L is indeed a bounded request set.

Let d be a coding constant for L (see [23, Sect. 2.3]). Choose e_0 such that $h(e) + u + d < 2h(e)$ for each $e \geq e_0$. Choose $s_0 \geq e_0$ such that $Y \upharpoonright_{e_0}$ is stable from stage s_0 on.

To show $A \leq_{ibT} Y$, given an input $x \geq s_0$, using Y as an oracle, compute $t > x$ such that $Y_t \upharpoonright_x = Y \upharpoonright_x$. We claim that $A(x) = A_t(x)$. Otherwise $A_s(x) \neq A_{s-1}(x)$ for some $s > t$. Let $e \leq x$ be the largest number such that $Y_r \upharpoonright_e = Y_t \upharpoonright_e$ for all r with $t < r \leq s$. If $e < x$ then $Y(e)$ changes in the interval $(t, s]$ of stages. Hence, by the choice of $t \geq s_0$, we cause $K(y) < 2h(e)$, where $y = Y_t \upharpoonright_e = Y \upharpoonright_e$, a contradiction. □

Example 5.6 For each order function h and constant d, the class

$$P_{h,d} = \{Y\colon\ \forall n\, 2h(n) \leq K(Y \upharpoonright_n) + d\}$$

is Π^0_1. Thus, by the foregoing proof, each strongly jump traceable c.e. set is *ibT* below each ω-c.e. member of $P_{h,d}$.

We discuss the foregoing Theorem 5.5, and relate it to results in [10, 13].

1. In [13, Theorem 2.9] it is shown that given a non-empty Π^0_1 class P, each jump traceable set A Turing below each superlow member of P is already strongly jump traceable. In particular, this applies to superlow c.e. sets A, since such sets are jump traceable [21]. For many non-empty Π^0_1 classes, such a set is in fact computable. For instance, it could be a class where any two distinct members form a minimal pair. In contrast, the nonempty among the Π^0_1 classes $P = P_{h,d}$ are examples of where being below each superlow (or ω-c.e.) member characterizes strong jump traceability for c.e. sets.
2. Each superlow set A is weak truth-table below *some* superlow set Y as in the hypothesis of Theorem 5.5. For let P be the class of $\{0,1\}$-valued d.n.c. functions. By [23, 1.8.41], there is a set $Z \in P$ such that $(Z \oplus A)' \leq_{tt} A'$. Now let $Y = Z \oplus A$. This contrasts with the case of ML-random covers: if a c.e. set A is not K-trivial, then each ML-random set Turing above A is already Turing above $\emptyset'$ by [14]. Thus, in the case of *ibT* reductions, Theorem 5.5 applies to more oracle sets Y than [10, Proposition 5.2].

3. Greenberg and Nies [10, Proposition 5.2] have shown that for each order function p, each strongly jump traceable c.e. set is Turing below each ω-c.e. ML-random set, via a reduction with use bounded by p. We could also strengthen Theorem 5.5 to yield such a "p-bounded" Turing reduction.

5.3 A Proper Implication Between Cost Functions

In this subsection we study a weakening of K-triviality using the monotonic cost function

$$\mathbf{c}_{\max}(x, s) = \max\{2^{-K_s(w)} : x < w \leq s\}.$$

Note that $\mathbf{c}_{\max}$ satisfies the limit condition, because

$$\underline{\mathbf{c}}_{\max}(x) = \max\{2^{-K(w)} : x < w\}.$$

Clearly, $\mathbf{c}_{\max}(x, s) \leq \mathbf{c}_{\mathcal{K}}(x, s)$, whence $\mathbf{c}_{\mathcal{K}} \rightarrow \mathbf{c}_{\max}$. We will show that this implication of cost functions is proper. Thus, some set obeys $\mathbf{c}_{\max}$ that is not K-trivial.

Firstly, we investigate sets obeying $\mathbf{c}_{\max}$. For a string α, let $g(\alpha)$ be the longest prefix of α that ends in 1, and $g(\alpha) = \emptyset$ if there is no such prefix.

Definition 5.7 We say that a set A is *weakly K-trivial* if

$$\forall n\, [K(g(A\upharpoonright_n)) \leq^+ K(n)].$$

Clearly, every K-trivial set is weakly K-trivial. By the following, every *c.e.* weakly K-trivial set is already K-trivial.

Fact 5.8 *If A is weakly K-trivial and not h-immune, then A is K-trivial.*

Proof By the second hypothesis, there is an increasing computable function p such that $[p(n), p(n+1)) \cap A \neq \emptyset$ for each n. Then

$$K(A\upharpoonright_{p(n)}) \leq^+ K(g(A\upharpoonright_{p(n+1)})) \leq^+ K(p(n+1)) \leq^+ K(p(n)).$$

This implies that A is K-trivial by [23, Exercise 5.2.9]. □

We say that a computable approximation $\langle A_s \rangle_{s\in\mathbb{N}}$ is *erasing* if for each x and each $s > 0$, $A_s(x) \neq A_{s-1}(x)$ implies $A_s(y) = 0$ for each y such that $x < y \leq s$. For instance, the computable approximation built in the proof of the implication "⇒" of Theorem 3.4 is erasing by the construction.

Proposition 5.9 *Suppose $\langle A_s \rangle_{s\in\mathbb{N}}$ is an erasing computable approximation of a set A, and $\langle A_s \rangle \models \mathbf{c}_{\max}$. Then A is weakly K-trivial.*

Proof This is a modification of the usual proof that every set A obeying $\mathbf{c}_{\mathcal{K}}$ is K-trivial (see, for instance, [23, Theorem 5.3.10]).

To show that A is weakly K-trivial, one builds a bounded request set W. When at stage $s > 0$ we have $r = K_s(n) < K_{s-1}(n)$, we put the request $\langle r + 1, g(A \upharpoonright_n)\rangle$ into W. When $A_s(x) \neq A_{s-1}(x)$, let r be the number such that $\mathbf{c}_{\max}(x, s) = 2^{-r}$, and put the request $\langle r + 1, g(A \upharpoonright_{x+1})\rangle$ into W.

Since the computable approximation $\langle A_s\rangle_{s\in\mathbb{N}}$ obeys $\mathbf{c}_{\max}$, the set W is indeed a bounded request set; since $\langle A_s\rangle_{s\in\mathbb{N}}$ is erasing, this bounded request set shows that A is weakly K-trivial. □

We now prove that $\mathbf{c}_{\max} \not\to \mathbf{c}_{\mathcal{K}}$. We do so by proving a reformulation that is of interest by itself.

Theorem 5.10 *For every $b \in \mathbb{N}$, there is an x such that $\underline{\mathbf{c}}_{\mathcal{K}}(x) \geq 2^b \underline{\mathbf{c}}_{\max}(x)$. In other words,*

$$\sum\{2^{-K(w)}\colon x < w\} \geq 2^b \max\{2^{-K(w)}\colon x < w\}.$$

By Theorem 3.4, the statement of the foregoing theorem is equivalent to $\mathbf{c}_{\max} \not\to \mathbf{c}_{\mathcal{K}}$. Thus, as remarked above, some set A obeys $\mathbf{c}_{\max}$ via an erasing computable approximation, and does not obey $\mathbf{c}_{\mathcal{K}}$. By Proposition 5.9 we obtain a separation.

Corollary 5.11 *Some weakly K-trivial set fails to be K-trivial.*

Melnikov and Nies [20, Proposition 3.7] have given an alternative proof of the preceding result by constructing a weakly K-trivial set that is Turing complete.

Proof of Theorem 5.10 Assume that there is a $b \in \mathbb{N}$ such that

$$\forall x\, [\underline{\mathbf{c}}_{\mathcal{K}}(x) < 2^b \underline{\mathbf{c}}_{\max}(x)].$$

To obtain a contradiction, the idea is that $\mathbf{c}_{\mathcal{K}}(x, s)$, which is defined as a sum, can be made large in many small bits; in contrast, $\mathbf{c}_{\max}(x, s)$, which depends on the value $2^{-K_s(w)}$ for a single w, cannot.

We will define a sequence $0 = x_0 < x_1 < \ldots < x_N$ for a certain number N. When x_v has been defined for $v < N$, for a certain stage $t > x_v$ we cause $\mathbf{c}_{\mathcal{K}}(x_v, t)$ to exceed a fixed quantity proportional to $1/N$. We wait until the opponent responds at a stage $s > t$ with some $w > x_v$ such that $2^{-K_s(w)}$ corresponds to that quantity. Only then do we define $x_{v+1} = s$. For us, the cost $\mathbf{c}_{\mathcal{K}}(x_i, x_j)$ will accumulate for $i < j$, while the opponent has to provide a new w each time. This means that eventually he will run out of space in the domain of the prefix-free machine, giving short descriptions of such w's.

In the formal construction, we will build a bounded request set L with the purpose to cause $\mathbf{c}_{\mathcal{K}}(x, s)$ to be large when it is convenient to us. We may assume by the recursion theorem that the coding constant for L is given in advance (see [23, Remark 2.2.21] for this standard argument). Thus, if we put a request $\langle n, y + 1\rangle$

into L at a stage y, there will be a stage $t > y$ such that $K_t(y+1) \leq n+d$, and hence $\mathbf{c}_{\mathcal{K}}(x,t) \geq \mathbf{c}_{\mathcal{K}}(x,y) + 2^{-n-d}$.

Let $k = 2^{b+d+1}$. Let $N = 2^k$.

Construction of L and a Sequence $0 = x_0 < x_1 < \ldots < x_N$ *of Numbers*

Suppose $v < N$, and x_v has already been defined. Put $\langle k, x_v + 1\rangle$ into L. As remarked above, we may wait for a stage $t > x_v$ such that $\mathbf{c}_{\mathcal{K}}(x_v, t) \geq 2^{-k-d}$. Now, by our assumption, we have $\underline{\mathbf{c}}_{\mathcal{K}}(x_i) < 2^b \underline{\mathbf{c}}_{\max}(x_i)$ for each $i \leq v$. Hence we can wait for a stage $s > t$ such that

$$\forall i \leq v\, \exists w \left[x_i < w \leq s \ \wedge\ \mathbf{c}_{\mathcal{K}}(x_i, s) \leq 2^{b-K_s(w)}\right]. \tag{5}$$

Let $x_{v+1} = s$. This ends the construction.

Verification Note that L is indeed a bounded request set. Clearly, we have $\mathbf{c}_{\mathcal{K}}(x_i, x_{i+1}) \geq 2^{-k-d}$ for each $i < N$.

Claim 5.12 Let $r \leq k$. Write $R = 2^r$. Suppose $p + R \leq N$. Let $s = x_{p+R}$. Then we have

$$\sum_{w=x_p+1}^{x(p+R)} \min(2^{-K_s(w)}, 2^{-k-b-d+r}) \geq (r+1)2^{-k-b-d+r-1}. \tag{6}$$

For $r = k$, the right-hand side equals $(k+1)2^{-(b+d+1)} > 1$, which is a contradiction because the left-hand side is at most $\Omega \leq 1$.

We prove the claim by induction on r. To verify the case $r = 0$, note that by (5) there is a $w \in (x_p, x_{p+1}]$ such that $\mathbf{c}_{\mathcal{K}}(x_p, x_{p+1}) \leq 2^{b-K_s(w)}$. Since $2^{-k-d} \leq \mathbf{c}_{\mathcal{K}}(x_p, x_{p+1})$, we obtain

$$2^{-k-b-d} \leq 2^{-K_s(w)} (\text{where } s = x_{p+1}).$$

Thus the left-hand side in the inequality (6) is at least 2^{-k-b-d}, while the right-hand side equals $2^{-k-b-d-1}$, and the claim holds for $r = 0$.

In the following, for $i < j \leq N$, we will write $\mathcal{S}(x_i, x_j)$ for a sum of the type occurring in (6), where w ranges from $x_i + 1$ to x_j.

Suppose inductively the claim has been established for $r < k$. To verify the claim for $r + 1$, suppose that $p + 2R \leq N$ where $R = 2^r$ as before. Let $s = x_{p+2R}$. Since $\mathbf{c}_{\mathcal{K}}(x_i, x_{i+1}) \geq 2^{-k-d}$, we have

$$\mathbf{c}_{\mathcal{K}}(x_p, s) \geq 2R2^{-k-d} = 2^{-k-d+r+1}.$$

By (5), this implies that there is w, $x_p < w \leq s$, such that

$$2^{-k-b-d+r+1} \leq 2^{-K_s(w)}. \tag{7}$$

Now, in sums of the form $\mathcal{S}(x_q, x_{q+R})$, because of taking the minimum, the "cut-off" for how much w can contribute is at $2^{-k-b-d+r}$. Hence we have

$$\mathcal{S}(x_p, x_{p+2R}) \geq 2^{-k-b-d+r} + \mathcal{S}(x_p, x_{p+R}) + \mathcal{S}(x_{p+R}, x_{p+2R}).$$

The additional term $2^{-k-b-d+r}$ is due to the fact that w contributes at most $2^{-k-b-d+r}$ to $\mathcal{S}(x_p, x_{p+R}) + \mathcal{S}(x_{p+R}, x_{p+2R})$, but by (7), w contributes $2^{-k-b-d-r+1}$ to $\mathcal{S}(x_p, x_{p+2R})$. By the inductive hypothesis, the right-hand side is at least

$$2^{-k-b-d+r} + 2 \cdot (r+1)2^{-k-b-d+r-1} = (r+2)2^{-k-b-d+r},$$

as required. □

6 A Cost Function-Related Basis Theorem for Π^0_1 Classes

The following strengthens [13, Theorem 2.6], which relied on the extra assumption that the Π^0_1 class is contained in the ML-randoms.

Theorem 6.1 *Let $\mathcal{P}$ be a nonempty Π^0_1 class, and let $\mathbf{c}$ be a monotonic cost function with the limit condition. Then there is a Δ^0_2 set $Y \in \mathcal{P}$ such that each c.e. set $A \leq_T Y$ obeys $\mathbf{c}$.*

Proof We may assume that $\mathbf{c}(x,s) \geq 2^{-x}$ for each $x \leq s$, because any c.e. set that obeys $\mathbf{c}$ also obeys the cost function $\mathbf{c}(x,s) + 2^{-x}$.

Let $\langle A_e, \Psi_e \rangle_{e \in \mathbb{N}}$ be an effective listing of all pairs consisting of a c.e. set and a Turing functional. We will define a Δ^0_2 set $Y \in \mathcal{P}$ via a computable approximation, $Y_{s s \in \mathbb{N}}$, where Y_s is a binary string of length s. We meet the requirements

$$N_e\colon\ A_e = \Psi_e(Y) \ \Rightarrow\ A_e \text{ obeys } \mathbf{c}.$$

We use a standard tree construction at the $\emptyset''$ level. Nodes on the tree $2^{<\omega}$ represent the strategies. Each node α of length e is a strategy for N_e. At stage s we define an approximation δ_s to the true path. We say that s is an *α-stage* if $\alpha \prec \delta_s$.

Suppose that a strategy α is on the true path. If $\alpha 0$ is on the true path, then strategy α is able to build a computable enumeration of A_e via which A_e obeys $\mathbf{c}$. If $\alpha 1$ is on the true path, the strategy shows that $A_e \neq \Psi_e(Y)$.

Let $\mathcal{P}^{\emptyset}$ be the given class $\mathcal{P}$. A strategy α has as an environment a Π^0_1 class $\mathcal{P}^\alpha$. It defines $\mathcal{P}^{\alpha 0} = \mathcal{P}^\alpha$, but usually lets $\mathcal{P}^{\alpha 1}$ be a proper refinement of $\mathcal{P}^\alpha$.

Let $|\alpha| = e$. The length of agreement for e at a stage t is $\min\{y\colon A_{e,t}(y) \neq \Psi_{e,t}(Y_t)\}$. We say that an α-stage s is *α-expansionary* if the length of agreement for e at stage s is larger than at u for all previous α-stages u.

Let $w^n_0 = n$, and

$$w^n_{i+1} \simeq \mu v > w^n_i.\, \mathbf{c}(w^n_i, v) \geq 4^{-n}. \tag{8}$$

Since $\mathbf{c}$ satisfies the limit condition, for each n this sequence breaks off.

Let $a = w_i^n$ be such a value. The basic idea is to *certify* $A_{e,s} \upharpoonright_w$, which means to ensure that all $X \succ Y_s \upharpoonright_{n+d}$ on $\mathcal{P}^\alpha$ compute $A_{e,s} \upharpoonright_w$. If $A \upharpoonright_w$ changes later, then also $Y \upharpoonright_{n+d}$ has to change. Since $Y \upharpoonright_{n+d}$ can only move to the right (as long as α is not initialized), this type of change for n can only contribute a cost of $4^{-n+1}2^{n+d} = 2^{-n+d+2}$.

By [23, p. 55], from an index $\mathcal{Q}$ for a Π^0_1 class in 2^ω we can obtain a computable sequence $(\mathcal{Q}_s)_{s\in\mathbb{N}}$ of clopen classes such that $\mathcal{Q}_s \supseteq \mathcal{Q}_{s+1}$ and $\mathcal{Q} = \bigcap_s \mathcal{Q}_s$. In the construction below we will have several indices for Π^1_1 classes $\mathcal{Q}$ that change over time. At stage s, as usual, by $\mathcal{Q}[s]$ we denote the value of the index at stage s. Thus $(\mathcal{Q}[s])_s$ is the clopen approximation of $\mathcal{Q}[s]$ at stage s.

Construction of Y

Stage 0. *Let* $\delta_0 = \emptyset$ *and* $\mathcal{P}^\emptyset = \mathcal{P}$. *Let* $Y_0 = \emptyset$.

Stage $s > 0$. Let $\mathcal{P}^\emptyset = \mathcal{P}$.

For each β such that $\delta_{s-1} <_L \beta$, we initialize strategy β. We let Y_s be the leftmost path on the current approximation to $\mathcal{P}^{\delta_{s-1}}$, i.e., the leftmost string y of length $s-1$ such that $[y] \cap (\mathcal{P}^{\delta_{s-1}}[s-1])_s \neq \emptyset$. For each α, n, if $Y_s \upharpoonright_{n+d} \neq Y_{s-1} \upharpoonright_{n+d}$, where $d = \mathsf{init}_s(\alpha)$, then we declare each existing value w_i^n to be (α, n)-*unsatisfied*.

Substage k, $0 \leq k < s$. Suppose we have already defined $\alpha = \delta_s \upharpoonright_k$. Run strategy α (defined below) at stage s, which defines an outcome $r \in \{0, 1\}$ and a Π^0_1 class $\mathcal{P}^{\alpha r}$. Let $\delta_s(k) = r$.

We now describe the strategies α and the procedures $\mathcal{S}^\alpha_n$ they call. To initialize a strategy α means to cancel the run of this procedure. Let

$$d = \mathsf{init}_s(\alpha) = |\alpha| + \text{thelaststagewhen } \alpha was initialized.$$

Strategy α *at an* α*-Stage* s

(a) If no procedure for α is running, call procedure $\mathcal{S}^\alpha_n$ with parameter w, where n is least, and i is chosen least for n, such that $w = w_i^n \leq s$ is not (α, n)-satisfied. Note that n exists because $w_0^s = s$ and this value is not (α, n)-satisfied at the beginning of stage s. By calling this procedure, we attempt to certify $A_{e,s} \upharpoonright_w$ as discussed above.

(b) While such a procedure $\mathcal{S}^\alpha_n$ is running, give outcome 1.
(This procedure will define the current class $\mathcal{P}^{\alpha 1}$.)

(c) If a procedure $\mathcal{S}^\alpha_n$ returns at this stage, goto (d).

(d) If s is α-expansionary, give outcome 0, let $\mathcal{P}^{\alpha 0} = \mathcal{P}^\alpha$, and continue at (a) at the next α-stage. Otherwise, give outcome 1, let $\mathcal{P}^{\alpha 1} = \mathcal{P}^\alpha$, and stay at (d).

Procedure $\mathcal{S}^\alpha_n$ *with Parameter* w *at a Stage* s If $n + d \geq s - 1$, let $\mathcal{P}^{\alpha 1} = \mathcal{P}^\alpha$. Otherwise, let

$$\mathcal{Q} = \mathcal{P}^\alpha \cap \{X \succ z \colon \Psi_e^X \not\succ A_{e,s} \upharpoonright_w\}, \tag{9}$$

where $z = Y_s \upharpoonright_{n+d}$. (Note that each time $Y \upharpoonright_{n+d}$ or $A_e \upharpoonright_w$ has changed, we update this definition of $\mathcal{Q}$.)

(e) If $\mathcal{Q}_s \neq \emptyset$, let $\mathcal{P}^{\alpha 1} = \mathcal{Q}$. If the definition of $\mathcal{P}^{\alpha 1}$ has changed since the last α-stage, then each β such that $\alpha 1 \preceq \beta$ is initialized.

(f) If $\mathcal{Q}_s = \emptyset$, declare w to be (α, n)-*satisfied* and return. ($A_{e,s} \upharpoonright_w$ is certified, as every $X \in \mathcal{P}^\alpha$ extending z computes $A_{e,s} \upharpoonright_w$ via Ψ_e. If $A_e \upharpoonright_w$ changes later, then necessarily $z \not\prec Y$.)

Claim 6.2 Suppose a strategy α is no longer initialized after stage s_0. Then for each n, a procedure $\mathcal{S}_n^\alpha$ is only called finitely many times after s_0.

There are only finitely many values $w = w_i^n$ because $\mathbf{c}$ satisfies the limit condition. Since α is not initialized after s_0, $\mathcal{P}^\alpha$ and $d = \mathsf{init}_s(\alpha)$ do not change. When a run of $\mathcal{S}_n^\alpha$ is called at a stage s, the strategies $\beta \succeq \alpha 1$ are initialized; hence $\mathsf{init}_t(\beta) \geq s > n + d$ for all $t \geq s$. So the string $Y_s \upharpoonright_{n+d}$ is the leftmost string of length $n + d$ on $\mathcal{P}^\alpha$ at stage s. This string has to move to the right between the stages when $\mathcal{S}_n^\alpha$ is called with the same parameter w, because w is declared (α, n)-unsatisfied before $\mathcal{S}_n^\alpha$ is called again with parameter w. Thus, procedure $\mathcal{S}_n^\alpha$ can only be called 2^{n+d} times with parameter w.

Claim 6.3 $\langle Y_s \rangle_{s \in \mathbb{N}}$ is a computable approximation of a Δ_2^0 set $Y \in \mathcal{P}$.

Fix $k \in \mathbb{N}$. For a stage s, if $Y_s \upharpoonright_k$ is to the left of $Y_{s-1} \upharpoonright_k$, then there are α, n with $n + \mathsf{init}_s(\alpha) \leq k$ such that $\mathcal{P}^\alpha[s] \neq \mathcal{P}^\alpha[s-1]$ because of the action of a procedure $\mathcal{S}_n^\alpha$ at (e) or (f).

There are only finitely many pairs α, s such that $\mathsf{init}_s(\alpha) \leq k$. Thus, by Claim 6.2, there is a stage s_0 such that at all stages $s \geq s_0$, for no α and n with $n + \mathsf{init}_s(\alpha) \leq k$, a procedure $\mathcal{S}_n^\alpha$ is called.

While a procedure $\mathcal{S}_n^\alpha$ is running with a parameter w, it changes the definition of $\mathcal{P}^{\alpha 1}$ only if $A_e \upharpoonright_w$ changes ($e = |\alpha|$), so at most w times. Thus there are only finitely many s such that $Y_s \upharpoonright_k \neq Y_{s-1} \upharpoonright_k$.

By the definition of the computable approximation $\langle Y_s \rangle_{s \in \mathbb{N}}$, we have $Y \in \mathcal{P}$. This completes Claim 6.3.

As usual, we define the true path f by $f(k) = \liminf_s \delta_s(k)$. By Claim 6.2, each $\alpha \prec f$ is only initialized finitely often, because each β such that $\beta 1 \prec \alpha$ eventually is stuck with a single run of a procedure $\mathcal{S}_m^\beta$.

Claim 6.4 If $e = |\alpha|$ and $\alpha 1 \prec f$, then $A_e \neq \Psi_e^Y$.

Some procedure $\mathcal{S}_n^\alpha$ was called with parameter w, and is eventually stuck at (e) with the final value $A_e \upharpoonright_w$. Hence the definition $\mathcal{Q} = \mathcal{P}^{\alpha 1}$ eventually stabilizes at α-stages s. Since $Y \in \mathcal{Q}$, this implies $A_e \neq \Psi_e^Y$.

Claim 6.5 If $e = |\alpha|$ and $\alpha 0 \prec f$, then A_e obeys $\mathbf{c}$.

Let $A = A_e$. We define a computable enumeration $(\hat{A}_p)_{p \in \mathbb{N}}$ of A via which A obeys $\mathbf{c}$.

Since $\alpha 0 \prec f$, each procedure $\mathcal{S}_n^\alpha$ returns. In particular, since $\mathbf{c}$ has the limit condition and by Claims 6.2 and 6.3 each value $w = w_i^n$ becomes permanently

(α, n)-satisfied. Let $d = \mathsf{init}_s(\alpha)$. Let s_0 be the least $\alpha 0$-stage such that $s_0 \geq d$, and let

$$s_{p+1} = \mu s \geq s_p + 2\,[s \text{ is} \alpha 0 - \text{stage } \wedge$$

$$\forall n, i\,(w = w_i^n < s_p \rightarrow w \text{ is}(\alpha, \text{n}) - \text{satisfiedat } s)].$$

As in similar constructions such as [23], for $p \in \mathbb{N}$ we let

$$\hat{A}_p = A_{s_{p+2}} \cap [0, p).$$

Consider the situation where $p > 0$ and $x \leq p$ is least such that $\hat{A}_p(x) \neq \hat{A}_{p-1}(x)$. We call this situation an *n-change* if n is least such that $x < w_i^n < s_p$ for some i. (Note that $n \leq p + 1$ because $w_0^{p+1} = p + 1$.) Thus (x, s_p) contains no value of the form w_j^{n-1}, whence $\mathbf{c}(x, p) \leq \mathbf{c}(x, s_p) \leq 4^{-n+1}$. We are done if we can show there are at most 2^{n+d} many n-changes, for in that case the total cost $\mathbf{c}\langle \hat{A}_p \rangle$ is bounded by $\sum_n 4^{-n+1} 2^{n+d} = O(2^d)$.

Recall that $\mathcal{P}^\alpha$ is stable by stage s_0. Note that $Y \upharpoonright_{n+d}$ can only move to the right after the first run of $\mathcal{S}_n^\alpha$, as observed in the proof of Claim 6.2.

Consider n-changes at stages $p < q$ via parameters $w = w_i^n$ and $w' = w_k^n$ (where possibly $k < i$). Suppose the last run of $\mathcal{S}_n^\alpha$ with parameter w that was started before s_{p+1} has returned at stage $t \leq s_{p+2}$, and similarly, the last run of $\mathcal{S}_n^\alpha$ with parameter w' that was started before s_{q+1} has returned at stage t'. Let $z = Y_t \upharpoonright_{n+d}$ and $z' = Y_{t'} \upharpoonright_{n+d}$. We show $z <_L z'$; this implies that there are at most 2^{n+d} many n-changes.

At stage t, by definition of returning at (f) in the run of $\mathcal{S}_n^\alpha$, we have $\mathcal{Q} = \emptyset$. Therefore $\Psi_{e,t}^X \succ A_{e,t} \upharpoonright_w$ for each X on $\mathcal{P}_t^\alpha$ such that $X \succ z$. Now

$$\hat{A}_p(x) \neq \hat{A}_{p-1}(x), x < w \text{ and } t \leq s_{p+1},$$

so $A_{s_{p+2}} \upharpoonright_w \neq A_t \upharpoonright_w$. The stage s_{p+2} is $\alpha 0$-expansionary, and $Y_{s_{p+2}}$ is on $\mathcal{P}_t^\alpha$. Therefore

$$Y_{r-1} \upharpoonright_{n+d} <_L Y_r \upharpoonright_{n+d}$$

for some stage r such that $t < r \leq s_{p+2}$. Thus, at stage r, the value w' was declared (α, n)-unsatisfied. Hence a new run of $\mathcal{S}_n^\alpha$ with parameter w' is started after r, which has returned by stage $s_{q+1} \geq s_{p+2}$. Thus $r < t'$. So $z \leq_L Y_{r-1} \upharpoonright_{n+d} <_L Y_r \upharpoonright_{n+d} \leq_L z'$, whence $z <_L z'$, as required. This concludes Claim 6.5 and the proof. □

7 A Dual Cost Function Construction

Given a relativizable cost function $\mathbf{c}$, let $D \rightarrow W^D$ be the c.e. operator given by the cost function construction in Theorem 2.7 relative to the oracle D. By pseudo-jump inversion, there is a c.e. set D such that $W^D \oplus D \equiv_T \emptyset'$, which implies $D <_T \emptyset'$.

Here, we give a direct construction of a c.e. set $D <_T \emptyset'$ so that the total cost of $\emptyset'$-changes as measured by $\mathbf{c}^D$ is finite. More precisely, there is a D-computable enumeration of $\emptyset'$ obeying $\mathbf{c}^D$.

If $\mathbf{c}$ is sufficiently strong, then the usual cost function construction builds an incomputable c.e. set A that is close to being computable. The dual cost function construction then builds a c.e. set D that is close to being Turing complete.

7.1 Preliminaries on Cost Functionals

Firstly we clarify how to relativize cost functions, and the notion of obedience to a cost function. Secondly we provide some technical details needed for the main construction.

Definition 7.1

(i) A *cost functional* is a Turing functional $\mathbf{c}^Z(x, t)$ such that, for each oracle Z, $\mathbf{c}^Z$ either is partial, or is a cost function relative to Z. We say that $\mathbf{c}$ is non-increasing in the main argument if this holds for each oracle Z such that $\mathbf{c}^Z$ is total. Similarly, $\mathbf{c}$ is non-decreasing in the stage argument if this holds for each oracle Z such that $\mathbf{c}^Z$ is total. If both properties hold, we say that $\mathbf{c}$ is monotonic.

(ii) Suppose $A \leq_T Z'$. Let $\langle A_s \rangle$ be a Z-computable approximation of A.
We write $\langle A_s \rangle \models^Z \mathbf{c}^Z$ if
$\mathbf{c}^Z \langle A_s \rangle = \sum_{x,s} \mathbf{c}^Z(x, s)$
$[\![x < s \wedge \mathbf{c}^Z(x, s) \downarrow \wedge x \text{ is least s.t. } A_{s-1}(x) \neq A_s(x)]\!]$
is finite. We write $A \models^Z c^Z$ if $\langle A_s \rangle \models^Z c^Z$ for some Z-computable approximation $\langle A_s \rangle$ of A.

For example, $\mathbf{c}^Z_{\mathcal{K}}(x, s) = \sum_{x<w\leq s} 2^{-K^Z_s(w)}$ is a total monotonic cost functional. We have $A \models^Z \mathbf{c}^Z_{\mathcal{K}}$ iff A is K-trivial relative to Z.

We may convert a cost functional $\mathbf{c}$ into a total cost functional $\tilde{\mathbf{c}}$ such that $\tilde{\mathbf{c}}^Z(x) = \mathbf{c}^Z(x)$ for each x with $\forall t\, \mathbf{c}^Z(x, t) \downarrow$, and, for each Z, x, t, the computation $\tilde{\mathbf{c}}^Z(x, t)$ converges in t steps. Let

$$\tilde{\mathbf{c}}^Z(x, s) = \mathbf{c}^Z(x, t), \text{ where } t \leq s \text{ is largest such that } \mathbf{c}^Z(x, t)[s] \downarrow .$$

Clearly, if $\mathbf{c}$ is monotonic in the main/stage argument, then so is $\tilde{\mathbf{c}}$.

Suppose that D is c.e. and we compute $\mathbf{c}^D(x, t)$ via hat computations [25, p. 131]: the use of a computation $\mathbf{c}^D(x, t)[s] \downarrow$ is no larger than the least number entering D at stage s. Let N_D be the set of non-deficiency stages; that is, $s \in N_D$ iff there is an $x \in D_s - D_{s-1}$ such that $D_s \upharpoonright_x = D \upharpoonright_x$. Any hat computation existing at a non-deficiency stage is final. We have

$$\mathbf{c}^D(x) = \sup_{s \in N_D} \tilde{\mathbf{c}}^{D_s}(x, s). \quad (10)$$

For, if $\mathbf{c}^D(x,t)[s_0] \downarrow$ with D stable below the use, then $\mathbf{c}^D(x,t) \le \tilde{\mathbf{c}}^{D_s}(x,s)$ for each $s \in N_D$. Therefore $\mathbf{c}^D(x) \le \sup_{s \in N_D} \tilde{\mathbf{c}}^{D_s}(x,s)$. For the converse inequality, note that for $s \in N_D$, we have $\tilde{\mathbf{c}}^{D_s}(x,s) = \mathbf{c}^D(x,t)$ for some $t \le s$ with D stable below the use.

7.2 The Dual Existence Theorem

Theorem 7.2 *Let* $\mathbf{c}$ *be a total cost functional that is nondecreasing in the stage component and satisfies the limit condition for each oracle. Then there is a Turing incomplete c.e. set D such that* $\emptyset' \models^D \mathbf{c}^D$.

Proof We define a cost functional $\Gamma^Z(x,s)$ that is nondecreasing in the stage. We will have $\underline{\Gamma}^D(x) = \mathbf{c}^D(x)$ for each x, where $\underline{\Gamma}^D(x) = \lim_t \Gamma^D(x,t)$, and $\emptyset'$ with its given computable enumeration obeys Γ^D. Then $\emptyset' \models^D \mathbf{c}^D$ by the easy direction '$\Leftarrow$' of Theorem 3.4 relativized to D.

Towards $\Gamma^D(x) \ge \mathbf{c}^D(x)$, when we see a computation $\tilde{\mathbf{c}}^{D_s}(x,s) = \alpha$, we attempt to ensure that $\Gamma^D(x,s) \ge \alpha$. To do so, we enumerate relative to D a set G of "wishes" of the form

$$\rho = \langle x, \alpha \rangle^u,$$

where $x \in \mathbb{N}$, α is a nonnegative rational, and $u+1$ is the use. We say that ρ is a *wish about* x. If such a wish is enumerated at a stage t and $D_t \upharpoonright_u$ is stable, then the wish is granted, namely, $\Gamma^D(x,t) \ge \alpha$. The converse inequality $\Gamma^D(x) \le \mathbf{c}^D(x)$ will hold automatically.

To ensure $D <_T \emptyset'$, we enumerate a set F, and meet the requirements

$$N_e\colon\ F \neq \Phi_e^D.$$

Suppose we have put a wish $\rho = \langle x, \alpha \rangle^u$ into G^D. To keep the total Γ^D-cost of the given computable enumeration of $\emptyset'$ down, when x enters $\emptyset'$, we want to remove ρ from G^D by putting u into D. However, sometimes D is preserved by some N_e. This will generate a *preservation cost.* N_e starts a run at a stage s via some parameter v, and "hopes" that $\emptyset'_s \upharpoonright_v$ is stable. If $\emptyset' \upharpoonright_v$ changes after stage s, then this run of N_e is cancelled. On the other hand, if $x \ge v$ and x enters $\emptyset'$, then the ensuing preservation cost can be afforded. This is so because we choose v such that $\tilde{c}_s^{D_s}(v,s)$ is small. Since $\tilde{\mathbf{c}}^D$ has the limit condition, eventually there is a run $N_e(v)$ with such a low-cost v where $\emptyset' \upharpoonright_v$ is stable. Then the diagonalization of N_e will succeed.

Construction of c.e. Sets F, D and a D-c.e. Set G of Wishes *Stage* $s > 0$. We may suppose that there is a unique $n \in \emptyset'_s - \emptyset'_{s-1}$.

1. *Canceling* N_e*'s.* Cancel all currently active $N_e(v)$ with $v > n$.
2. *Removing wishes.* For each $\rho = \langle x, \alpha \rangle^u \in G^D[s-1]$ put in at a stage $t < s$, if $\emptyset'_s \upharpoonright_{x+1} \neq \emptyset'_t \upharpoonright_{x+1}$ and ρ is not held by any $N_e(v)$, then put $u-1$ into D_s, thereby removing ρ from G^D.

3. *Adding wishes.* For each $x < s$, pick a large u (in particular, $u \notin D_s$) and put a wish $\langle x, \alpha \rangle^u$ into G, where $\alpha = \tilde{\mathbf{c}}^{D_s}(x, s)$. The set of queries to the oracle D for this enumeration into G is contained in $[0, r) \cup \{u\}$, where r is the use of $\tilde{\mathbf{c}}^{D_s}(x, s)$ (which may be much smaller than s). Then, from now on this wish is kept in G^D unless (a) $D \upharpoonright_r$ changes , or (b) u enters D.
4. *Activating $N_e(v)$.* For each $e < s$ such that N_e is not currently active, see if there is a v, $e \leq v \leq n$, such that
   - $\tilde{\mathbf{c}}^{D_s}(v, s) \leq 3^{-e}/2$,
   - $v > w$ for each w such that $N_i(w)$ is active for some $i < e$, and
   - $\Phi_e^D \upharpoonright_{x+1} = F \upharpoonright_{x+1}$ where $x = \langle e, v, |\emptyset' \cap [0, v)| \rangle$,

If so, choose v least and activate $N_e(v)$. Put x into F. Let N_e *hold* all wishes for some $y \geq v$ that are currently in G^D. Declare that such wishes are no longer held by any $N_i(w)$ for $i \neq e$. (We also say that N_e *takes over* the wishes.)

Go to stage s', where s' is larger than any number mentioned so far.

Claim 1 Each requirement N_e is activated only finitely often, and met. Hence $F \not\leq_T D$.

Inductively suppose that N_i for $i < e$ is no longer activated after stage t_0. Assume for a contradiction that $F = \Phi_e^D$. Since $\mathbf{c}^D$ satisfies the limit condition, by (10) there is a least v such that $\tilde{\mathbf{c}}^{D_s}(v, s) \leq 3^{-e}/2$ for infinitely many $s > t_0$. Furthermore, $v > w$ for any w such that some $N_i(w)$, $i < e$, is active at t_0. Once $N_e(v)$ is activated, it can only be canceled by a change of $\emptyset' \upharpoonright_v$. Then there is a stage $s > t_0$, $\tilde{\mathbf{c}}^{D_s}(v, s) \leq 3^{-e}/2$, such that $\emptyset' \upharpoonright_v$ is stable at s and $\Phi_e^D \upharpoonright_{x+1} = F \upharpoonright_{x+1}$, where $x = \langle e, v, |\emptyset' \cap [0, v)| \rangle$. If some $N_e(v')$ for $v' \leq v$ is active after (1.) of stage s, then it remains active, and N_e is met. Now suppose otherwise.

Since we do not activate $N_e(v)$ in (4) of stage s, some $N_e(w)$ is active for $w > v$. Say it was activated last at a stage $t < s$ via $x = \langle e, w, |\emptyset'_t \cap [0, w]|$. Then $x' = \langle e, v, |\emptyset'_t \cap [0, v)| \rangle$ was available to activate $N_e(v)$ as $x' \leq x$ and hence $\Phi_e^D \upharpoonright_{x'+1} = F \upharpoonright_{x'+1} [t]$. Since w was chosen minimal for e at stage t, we had $\tilde{\mathbf{c}}^{D_t}(v, t) > 3^{-e}/2$. On the other hand, $\tilde{\mathbf{c}}^{D_s}(v, s) \leq 3^{-e}/2$; hence $D_t \upharpoonright_t \neq D_s \upharpoonright_t$. When $N_e(w)$ became active at t, it tried to preserve $D \upharpoonright_t$ by holding all wishes about some $y \geq w$ that were in $G^D[t]$. Since $N_e(w)$ did not succeed, it was cancelled by a change $\emptyset'_t \upharpoonright_w \neq \emptyset'_s \upharpoonright_w$. Hence $N_e(w)$ is not active at stage s, a contradiction. ◇

We now define $\Gamma^Z(x, t)$ for an oracle Z (we are interested only in the case $Z = D$). Let s be least such that $D_s \upharpoonright_t = Z \upharpoonright_t$. Output the maximum α such that some wish $\langle x, \alpha \rangle^u$ for $u \leq t$ is in $G^D[s]$.

Claim 2 (i) $\Gamma^D(x, t)$ is nondecreasing in t. (ii) $\forall x \, \underline{\Gamma}^D(x) = \underline{\mathbf{c}}^D(x)$.

(i) Suppose $t' \geq t$. As above, let s be least such that $D_s \upharpoonright_t$ is stable. Let s' be least such that $D_{s'} \upharpoonright_{t'}$ is stable. Then $s' \geq s$, so a wish as in the definition of $\Gamma^D(x, t)$ above is also in $G^D[s']$. Hence $\Gamma^D(x, t') \geq \Gamma^D(x, t)$.
(ii) Given x, to show that $\Gamma^D(x) \geq \mathbf{c}^D(x)$, pick t_0 such that $\emptyset' \upharpoonright_{x+1}$ is stable at t_0. Let $s \in N_D$ and $s > t_0$. At stage s we put a wish $\langle x, \alpha \rangle^u$ into G_D, where $\alpha = \tilde{\mathbf{c}}^{D_s}(x, s)$. This wish is not removed later, so $\Gamma^D(x) \geq \alpha$.

For $\Gamma^D(x) \leq \mathbf{c}^D(x)$, note that for each $s \in N_D$ we have $\tilde{\mathbf{c}}^{D_s}(x,s) \geq \Gamma^{D_s}(x,s)$ by the removal of a wish in 3(a) of the construction when the reason the wish was there disappears. ◇

Claim 3 The given computable enumeration of $\emptyset'$ obeys Γ^D.

First we show by induction on stages s that N_e holds in total at most 3^{-e} at the end of stage s, namely,

$$3^{-e} \geq \sum_x \max\{\alpha : N_e \text{ holdsawish } \langle x, \alpha \rangle^u\} \tag{11}$$

Note that once $N_e(v)$ is activated and holds some wishes, it will not hold any further wishes later, unless it is cancelled by a change of $\emptyset' \upharpoonright_v$ (in which case the wishes it holds are removed).

We may assume that $N_e(v)$ is activated at (3.) of stage s. Wishes held at stage s by some $N_i(w)$ where $i < e$ will not be taken over by $N_e(v)$ because $w < v$. Now consider wishes held by an $N_i(w)$ where $i > e$. By inductive hypothesis, the total of such wishes is at most $\sum_{i>e} 3^{-i} = 3^{-e}/2$ at the beginning of stage s. The activation of $N_e(v)$ adds at most another $3^{-e}/2$ to the sum in (11).

To show that $\Gamma^D\langle \emptyset'_s \rangle < \infty$, note that any contribution to this quantity due to n's entering $\emptyset'$ at stage s is because a wish $\langle n, \delta \rangle^u$ is eventually held by some $N_e(v)$. The total is at most $\sum_e 3^{-e}$. □

The study of non-monotonic cost functions is left for the future. For instance, we conjecture that there are cost functions $\mathbf{c}, \mathbf{d}$ with the limit condition that for any Δ^0_2 sets A, B,

$$A \models \mathbf{c} \text{ and } B \models \mathbf{d} \Rightarrow A, B \text{ formaminimalpair.}$$

It is not hard to build cost functions $\mathbf{c}, \mathbf{d}$ such that only computable sets obey both of them. This provides some evidence for the conjecture.

Acknowledgements Research partially supported by the Marsden Fund of New Zealand, grant no. 08-UOA-187, and by the Hausdorff Institute of Mathematics, Bonn.

References

1. L. Bienvenu, N. Greenberg, A. Kučera, A. Nies, D. Turetsky, Coherent randomness tests and computing the K-trivial sets. J. Eur. Math. Soc. **18**, 773–812 (2016)
2. L. Bienvenu, A. Day, N. Greenberg, A. Kučera, J. Miller, A. Nies, D. Turetsky, Computing *K*-trivial sets by incomplete random sets. Bull. Symb. Log. **20**, 80–90 (2014)
3. L. Bienvenu, R. Downey, N. Greenberg, W. Merkle, A. Nies, Kolmogorov complexity and Solovay functions. J. Comput. Syst. Sci. **81**(8), 1575–1591 (2015)
4. C. Calude, A. Grozea, The Kraft-Chaitin theorem revisited. J. Univ. Comput. Sci. **2**, 306–310 (1996)

5. A.R. Day, J.S. Miller, Density, forcing and the covering problem. Math. Res. Lett. **22**(3), (2015). http://arxiv.org/abs/1304.2789
6. D. Diamondstone, N. Greenberg, D. Turetsky, Inherent enumerability of strong jump-traceability. http://arxiv.org/abs/1110.1435 (2012)
7. R. Downey, D. Hirschfeldt, *Algorithmic Randomness and Complexity* (Springer, Berlin, 2010)
8. R. Downey, D. Hirschfeldt, A. Nies, F. Stephan, Trivial reals, in *Proceedings of the 7th and 8th Asian Logic Conferences* (Singapore University Press, Singapore, 2003), pp. 103–131
9. S. Figueira, A. Nies, F. Stephan, Lowness properties and approximations of the jump. Ann. Pure Appl. Log. **152**, 51–66 (2008)
10. N. Greenberg, A. Nies, Benign cost functions and lowness properties. J. Symb. Log. **76**, 289–312 (2011)
11. N. Greenberg, D. Turetsky, Strong jump-traceability and Demuth randomness. Proc. Lond. Math. Soc. **108**, 738–779 (2014)
12. N. Greenberg, J. Miller, A. Nies, Computing from projections of random sets (submitted)
13. N. Greenberg, D. Hirschfeldt, A. Nies, Characterizing the strongly jump-traceable sets via randomness. Adv. Math. **231**(3–4), 2252–2293 (2012)
14. D. Hirschfeldt, A. Nies, F. Stephan, Using random sets as oracles. J. Lond. Math. Soc. (2) **75**(3), 610–622 (2007)
15. B. Kjos-Hanssen, W. Merkle, F. Stephan, Kolmogorov complexity and the Recursion Theorem, in *STACS 2006*. Lecture Notes in Computer Science, vol. 3884 (Springer, Berlin, 2006), pp. 149–161
16. A. Kučera, An alternative, priority-free, solution to Post's problem, in *Mathematical Foundations of Computer Science, 1986 (Bratislava, 1986)*. Lecture Notes in Computer Science, vol. 233 (Springer, Berlin, 1986), pp. 493–500
17. A. Kučera, A. Nies, Demuth randomness and computational complexity. Ann. Pure Appl. Log. **162**, 504–513 (2011)
18. A. Kučera, T. Slaman, Low upper bounds of ideals. J. Symb. Log. **74**, 517–534 (2009)
19. A. Kučera, S. Terwijn, Lowness for the class of random sets. J. Symb. Log. **64**, 1396–1402 (1999)
20. A. Melnikov, A. Nies, K-triviality in computable metric spaces. Proc. Am. Math. Soc. **141**(8), 2885–2899 (2013)
21. A. Nies, Reals which compute little, in *Logic Colloquium '02*. Lecture Notes in Logic (Springer, Heidelberg, 2002), pp. 260–274
22. A. Nies, Lowness properties and randomness. Adv. Math. **197**, 274–305 (2005)
23. A. Nies, *Computability and Randomness*. Oxford Logic Guides, vol. 51 (Oxford University Press, Oxford, 2009)
24. A. Nies, Interactions of computability and randomness, in *Proceedings of the International Congress of Mathematicians* (World Scientific, Singapore, 2010), pp. 30–57
25. R.I. Soare, *Recursively Enumerable Sets and Degrees*. Perspectives in Mathematical Logic, Omega Series (Springer, Heidelberg, 1987)
26. R. Solovay, *Handwritten Manuscript Related to Chaitin's Work* (IBM Thomas J. Watson Research Center, Yorktown Heights, 1975), 215 pp.

Part V
The Mathematics of Emergence and Morphogenesis

Turing's Theory of Morphogenesis: Where We Started, Where We Are and Where We Want to Go

Thomas E. Woolley, Ruth E. Baker, and Philip K. Maini

Abstract Over 60 years have passed since Alan Turing first postulated a mechanism for biological pattern formation. Although Turing did not have the chance to extend his theories before his unfortunate death two years later, his work has not gone unnoticed. Indeed, many researchers have since taken up the gauntlet and extended his revolutionary and counter-intuitive ideas. Here, we reproduce the basics of his theory as well as review some of the recent generalisations and applications that have led our mathematical models to be closer representations of the biology than ever before. Finally, we take a look to the future and discuss open questions that not only show that there is still much life in the theory, but also that the best may be yet to come.

1 Introduction

The initiation and maintenance of biological heterogeneity, known as morphogenesis, is an incredibly broad and complex issue. In particular, the mechanisms by which biological systems maintain robustness, despite being subject to numerous sources of noise, are shrouded in mystery. Although molecular genetic studies have led to many advances in determining the active species involved in patterning, simply identifying genes alone does not help our understanding of the mechanisms by which structures form. This is where the strengths of mathematical modelling lie. Not only are models able to complement experimental results by testing hypothetical relationships, they are also able to predict mechanisms by which populations interact, thus suggesting further experiments [1].

The patterns we are considering are thought to arise as the consequence of an observable population, e.g. skin cells, responding to diffusing signalling populations, known as morphogens, e.g. proteins. Specifically, the morphogens we consider are simply chemical reactants that do not sense their surroundings and freely diffuse. Through morphogen diffusion and interactions, non-uniform

T.E. Woolley (✉) • R.E. Baker • P.K. Maini
Wolfson Centre for Mathematical Biology, Mathematical Institute, University of Oxford, Woodstock Road, Oxford OX2 6GG, UK
e-mail: woolley@maths.ox.ac.uk

S.B. Cooper, M.I. Soskova (eds.), *The Incomputable*, Theory and Applications of Computability, DOI 10.1007/978-3-319-43669-2_13

patterns in concentration can emerge. The observable population is then thought to undergo concentration-dependent differentiation based on this heterogeneous morphogen distribution, thereby producing a corresponding heterogeneous pattern in the observable population [2, 3].

Many mathematical frameworks have been postulated to explain how such patterns arise. Here, we focus on one such paradigm mechanism: Alan Turing's diffusion-driven instability [4]. Turing conjectured that diffusion, normally known as a homogenising process, could destabilise a spatially homogeneous stable steady state of morphogen concentration. At its simplest, the instability can be characterised by interactions between two diffusing morphogen populations. There are two possible types of kinetics that can lead to instability, the better studied being the type where one of the species acts as an activator and the other behaves as an inhibitor [5, 6]. These names are derived from the fact that the activator promotes its own production in a positive feedback loop, which, in turn, is controlled by an inhibitor in a negative feedback loop. If the reaction domain is small enough such that the populations are well-mixed everywhere diffusion dominates the system, i.e. the product of the diffusion rate and reaction time scale is much greater than the domain size squared, the reactions will simply tend to a homogeneous stable steady state of concentration. However, as the domain size increases, diffusion can destabilise the homogeneous steady state. Explicitly, if the inhibitor diffuses faster than the activator, local growth in the activator is able to occur whilst the inhibitor prevents activator spreading [7]; thus, once the domain is large enough, spatial heterogeneity will arise.

Although we will be specifically thinking about Turing's theory in terms of biological pattern formation, the mathematical formalism is quite general and can be used to discuss any situation where the morphogen populations can be considered to be randomly moving reactive agents. Thus, the ideas of diffusion-driven instability are not restricted to biology. Indeed, the idea has been applied to such diverse areas as semiconductor physics [8], hydrodynamics [9] and even astrophysics [10].

2 Where We Started

We will be primarily concerned with multiple biochemical populations, U_i, where $i = 1, \dots, n$, which are collectively denoted by the vector $\boldsymbol{U} = (U_1, \dots, U_n)$. Further, the populations are identified with chemical concentrations $\boldsymbol{\phi} = (\phi_1, \dots, \phi_n)$. These populations are able to diffuse in a spatial domain $\mathcal{V}$ with boundary surface $\partial\mathcal{V}$. As the populations diffuse around the domain, individual particles will often collide with each other, allowing reactions to occur. The reactions that occur are either motivated through biological observations or are of mathematical interest, proposed to reflect general aspects of the underlying biology.

The system could be completely described deterministically by Newton's laws of motion, treating individual morphogen particles as point masses that can collide and bounce off one another. In this framework, reactions are defined to occur when particles collide with sufficient force. However, due to our ignorance of the initial positions and velocities of the active and solvent particles and our inability to cope computationally with the large number of particles involved (which can easily be of the order of 10^7 particles and higher [11]), we instead choose to assume that the discrete populations, U_i, are large enough to be approximated by the continuous chemical concentrations, ϕ_i, which are described using deterministic differential equations. Immediately, we see that this assumption has produced an error as the populations can only physically take integer values, whereas, once we take the continuum limit, the concentrations can take any continuous value. However, it has been shown that stochastic influences that arise due to the discrete nature of the particles scale as the reciprocal of the square root of the population size [12]. Thus, if in a specific biological application the chemical population of interest can be justified to be large, then a deterministic description is, in general, valid [13].

Since we are dealing with biochemical species, the populations will not be able to sense their surroundings and, thus, in the absence of some external force producing directionality, e.g. an electric field, their movement will be a simple random walk down concentration gradients, deterministically modelled by Fick's Law of Diffusion [14]. This law postulates that the chemicals move from regions of high concentration to regions of low concentration, with a magnitude that is proportional to the size of the concentration gradient. Although the framework is described in the context of molecular particles, it is in fact more general and can be applied to any system where motility is considered to be governed by an unbiased random walk [15–17].

The evolution of the concentrations ϕ_i at position $\boldsymbol{x} \in \mathcal{V}$ and time $t \geq 0$ is defined by the coupled system of partial differential equations (PDEs)

$$\frac{\partial \boldsymbol{\phi}(\boldsymbol{x}, t)}{\partial t} = \boldsymbol{D}\nabla^2\boldsymbol{\phi}(\boldsymbol{x}, t) + \boldsymbol{F}(\boldsymbol{\phi}(\boldsymbol{x}, t)), \tag{1}$$

$$\boldsymbol{\phi}(\boldsymbol{x}, 0) = \boldsymbol{\phi}_0(\boldsymbol{x}) \quad \forall \boldsymbol{x} \in \mathcal{V},$$

$$\boldsymbol{G}(\boldsymbol{\phi}(\boldsymbol{x}, t)) = 0 \quad \forall \boldsymbol{x} \in \partial\mathcal{V} \text{ and } t > 0,$$

where ∇^2 denotes the Laplacian operator and represents diffusion. The term

$$\boldsymbol{F} = (F_1(\boldsymbol{\phi}), \ldots, F_n(\boldsymbol{\phi})) \tag{2}$$

defines the (usually non-linear and highly coupled) interactions between the populations whilst $\boldsymbol{D} = [d_{ii}]$ is a diagonal matrix of diffusivities that is generally constant in space and time. The diffusivity constants control how quickly the chemicals spread throughout the domain. Finally, the functional form of $\boldsymbol{G}$ specifies how the chemicals behave on the boundary of the spatial domain that we are considering and $\boldsymbol{\phi}_0(\boldsymbol{x})$ is the initial concentrations of the chemicals [18].

Numerous different types of boundary conditions are possible: for example, homogeneous Neumann, or zero-flux boundary conditions, i.e.

$$\boldsymbol{G} = \frac{\partial \boldsymbol{\phi}}{\partial \boldsymbol{n}} = \boldsymbol{0}, \tag{3}$$

where $\boldsymbol{n}$ is the outward pointing normal of $\partial\mathcal{V}$. This simply states that no material may leave the domain; effectively, the domain is insulated. An alternative type of boundary condition is known as Dirichlet, or fixed concentration boundary condition, which, as the name suggests, simply fixes the concentration of the chemical on the boundary,

$$\boldsymbol{G} = \boldsymbol{\phi} - \boldsymbol{C} = \boldsymbol{0}, \tag{4}$$

where $\boldsymbol{C}$ is normally a constant. Other boundary conditions, e.g. reactive boundary conditions [19] or periodic boundary conditions [20, 21], also can be used although they are not considered here.

Systems such as Eq. (1) are known as 'reaction-diffusion' equations. They are able to produce a large variety of stationary and temporally varying patterns, such as stationary gradients, travelling waves and moving fronts [22], even without the Turing instability. Thus, unless biologically motivated to add further components to capture relevant dynamics, we concentrate on capturing the maximum amount of complexity through the simplest forms of reaction-diffusion equations.

2.1 Turing Instability

For clarity the current formulae are quoted for reaction-diffusion systems of two concentrations (ϕ, ψ), with Neumann (zero flux) boundary conditions in a one-dimensional domain, $[0, L]$. Extensions to higher dimensions and various other boundary conditions are possible [18, 23]. In full generality the equations are

$$\phi_t = D_\phi \phi_{xx} + f(\phi, \psi), \tag{5}$$

$$\psi_t = D_\psi \psi_{xx} + g(\phi, \psi), \tag{6}$$

where the subscripts x and t denote partial derivatives, and suitable initial conditions are defined to close the system. Usually the initial conditions are taken simply to be random perturbations around a spatially uniform steady state, as it is the final pattern that is evolved that is important, not the initialisation of the system.

The first requirement of a diffusion-driven instability is that there exists a spatially homogeneous, linearly stable steady state, i.e. there exists (ϕ_0, ψ_0) such that $f(\phi_0, \psi_0) = g(\phi_0, \psi_0) = 0$ and all eigenvalues of the Jacobian (evaluated at the homogeneous steady state),

$$J(\phi_0, \psi_0) = \begin{pmatrix} \frac{\partial f}{\partial \phi}(\phi_0, \psi_0) & \frac{\partial f}{\partial \psi}(\phi_0, \psi_0) \\ \frac{\partial g}{\partial \phi}(\phi_0, \psi_0) & \frac{\partial g}{\partial \psi}(\phi_0, \psi_0) \end{pmatrix}, \tag{7}$$

have a negative real part. The second requirement is that the steady state becomes linearly unstable in the presence of diffusion. Note that although we derive conditions that will allow a reaction-diffusion system to realise a Turing pattern, as we will see in the biological applications section, the solution domain also has to be bigger than a critical size in order for the patterns to exist.

To derive necessary conditions for pattern formation to occur, the steady state is perturbed using functions that also satisfy the boundary conditions. Since we are using zero flux boundary conditions, we use a Fourier cosine expansion of the form $(\phi(x,t), \psi(x,t)) = (\phi_0 + \hat{\phi}(x,t), \psi_0 + \hat{\psi}(x,t))$, where

$$\begin{pmatrix} \hat{\phi} \\ \hat{\psi} \end{pmatrix} = \sum_{m=0}^{\infty} \begin{pmatrix} a_m \\ b_m \end{pmatrix} e^{\lambda_m t} \cos(k_m x), \tag{8}$$

and $k_m = m\pi/L$, $m = 0, 1, 2, \ldots$. Explicitly, the cosine function allows us to satisfy the zero flux boundary conditions since at the boundaries of the solution domain its spatial derivative will take the form of a sine function, which evaluates to zero when $x = 0$ or L. Moreover, because the cosine functions, $\{\cos(k_m x)\}_{m=0}^{\infty}$, form a complete orthogonal set, any solution of the linearised equation system can be decomposed into a series solution of superpositions.

The growth rate λ_m informs us about the stability of the homogeneous steady state with respect to the wave mode, k_m. If the real part of λ_m is negative for all m, then any perturbations will tend to decay exponentially quickly. However, in the case that the real part of λ_m is positive for any non-zero value of m, our expansion solution suggests that the amplitude of these modes will grow exponentially quickly and so the homogeneous steady state is now linearly unstable. Moreover, in the case where there are multiple $\cos(k_m x)$ terms growing, small alterations in the initial conditions (which are bound to occur, since we are assuming that initial conditions are random perturbations around the homogeneous steady state) can lead to completely different final outcomes. Critically, the integer values of m for which λ_m has a positive real part then indicate how many pattern peaks we will see in the final solution. For example, if λ_5 is the only growth rate with positive real part, then we expect that the system will tend to a solution in which a $\cos(5\pi x/L)$ function is dominant, so the final pattern will have the corresponding number of peaks. However, if a range of growth rates is positive, then multiple cosine modes will fight for dominance and we will be unable to predict with certainty which mode will dominate in the final solution, because of the initial random perturbations and nonlinear interactions. This is the robustness problem. When dealing with animal pigmentation patterns, this dependence on initial conditions can be a useful property; for example zebra stripes are as individual as fingerprints [24]. However, such variability is problematic when we apply Turing's theory to more robust forms of biological development. Fortunately, as we will see later, this robustness problem is surmountable.

Substituting Eq. (8) into the linearised form of Eqs. (5) and (6), we obtain

$$\mathbf{0} = \begin{pmatrix} \lambda_m + D_\phi k_m^2 - f_\phi & -f_\psi \\ -g_\phi & \lambda_m + D_\psi k_m^2 - g_\psi \end{pmatrix} \begin{pmatrix} a_m \\ b_m \end{pmatrix}. \tag{9}$$

This matrix equation has a non-trivial solution ($(a_m, b_m) \neq (0,0)$) if and only if the determinant is zero:

$$\lambda_m^2+\lambda_m((D_\phi+D_\psi)k_m^2-f_\phi-g_\psi)+D_\phi D_\psi k_m^4-k_m^2(D_\phi g_\psi+D_\psi f_\phi)+f_\phi g_\psi-f_\psi g_\phi=0. \tag{10}$$

Letting $h(k^2)=D_\phi D_\psi k^4-k^2(D_\phi g_\psi+D_\psi f_\phi)+f_\phi g_\psi-f_\psi g_\phi$, the linear stability of the homogeneous steady state is now governed by the signs of the real parts of

$$\lambda_{m\pm}=\frac{f_\phi+g_\psi-(D_\phi+D_\psi)k_m^2\pm\sqrt{(f_\phi+g_\psi-(D_\phi+D_\psi)k_m^2)^2-4h(k_m^2)}}{2}. \tag{11}$$

First, we consider the linear stability in the case where there is no diffusion, $D_\phi = D_\psi = k_m = 0$. For the homogeneous steady state to be linearly stable, the real parts of both eigenvalues need to be negative. Thus

$$f_\phi+g_\psi<0, \tag{12}$$

and

$$h(0)=f_\phi g_\psi-f_\psi g_\phi>0. \tag{13}$$

Diffusion is now included and we derive conditions to ensure that at least one of the eigenvalues has positive real part. Since $f_\phi+g_\psi<0$ by inequality (12), it follows that $f_\phi+g_\psi-(D_\phi+D_\psi)k_m^2<0$; thus the real part of λ_{m-} is always negative. The only way to obtain an instability is if the real part of λ_{m+} is positive. From (11), this occurs if $h(k_m^2)<0$. Explicitly,

$$D_\phi D_\psi k_m^4-k_m^2(D_\phi g_\psi+D_\psi f_\phi)+f_\phi g_\psi-f_\psi g_\phi<0, \tag{14}$$

$$\Rightarrow k_-^2<k_m^2<k_+^2, \tag{15}$$

where

$$2D_\phi D_\psi k_\pm^2=D_\phi g_\psi+D_\psi f_\phi\pm\sqrt{(D_\phi g_\psi+D_\psi f_\phi)^2-4D_\phi D_\psi(f_\phi g_\psi-f_\psi g_\phi)}. \tag{16}$$

For inequality (15) to be realised, k_+^2 needs to be real and positive, implying

$$D_\phi g_\psi+D_\psi f_\phi>0, \tag{17}$$

$$(D_\phi g_\psi+D_\psi f_\phi)^2-4D_\phi D_\psi(f_\phi g_\psi-f_\psi g_\phi)>0. \tag{18}$$

Since $f_\phi g_\psi - f_\psi g_\phi > 0$, from inequality (13), these two inequalities yield one condition,

$$D_\phi g_\psi+D_\psi f_\phi>2\sqrt{D_\phi D_\psi}\sqrt{(f_\phi g_\psi-f_\psi g_\phi)}>0. \tag{19}$$

Thus inequalities (12), (13), (15) and (19) form the conditions needed for a Turing instability in a reaction-diffusion system.

$$f_\phi + g_\psi < 0,$$

$$f_\phi g_\psi - f_\psi g_\phi > 0,$$

$$D_\phi g_\psi + D_\psi f_\phi > 2\sqrt{D_\phi D_\psi}\sqrt{(f_\phi g_\psi - f_\psi g_\phi)} > 0,$$

$$k_-^2 < \left(\frac{m\pi}{L}\right)^2 < k_+^2.$$

Turing's computer science background ideally suited this problem, as not only did he possess the mathematical skills to create the theoretical framework, but he was perfectly situated to numerically simulate the equations and, hence, visualise coarse-grained versions of the patterns (Fig. 1a, b). Due to the dramatic increase in computational speed and numerical algorithms, we are able to revisit the calculations (Fig. 1c, d) and see just how good Turing's first simulations were. Clearly, although his simulations were very coarse approximations to the equations,

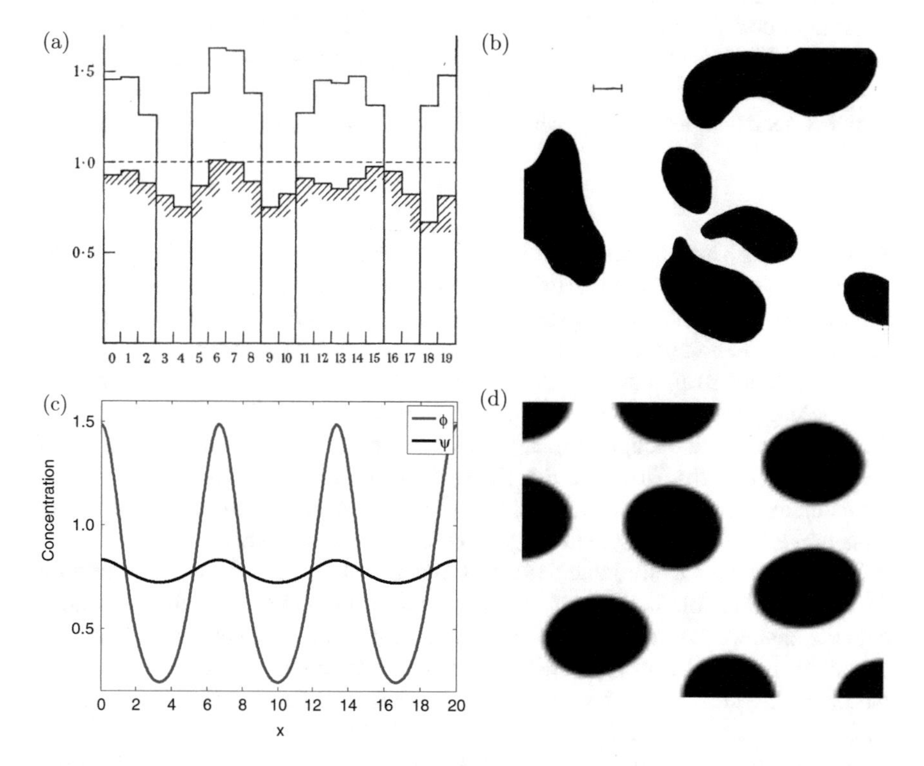

Fig. 1 Heterogeneous patterns visualised in (**a**) one and (**b**) two dimensions, originally created by Turing himself. Modern versions of the Turing pattern in (**c**) one and (**d**) two dimensions. Figures (**a**) and (**b**) is reproduced from [4] by permission of the Royal Society

the basic patterns are still visible and very close to those we are now able to generate, illustrating his impressive computational abilities.

3 Where We Are

Turing's research was ahead of its time and so, for a while, his ideas lay dormant. However, the fast pace of theoretical, numerical and biological development that occurred towards the end of the twentieth century meant that it was the perfect time for Turing's theory to enjoy a successful renaissance [25, 26]. Here, we review just a few of the convincing biological applications, as well as some of the theoretical extensions, illustrating the richness of the original theory.

3.1 Biological Applications

Perhaps the most colourful application is to pigmentation patterns. Importantly, this is not restricted to coat markings and animal skin. The Turing instability has also been suggested to be the mechanism behind the patterns on many seashells [27].

One prediction that immediately springs from the theory concerns tapered domains, for example a tail. By rearranging inequality (15), we obtain a bound on m,

$$\frac{Lk_-}{\pi} < m < \frac{Lk_+}{\pi}. \tag{20}$$

Since k_+ and k_- are constants, this means that as the domain size, L, decreases, the window of viable wave modes shrinks, eventually disappearing. This means that as a domain becomes smaller, we should see a simplification of the pattern, e.g. from peak patterns to homogeneity. This result can be extended to the second dimension, where spot and stripe patterns are available. Once again, as the domain shrinks, we would expect a transition from spots to stripes, and finally to homogeneity, if the domain is small enough (Fig. 2a). This is excellently exemplified on the tail of the cheetah (Fig. 2b). However, the biological world does not always have respect for mathematics, as illustrated in Fig. 2c, where we observe that the lemur's pattern transition goes from a simple homogeneous colour on the body to a more complex striped pattern on the tail. Potentially, this means that Turing's theory does hold for the lemur's skin. Alternatively, if Turing's theory is used to account for the lemur's patterns, then we have to postulate either that the parameter values for the body and the tail are different, causing the difference in pattern, or that the patterns arise from the highly nonlinear regime of the kinetics, where our linear theory breaks down and, hence, we can no longer use the above predictions.

Importantly, we are not restricted to stationary domains, and these predictions were extended by Kondo and Asai [28] to pattern transitions on growing angelfish. As angelfish age, their bodies grow in size and more stripes are included in the pattern. Critically, the evolving patterns maintain a near-constant stripe spacing, which is one of the crucial features of a Turing pattern.

Fig. 2 (**a**) Turing pattern on a tapered domain. Pattern transitions on (**b**) a cheetah and (**c**) a lemur

Turing patterns have also been postulated to underlie formation of the precursor patterns of many developmental systems, for example in mice, where it has been suggested that molar placement can be described by a diffusion-driven instability. Critically, not only can the normal molar placement be predicted by the model, but, by altering the model parameters to mimic biological perturbations, fused molar prepatterns are predicted, thereby reproducing experimental results [29]. Sheth et al. [30] further showed that Turing systems could underlie mouse digit development. In particular, experimental perturbations produced paws that did not change in size, but the number of digits did increase, leading to a reduction in digit spacing. Like the stripes on the angelfish, this new digit spacing in the treated mice was constant, consistent with a Turing-like mechanism. Critically, the reduction in wavelength could be linked to changes of parameters in a general Turing model.

3.2 *Theoretical Extensions*

As already discussed, growth is an essential and readily observed process in development that has been identified as an important factor in the production of spatial heterogeneity since it can fundamentally change the observed dynamics of patterning mechanisms. Although growth had previously been included in an ad hoc manner [31], Crampin et al. [32] were the first to rigorously incorporate the effects of domain growth into the reaction-diffusion framework. This led to the discovery that uniform exponential domain growth can robustly generate persistent pattern doubling, even in the face of random initial conditions (Fig. 3a). This insensitivity to initial conditions is particularly significant in the context of biological development,

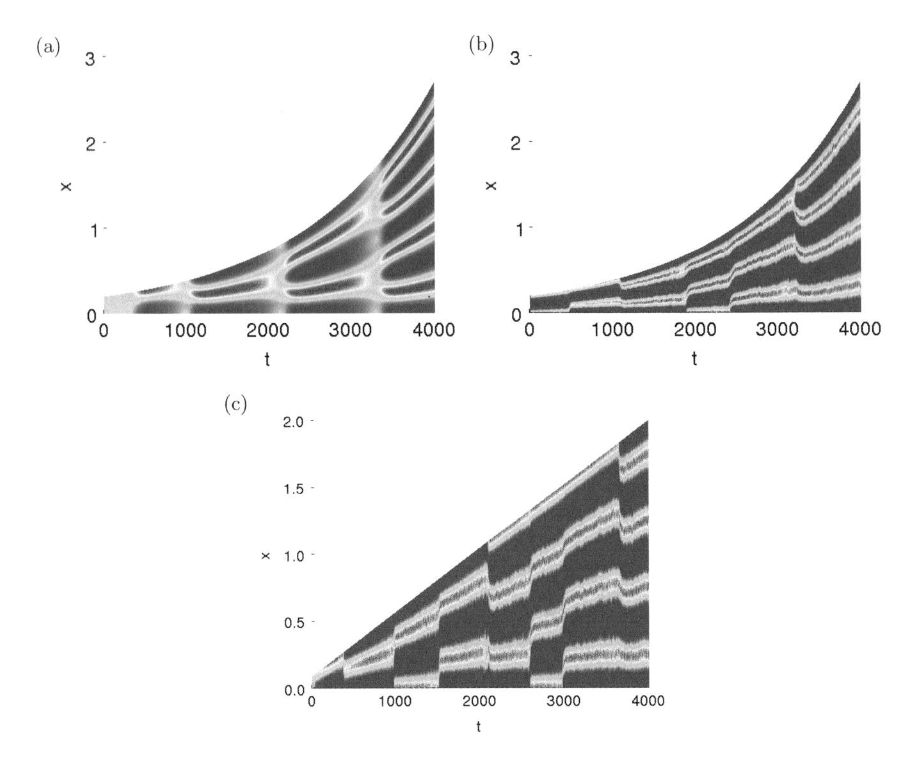

Fig. 3 (**a**) Deterministic Turing kinetics on an exponentially growing domain. (**b**) Stochastic Turing kinetics on an exponentially growing domain. (**c**) Stochastic Turing kinetics on an linearly, apically growing domain. Figure (**c**) is reproduced with permission from [35]. Copyright 2011 American Physical Society

as not only does heterogeneity need to form, but also, in many cases, it is imperative that the final pattern be reliably reproducible.

Continuing this idea of robustness, we note that biological systems are frequently subject to noisy environments, inputs and signalling, not to mention that important proteins may only appear in very small quantities. Fundamentally, we based the derived partial differential equation (PDE) framework on the assumption that each species was present in high concentration, which allowed us to use a continuous approximation of the chemical concentrations. In order to investigate the Turing mechanism's sensitivity to noise, stochastic formulations have been created and even extended to encompass descriptions of domain growth [33–35]. Although it is clear that the Turing instability is able to exist (Fig. 3b), even in the face of intrinsic randomness, we see that uniform domain growth is no longer able to support the robust peak splitting that Crampin et al. [32] demonstrated in the deterministic system. However, if growth is localized to one of the boundaries (known as apical growth), then we see that pattern peaks appear in the domain one at a time, creating a consistent consecutive increase in the pattern wavenumber (Fig. 3c). If apical domain growth and wavenumber were connected in some form of feedback loop, then, once the desired wavenumber was reached, growth would stop, leaving a stable pattern of exactly the desired wave mode. Thus, robust pattern generation can be recovered. It should be noted that noise does not need to be generated explicitly through stochastic reactions. Turing systems can also be chaotic, thus producing a deterministic form of noise [36].

A further relaxation of the fundamental assumptions behind the PDE formulation concerns the reaction rates as being defined by the Law of Mass Action. As originally stated, the law assumes that reactant products are created at the same moment that the reaction occurs. However, this may not always be the case. Reaction delays are particularly important when dealing with the production of important proteins as a cascade of time-consuming biological processes must occur in order for a single protein to be produced. Firstly, a linear polymeric ribonucleic acid (RNA) molecule is produced in a cell nucleus. This RNA molecule is an exact copy of the relevant gene sequence and is modified into a form called messenger RNA (mRNA). The mRNA is then transported into the nuclear membrane, where it is used as a blueprint for protein synthesis. In particular, the process of mRNA translation involves the polymerization of thousands to millions of amino acids. Given the complexity of this mechanism, it should not be surprising that a delay occurs between the initiation of protein translation and the point at which mature proteins are observed. The exact delay depends both on the length of the sequence being read and the sequence being created. However, typically the delay ranges from tens of minutes to as long as several hours [37]. Work has been done on including these gene-expression delays into both the deterministic and stochastic PDE formulations of the Turing instability, leading to observations of wildly different outcomes when compared to the non-delayed equations [38–40]. The potentially most worrying case is that of kinetic delays causing a catastrophic collapse of the pattern formation mechanism. Furthermore, such pathological dynamics occur consistently, regardless of domain growth profiles [41].

4 Computational Extensions

Of course, our simulations on one-dimensional lines and two-dimensional flat surfaces should always be questioned as to their accuracy in reproducing the effects of a real surface, which may have high curvatures. For example, pigmentation patterns are produced on skin surfaces that are stretched over skeletons that have highly non-trivial geometries. Turing mechanisms have been studied on simple regular surfaces, e.g. spheres, cones, etc. [42]. However, recent developments in numerical algorithms have allowed us to push our studies even further, allowing us to greatly generalise the geometries on which we numerically simulate the reaction-diffusion systems.

PDEs on surfaces are normally solved using finite element discretisations on a triangulation of the surface [43] or some other discretisation based on a suitable parameterisation of the surface [42]. An alternative approach to parameterizing the surface is to embed it in a higher-dimensional space [44]. The PDEs are then

Fig. 4 Examples of Turing patterns on general surfaces, computed using the closest point method [44]

solved in the embedding space, rather than just on the lower dimensional surface. Embedding methods have the attractive feature of being able to work using standard Cartesian coordinates and Cartesian grid methods [44]. Thus, it is within this class that the Closest Point Method was developed and analyzed [45]. Although we will not go into full details concerning the technique here, we do present the simple central idea of the embedding, which, as the name suggests, is the construction of the closest point function.

Definition 1 For a given surface $\mathcal{S}$, the closest point function $cp : R^d \rightarrow R^d$ takes a point $\boldsymbol{x} \in R^d$ and returns a unique point $cp(\boldsymbol{x}) \in \mathcal{S} \subset R^d$ which is closest in Euclidean distance to $\boldsymbol{x}$. Namely,

$$cp(\boldsymbol{x}) = \min_{q \in \mathcal{S}} ||\boldsymbol{x} - \boldsymbol{q}||_2. \tag{21}$$

If more than one $\boldsymbol{q}$ should fit this property, a single one is chosen arbitrarily.

From this definition, equations governing quantities on the surface can be extended to the embedding space. The equations are then solved more easily in the regular grid of the embedding space. This solution in the embedded space evaluated on the original surface will then agree with the solution that would have been generated if we had simply tried to solve the equation on just the surface. Examples of the impressive generality of this technique are given in Fig. 4.

5 Where We Want to Go

Now that we better understand the formation of Turing patterns on general two-dimensional surfaces, it is natural to want to extend to three and higher dimensions. Indeed, theoretical, experimental and computational work does exist heading in this direction [46–49]. However, by going to higher dimensions we start having problems of pattern degeneracy. In one dimension, we are guaranteed only discrete peaks. The only degeneracy is in the choice of polarity, i.e. for any pattern mode that is based on a $\cos(kx)$ form, $-\cos(kx)$ is also a possible solution with opposite polarity. In two dimensions, not only can we obtain stripes, spots and labrythine patterns, but the orientation of these patterns is also variable, because any wave vector $\boldsymbol{k}$, which is associated with a critical wave number $|\boldsymbol{k}| = \sqrt{k_{xm}^2 + k_{yn}^2} = k_c$, such that $Re(\lambda_m(k_c)) > 0$, defines a growing mode. This means that in the spatially bounded two-dimensional case a finite number of Fourier modes can have wave vectors that lie on the critical circle [50]. Thus, spots can be arranged in rectangular, hexagonal, or rhombic patterns amongst other, more varied templates. This degeneracy problem becomes even more complex in three dimensions, where lamellae, prisms, and various other cubic structures all exist, making prediction even more difficult [51]. Weakly nonlinear theory and equivariant bifurcation theory [7, 51–53] can be used to derive amplitude equations near a critical bifurcation point that separate the homogeneous and patterned stationary stable states.

However, analysis will only get us so far and thus we are depending more and more on numerical simulation in order to explore patterning parameter space. This illustrates the great need for three-dimensional PDE solvers that are not only able to efficiently approximate the solutions of stiff PDEs with fine spatial resolution, but also are flexible enough to incorporate various boundary conditions, geometries and spatial heterogeneities. Further, analogously to the above work, changing from continuous descriptions of the populations to individual-based stochastic simulations in three dimensions poses another computationally intensive task. There has been work done on speeding up stochastic simulation algorithms [54–56]; however, work has only just begun to consider the potential powerful use of parallel computing, which is a much underexplored territory [57].

Equally, the computational visualisation of Turing patterns in higher dimensions needs consideration, as the basic planiforms, discussed above, are much more complicated. Moreover, the ability to compare such visualisations with actual data is still in its infancy and there are, as yet, few metrics by which a simulation can be compared to an experiment. Currently, we depend on simply matching the general pattern and the ability of the kinetics to reproduce experimental perturbations. However, to rigorously compare such patterns we must be able to develop image segmentation software that is capable of extracting dominant features of numerical and experimental results and comparing them using statistical methods.

Importantly, we do not need to extend to a third spatial dimension to find new problems. There are many still unanswered questions in lower spatial dimensions, but with more than two chemical species [58]. To suggest that many complex biological phenomena occur because of the interactions of two chemical species is misguided, at best. In reality, a single developmental pathway can depend on many hundreds of gene products interacting through a complex network of non-linear kinetics. Moreover, living systems have numerous fail-safe mechanisms, such as multiple redundant pathways, that only activate when there is a problem with the main network. This means that even if we are able to produce a complete gene product interaction map for a given biological phenomenon, the phenomenon may still occur if the network is disrupted, making conclusions difficult.

Once again, analysis of such large systems can only lead us so far, before numerical simulations are required [59]. However, we are starting to see new branches of mathematical biology that seek to deal with these large networks, either through mass computer parallelisation of data processing [60, 61] or through rigorously and consistently identifying key features and time scales that allow the full system to be greatly reduced to a much smaller number of important species [62–66]. In either approach, efficient numerical algorithms are of paramount importance, and we hope to see more development in this direction in the future.

A rapidly growing research area is that of synthetic biology [67, 68]. In the future, no longer will we use mathematics to mimic a natural system's ability to produce patterns; instead, we will design tissues and cells that are able to reproduce mathematical predictions. Further, by utilising the large knowledge base surrounding the numerous extensions of Turing's theory, we may be able to customise such designs in order to produce patterns with specific properties.

6 Discussion

As can be clearly seen, Turing's theory for the chemical basis of morphogenesis has been applied to a wide range of patterning phenomena in developmental biology. The incredible richness in behaviour of the diffusion-driven instability has also allowed the theory to be extended dramatically from its humble beginnings of two chemicals deterministically reacting in a simple domain. Indeed, it is testament to Turing's genius that, not only did he discover such a counter-intuitive mechanism, it is still generating new ideas, even after 60 years of research. Importantly, our progress has significantly benefited from the recent rapid developments in computational software and hardware. Indeed, with the continued development of the biological techniques and computational visualisation abilities discussed in the last section, we could be at the dawn of a new age of Turing's theory, enabling us to further strengthen the links between experimental and theoretical researchers.

Acknowledgements TEW would like to thank St John's College Oxford for its financial support. This publication is based on work supported by Award No. KUK-C1-013-04, made by King Abdullah University of Science and Technology (KAUST). The cheetah and lemur photos were used under the Attribution-ShareAlike 2.0 license and were downloaded from http://www.flickr.com/photos/53936799@N05/ and http://www.flickr.com/photos/ekilby/.

References

1. C.J. Tomlin, J.D. Axelrod, Biology by numbers: mathematical modelling in developmental biology. Nat. Rev. Genet. **8**(5), 331–340 (2007)
2. L. Wolpert, Positional information and the spatial pattern of cellular differentiation. J. Theor. Biol. **25**(1), 1–47 (1969)
3. L. Wolpert, Positional information revisited. Development **107**(Suppl.), 3–12 (1989)
4. A.M. Turing, The chemical basis of morphogenesis. Philos. Trans. R. Soc. Lond. B **237**, 37–72 (1952)
5. A. Gierer, H. Meinhardt, A theory of biological pattern formation. Biol. Cybern. **12**(1), 30–39 (1972)
6. R. Kapral, K. Showalter, *Chemical Waves and Patterns* (Kluwer, Dordrecht, 1995)
7. P. Borckmans, G. Dewell, A. De wit, D. Walgraef, Turing bifurcations and pattern selection, in *Chemical Waves and Patterns*, Chap. 10 (Kluwer, Dordrecht, 1995), pp. 325–363
8. Y.I. Balkarei, A.V. Grigor'yants, Y.A. Rzhanov, M.I. Elinson, Regenerative oscillations, spatial-temporal single pulses and static inhomogeneous structures in optically bistable semiconductors. Opt. Commun. **66**(2–3), 161–166 (1988)
9. D.B. White, The planforms and onset of convection with a temperature-dependent viscosity. J. Fluid Mech. **191**(1), 247–286 (1988)
10. T. Nozakura, S. Ikeuchi, Formation of dissipative structures in galaxies. Astrophys. J. **279**, 40–52 (1984)
11. B. Futcher, G.I. Latter, P. Monardo, C.S. McLaughlin, J.I. Garrels, A sampling of the yeast proteome. Mol. Cell. Biol. **19**(11), 7357 (1999)
12. N.G. van Kampen, *Stochastic Processes in Physics and Chemistry*, 3rd edn. (North Holland, Amsterdam, 2007)
13. S. Cornell, M. Droz, B. Chopard, Role of fluctuations for inhomogeneous reaction-diffusion phenomena. Phys. Rev. A **44**, 4826–4832 (1991)
14. A. Fick, On liquid diffusion. Philos. Mag. J. Sci. **10**(1), 31–39 (1855)
15. J.D. Murray, E.A. Stanley, D.L. Brown, On the spatial spread of rabies among foxes. Proc. R. Soc. Lond. B. Biol. **229**(1255), 111–150 (1986)
16. A. Okubo, P.K. Maini, M.H. Williamson, J.D. Murray, On the spatial spread of the grey squirrel in Britain. Proc. R. Soc. Lond. B. Biol. **238**(1291), 113 (1989)
17. T.E. Woolley, R.E. Baker, E.A. Gaffney, P.K. Maini, How long can we survive? in *Mathematical Modelling of Zombies*, Chap. 6 (University of Ottawa Press, Ottawa, 2014)
18. J.D. Murray, *Mathematical Biology I: An Introduction*, vol. 1, 3rd edn. (Springer, Heidelberg, 2003)
19. R. Erban, S.J. Chapman, Reactive boundary conditions for stochastic simulations of reaction–diffusion processes. Phys. Biol. **4**, 16 (2007)
20. T.E. Woolley, R.E. Baker, P.K. Maini, J.L. Aragón, R.A. Barrio, Analysis of stationary droplets in a generic Turing reaction-diffusion system. Phys. Rev. E **82**(5), 051929 (2010)
21. T.E. Woolley, Spatiotemporal behaviour of stochastic and continuum models for biological signalling on stationary and growing domains. Ph.D. thesis, University of Oxford, 2011
22. R.A. Barrio, R.E. Baker, B. Vaughan Jr, K. Tribuzy, M.R. de Carvalho, R. Bassanezi, P.K. Maini, Modeling the skin pattern of fishes. Phys. Rev. E **79**(3), 31908 (2009)
23. R. Dillon, P.K. Maini, H.G. Othmer, Pattern formation in generalized Turing systems. J. Math. Biol. **32**(4), 345–393 (1994)
24. J.C.B. Petersen, An identification system for zebra (Equus burchelli, Gray). Afr. J. Ecol. **10**(1), 59–63 (1972)
25. P.K. Maini, T.E. Woolley, R.E. Baker, E.A. Gaffney, S.S. Lee, Turing's model for biological pattern formation and the robustness problem. Interface Focus **2**(4), 487–496 (2012)
26. T.E. Woolley, Mighty morphogenesis, in *50 Visions of Mathematics*, Chap. 48 (Oxford University Press, Oxford, 2014)

27. H. Meinhardt, P. Prusinkiewicz, D.R. Fowler, *The Algorithmic Beauty of Sea Shells* (Springer, Heidelberg, 2003)
28. S. Kondo, R. Asai, A reaction-diffusion wave on the skin of the marine angelfish Pomacanthus. Nature **376**, 765–768 (1995)
29. S.W. Cho, S. Kwak, T.E. Woolley, M.J. Lee, E.J. Kim, R.E. Baker, H.J. Kim, J.S. Shin, C. Tickle, P.K. Maini, H.S. Jung, Interactions between Shh, Sostdc1 and Wnt signaling and a new feedback loop for spatial patterning of the teeth. Development **138**, 1807–1816 (2011)
30. R. Sheth, L. Marcon, M.F. Bastida, M. Junco, L. Quintana, R. Dahn, M. Kmita, J. Sharpe, M.A. Ros, How genes regulate digit patterning by controlling the wavelength of a Turing-type mechanism. Science **338**(6113), 1476–1480 (2012)
31. P. Arcuri, J.D. Murray, Pattern sensitivity to boundary and initial conditions in reaction-diffusion models. J. Math. Biol. **24**(2), 141–165 (1986)
32. E.J. Crampin, E.A. Gaffney, P.K. Maini, Reaction and diffusion on growing domains: scenarios for robust pattern formation. Bull. Math. Biol. **61**(6), 1093–1120 (1999)
33. T.E. Woolley, R.E. Baker, E.A. Gaffney, P.K. Maini, Power spectra methods for a stochastic description of diffusion on deterministically growing domains. Phys. Rev. E **84**(2), 021915 (2011)
34. T.E. Woolley, R.E. Baker, E.A. Gaffney, P.K. Maini, Influence of stochastic domain growth on pattern nucleation for diffusive systems with internal noise. Phys. Rev. E **84**(4), 041905 (2011)
35. T.E. Woolley, R.E. Baker, E.A. Gaffney, P.K. Maini, Stochastic reaction and diffusion on growing domains: understanding the breakdown of robust pattern formation. Phys. Rev. E **84**(4), 046216 (2011)
36. J.L. Aragón, R.A. Barrio, T.E. Woolley, R.E. Baker, P.K. Maini, Nonlinear effects on Turing patterns: time oscillations and chaos. Phys. Rev. E **86**(2), 026201 (2012)
37. C.N. Tennyson, H.J. Klamut, R.G. Worton, The human dystrophin gene requires 16 hours to be transcribed and is cotranscriptionally spliced. Nat. Genet. **9**(2), 184–190 (1995)
38. T.E. Woolley, R.E. Baker, E.A. Gaffney, P.K. Maini, S. Seirin-Lee, Effects of intrinsic stochasticity on delayed reaction-diffusion patterning systems. Phys. Rev. E **85**(5), 051914 (2012)
39. E.A. Gaffney, N.A.M. Monk, Gene expression time delays and Turing pattern formation systems. Bull. Math. Biol. **68**(1), 99–130 (2006)
40. S.S. Lee, E.A. Gaffney, Aberrant behaviours of reaction diffusion self-organisation models on growing domains in the presence of gene expression time delays. Bull. Math. Biol. **72**, 2161–2179 (2010)
41. S.S. Lee, E.A. Gaffney, R.E. Baker, The dynamics of Turing patterns for morphogen-regulated growing domains with cellular response delays. Bull. Math. Biol. **73**(11), 2527–2551 (2011)
42. R.G. Plaza, F. Sanchez-Garduno, P. Padilla, R.A. Barrio, P.K. Maini, The effect of growth and curvature on pattern formation. J. Dyn. Differ. Equ. **16**(4), 1093–1121 (2004)
43. K.W. Morton, D.F. Mayers, *Numerical Solution of Partial Differential Equations: An Introduction* (Cambridge University Press, Cambridge, 2005)
44. C.B. Macdonald, S.J. Ruuth, The implicit closest point method for the numerical solution of partial differential equations on surfaces. SIAM J. Sci. Comput. **31**(6), 4330–4350 (2009)
45. S.J. Ruuth, B. Merriman, A simple embedding method for solving partial differential equations on surfaces. J. Comput. Phys. **227**(3), 1943–1961 (2008)
46. T.K. Callahan, E. Knobloch, Bifurcations on the fcc lattice. Phys. Rev. E **53**(4), 3559–3562 (1996)
47. T. Leppänen, M. Karttunen, K. Kaski, R.A. Barrio, L. Zhang, A new dimension to Turing patterns. Physica D **168**, 35–44 (2002)
48. E. Dulos, P. Davies, B. Rudovics, P. De Kepper, From quasi-2D to 3D Turing patterns in ramped systems. Physica D **98**(1), 53–66 (1996)
49. S. Muraki, E.B. Lum, K.-L. Ma, M. Ogata, X. Liu, A PC cluster system for simultaneous interactive volumetric modeling and visualization, in *Proceedings of the 2003 IEEE Symposium on Parallel and Large-Data Visualization and Graphics* (2003), p. 13

50. S.L. Judd, M. Silber, Simple and superlattice Turing patterns in reaction-diffusion systems: bifurcation, bistability, and parameter collapse. Physica D **136**(1–2), 45–65 (2000)
51. T.K. Callahan, E. Knobloch, Pattern formation in three-dimensional reaction–diffusion systems. Physica D **132**(3), 339–362 (1999)
52. T.K. Callahan, E. Knobloch, Symmetry-breaking bifurcations on cubic lattices. Nonlinearity **10**(5), 1179–1216 (1997)
53. T. Leppänen, M. Karttunen, R.A. Barrio, K. Kaski, Morphological transitions and bistability in Turing systems. Phys. Rev. E. **70**, 066202 (2004)
54. D.T. Gillespie, Approximate accelerated stochastic simulation of chemically reacting systems. J. Chem. Phys. **115**, 1716 (2001)
55. M. Rathinam, L.R. Petzold, Y. Cao, D.T. Gillespie, Stiffness in stochastic chemically reacting systems: the implicit tau-leaping method. J. Chem. Phys. **119**(24), 12784–12794 (2003)
56. Y. Yang, M. Rathinam, Tau leaping of stiff stochastic chemical systems via local central limit approximation. J. Comput. Phys. **242**, 581–606 (2013)
57. G. Klingbeil, R. Erban, M. Giles, P.K. Maini, STOCHSIMGPU: parallel stochastic simulation for the Systems Biology Toolbox 2 for Matlab. Bioinformatics **27**(8), 1170–1171 (2011)
58. R.A. Satnoianu, M. Menzinger, P.K. Maini, Turing instabilities in general systems. J. Math. Biol. **41**(6), 493–512 (2000)
59. V. Klika, R.E. Baker, D. Headon, E.A. Gaffney, The influence of receptor-mediated interactions on reaction-diffusion mechanisms of cellular self-organisation. B. Math. Biol. **74**(4), 935–957 (2012)
60. J. Dean, S. Ghemawat, MapReduce: simplified data processing on large clusters. Commun. ACM **51**(1), 107–113 (2008)
61. T. Ideker, T. Galitski, L. Hood, A new approach to decoding life: systems biology. Annu. Rev. Genomics Hum. Genet. **2**(1), 343–372 (2001)
62. M.W. Covert, B.O. Palsson, Constraints-based models: regulation of gene expression reduces the steady-state solution space. J. Theor. Biol. **221**(3), 309–325 (2003)
63. O. Cominetti, A. Matzavinos, S. Samarasinghe, D. Kulasiri, S. Liu, P.K. Maini, R. Erban, DifFUZZY: a fuzzy clustering algorithm for complex datasets. Int. J. Comput. Intel. Bioinf. Syst. Biol. **1**(4), 402–417 (2010)
64. H. Conzelmann, J. Saez-Rodriguez, T. Sauter, E. Bullinger, F. Allgöwer, E.D. Gilles, Reduction of mathematical models of signal transduction networks: simulation-based approach applied to EGF receptor signalling. Syst. Biol. **1**(1), 159–169 (2004)
65. O. Radulescu, A.N. Gorban, A. Zinovyev, A. Lilienbaum, Robust simplifications of multiscale biochemical networks. BMC Syst. Biol. **2**(1), 86 (2008)
66. L. Marcon, X. Dirego, J. Sharpe, P. Muller, High-throughput mathematical analysis identifies Turing networks for patterning with equally diffusing signals. eLife **5**, e14022 (2016)
67. W. Weber, J. Stelling, M. Rimann, B. Keller, M. Daoud-El Baba, C.C. Weber, D. Aubel, M. Fussenegger, A synthetic time-delay circuit in mammalian cells and mice. Proc. Natl. Acad. Sci. **104**(8), 2643–2648 (2007)
68. E. Fung, W.W. Wong, J.K. Suen, T. Bulter, S. Lee, J.C. Liao, A synthetic gene–metabolic oscillator. Nature **435**(7038), 118–122 (2005)

Construction Kits for Biological Evolution

Aaron Sloman

Abstract This is part of the Turing-inspired Meta-Morphogenesis project, which aims to identify transitions in information-processing since the earliest proto-organisms, in order to provide new understanding of varieties of biological intelligence, including the mathematical intelligence that produced Euclid's *Elements*. (Explaining evolution of mathematicians is much harder than explaining evolution of consciousness!) Transitions depend on "construction kits", including the initial "Fundamental Construction Kit" (FCK) based on physics and Derived Construction Kits (DCKs) produced by evolution, development, learning and culture.

Some construction kits (e.g. Lego, Meccano, plasticine, sand) are *concrete* using physical components and relationships. Others (e.g. grammars, proof systems and programming languages) are *abstract*, producing abstract entities, e.g. sentences, proofs, and new abstract construction kits. Mixtures of the two are *hybrid* kits. Some are meta-construction kits able to create, modify or combine construction kits. Construction kits are generative: they explain sets of possible construction processes, and possible products, with mathematical properties and limitations that are mathematical consequences of properties of the kit and its environment. Evolution and development both make new construction kits possible. Study of the FCK and DCKs can lead us to new answers to old questions, e.g. about the nature of mathematics, language, mind, science, and life, exposing deep connections between science and metaphysics. Showing how the FCK makes its derivatives, including all the processes and products of natural selection, possible is a challenge for science and philosophy. This is a long-term research programme with a good chance of being progressive in the sense of Lakatos. Later, this may explain how to overcome serious current limitations of AI (artificial intelligence), robotics, neuroscience and psychology.

A. Sloman (✉)
School of Computer Science, University of Birmingham, Birmingham, UK
http://www.cs.bham.ac.uk/~axs

S.B. Cooper, M.I. Soskova (eds.), *The Incomputable*, Theory and Applications of Computability, DOI 10.1007/978-3-319-43669-2_14

1 Background: What Is Science? Beyond Popper and Lakatos

How is it possible for very varied forms of life to evolve from lifeless matter, including a mathematical species able to make the discoveries presented in Euclid's *Elements*?[1] Explaining evolution of mathematical insight is much harder than explaining evolution of consciousness! (Even insects must be conscious of aspects of their surroundings.) An outline answer is based on construction kits that make other things (including new construction kits) possible. The need for science to include theories that explain how something is possible has not been widely acknowledged. Explaining how X is possible (e.g. how humans playing chess can produce a certain board configuration) need not provide a basis for *predicting* when X will be realised, so the theory used cannot be falsified by non-occurrence. Popper [33] labelled such theories "non-scientific"—at best metaphysics. His falsifiability criterion has been blindly followed by many scientists who ignore the history of science. E.g. the ancient atomic theory of matter was not falsifiable, but was an early example of a deep scientific theory. Later, Popper shifted his ground, e.g. in [35], and expressed great admiration for Darwin's theory of Natural Selection, despite its unfalsifiability.

Lakatos [25] extended Popper's philosophy of science, showing how to evaluate competing scientific research programmes over time, according to their progress. He offered criteria for distinguishing "progressive" from "degenerating" research programmes on the basis of their patterns of development, e.g. whether they systematically generate questions that lead to new empirical discoveries and new applications. It is not clear to me whether he understood that his distinction could also be applied to theories explaining how something is possible. Chapter 2 of [45][2] modified ideas of Popper and Lakatos to accommodate scientific theories about what is *possible*, e.g. types of plant, types of animal, types of reproduction, types of consciousness, types of thinking, types of learning, types of communication, types of molecule, types of chemical interaction, and types of biological information-processing. It presented criteria for evaluating theories about things that are possible and how they are possible, including theories that straddle science and metaphysics. Insisting on sharp boundaries between science and metaphysics harms both. Each can be pursued with rigour and openness to specific kinds of criticism. A separate paper[3] includes a section entitled "Why allowing non-falsifiable theories doesn't make science soft and mushy", and discusses the general concept of "explaining possibilities", its importance in science, the criteria for evaluating such explanations, and how this notion conflicts with the falsifiability requirement for scientific theories. Further examples are in [48]. The extremely ambitious Turing-inspired

[1] http://www.gutenberg.org/ebooks/21076.

[2] http://www.cs.bham.ac.uk/research/projects/cogaff/crp/#chap2.

[3] http://www.cs.bham.ac.uk/research/projects/cogaff/misc/explaining-possibility.html.

Meta-Morphogenesis project, proposed in [57],[4] depends on these ideas, and will be a test of their fruitfulness, in a combination of metaphysics and science.

This paper, straddling science and metaphysics, asks: *How is it possible for natural selection, starting on a lifeless planet, to produce billions of enormously varied organisms, in environments of many kinds, including mathematicians able to discover and prove geometrical theorems?* An outline answer is presented in terms of *construction kits*: the Fundamental (physical) Construction Kit (the FCK), and a variety of "concrete", "abstract" and "hybrid" Derived Construction Kits (DCKs), that together are conjectured to explain how evolution is possible, including evolution of mathematicians. The FCK and its relations to DCKs are crudely depicted later in Sect. 2.1. Inspired by Kant's ideas in [18], construction kits are also offered as providing Biological/Evolutionary foundations for core parts of mathematics, including parts used by evolution (but not consciously, of course) long before there were human mathematicians.

Note on "Making Possible" "X makes Y possible" as used here does not imply that if X does not exist then Y is impossible, only that *one* route to existence of Y is via X. Other things can also make Y possible, e.g., an alternative construction kit. So "makes possible" is a relation of sufficiency, not necessity. The exception is the case where X is the FCK—the *Fundamental Construction Kit*—since all concrete constructions must start from it (in this universe?). If Y is abstract, there need not be something like the FCK from which it must be derived. The space of abstract construction kits may not have a fixed "root". However, the abstract construction kits that can be thought about by physically implemented thinkers may be constrained by a future replacement for the Church-Turing thesis, based on later versions of ideas presented here. Although my questions about explaining possibilities arise in the overlap between philosophy and science [45, Chap. 2], I am not aware of any philosophical work that explicitly addresses the theses discussed here, though there seem to be examples of potential overlap, e.g. [7, 69].

2 Fundamental and Derived Construction Kits (FCK, DCKs)

Natural selection alone cannot explain how evolution happens, for it must have options to select from. What sorts of mechanisms can produce options that differ so much in so many ways, allowing evolution to produce microbes, fungi, oaks, elephants, octopuses, crows, new niches, ecosystems, cultures, etc.? Various sorts of construction kit, including evolved/derived construction kits, help to explain the emergence of new options. What explains the possibility of these construction kits? Ultimately features of fundamental physics including those emphasised by Schrödinger [37], discussed below. Why did it take so much longer for evolution

[4] Expanded in http://www.cs.bham.ac.uk/research/projects/cogaff/misc/meta-morphogenesis.html.

to produce baboons than bacteria? Not merely because baboons are more complex, but also because evolution had to produce more complex construction kits, to make baboon-building possible.

Construction kits are the "hidden heroes" of evolution. Life as we know it requires construction kits supporting construction of machines with many capabilities, including growing many types of material, many types of mechanism, many types of highly functional bodies, immune systems, digestive systems, repair mechanisms, reproductive machinery, and even mathematicians!

A kit needs more than basic materials. If all the atoms required for making a loaf of bread could somehow be put into a container, no loaf could emerge. Not even the best bread-making machine, with paddle and heater, could produce bread from atoms, since that would require atoms pre-assembled into the right amounts of flour, sugar, yeast, water, etc. Only different, separate, histories can produce the molecules and multi-molecule components, e.g. grains of yeast or flour. Likewise, no fish, reptile, bird, or mammal could be created simply by bringing together enough atoms of all the required sorts; and no machine, not even an intelligent human designer, could assemble a functioning airliner, computer, or skyscraper directly from the required atoms. Why not, and what are the alternatives? We first state the problem of constructing very complex working machines in very general terms and indicate some of the variety of strategies produced by evolution, followed later by conjectured features of a very complex, but still incomplete, explanatory story.

2.1 Combinatorics of Construction Processes

Reliable construction of a living entity requires: appropriate types of matter, machines that manipulate matter, physical assembly processes, stores of energy for use during construction, and usually *information*, e.g. about which components to assemble at each stage, how to assemble them, and how to decide in what order to do so. This requires, at every stage, at least: (1) components available for the remaining stages, (2) mechanisms capable of assembling the components, (3) mechanisms able to decide what should happen next.

If there are N types of basic component and a task requires an object of type O composed of K basic components, the size of a blind exhaustive search for a sequence of types of basic component to assemble an O is up to N^K sequences, a number that rapidly grows astronomically large as K increases. If, instead of starting from the N types of *basic* components, the construction uses M types of *pre-assembled* component, each containing P basic components, then an O will require only K/P pre-assembled parts. The search space for a route to O is reduced in size to $M^{(K/P)}$.

Compare assembling an essay of length 10,000 characters (a) by systematically trying elements of a set of about 30 possible characters (including punctuation and spaces) with (b) choosing from a set of 1000 useful words and phrases, of average length 50 characters. In the first case each choice has 30 options but 10,000 choices are required. In the second case there are 1000 options per choice, but far fewer stages: 200 instead of 10,000. So the size of the (exhaustive) search space is reduced from $30^{10,000}$, a number with 14,773 digits, to about 1000^{200}, a number with only 602 digits: a very much smaller number. Therefore trying only good pre-built substructures at each stage of a construction process can make a huge reduction to the search space for solutions of a given size, though some solutions may be missed.

So, learning from experience by storing useful subsequences can achieve dramatic reductions, analogous to a house designer moving from thinking about how to assemble atoms, to thinking about assembling molecules, then bricks, planks, tiles, then pre-manufactured house sections. The reduced search space contains fewer samples from the original possibilities, but the original space has a much larger proportion of useless options. As sizes of pre-designed components increase so does the variety of pre-designed options to choose from at each step, though far, far fewer search steps are required for a working solution: a very much shorter evolutionary process. The cost may be exclusion of some design options.

This indicates intuitively, but very crudely, how using increasingly large, already tested useful part-solutions can enormously reduce the search for viable solutions. The technique is familiar to many programmers, in the use of "memo-functions" ("memoization") to reduce computation time, e.g. when computing Fibonacci numbers. The family of computational search techniques known as "Genetic Programming"[5] makes use of related ideas. The use of "crossover" in evolution (and in Genetic Algorithms) allows parts of each parent's design specification to be used in new combinations.

In biological evolution, instead of previous *solutions* being stored for future re-use, *information about how to build components of previous solutions* is stored in genomes. Evolution, the Great Blind Mathematician, discovered memoization long before we did. A closely related strategy is to record fragments that cannot be useful in certain types of problem, in order to prevent wasteful attempts to use such fragments. Expert mathematicians learn from experience which options are useless (e.g. dividing by zero). This could be described as "negative-memoization". Are innate aversions examples of evolution doing something like that?

Without prior information about useful components and combinations of pre-built components, random assembly processes can be used. If mechanisms are available for recording larger structures that have been found to be useful or useless, the search space for new designs can be shrunk. By doing the searching and experimentation using *information* about how to build things rather than directly recombining the built physical structures themselves, evolution reduces the problem of *recording* what has been learnt.

The Fundamental Construction Kit (FCK) provided by the physical universe made possible all the forms of life that have so far evolved on earth, and also possible but still unrealised forms of life, in possible types of physical environment.

[5]https://en.wikipedia.org/wiki/Genetic_programming.

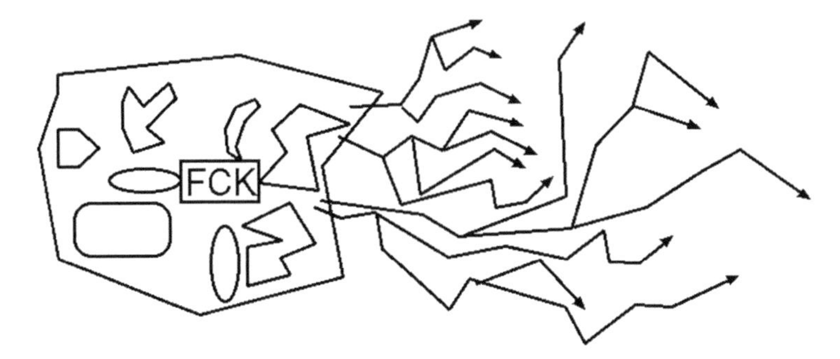

Fig. 1 This is a crude representation of the Fundamental Construction Kit (FCK) (on *left*) and (on *right*) a collection of trajectories from the FCK through the space of possible trajectories to increasingly complex mechanisms

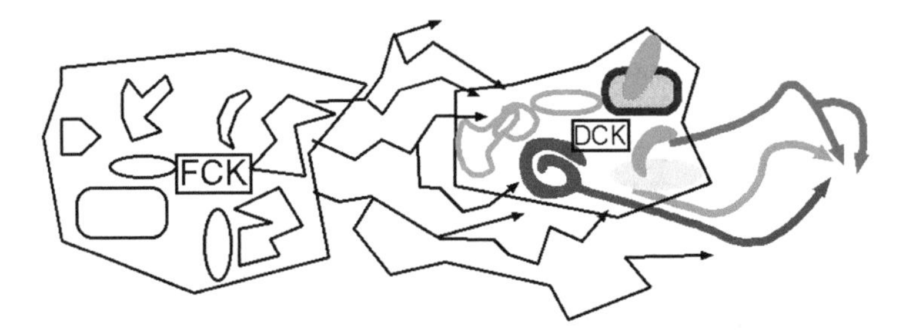

Fig. 2 Further transitions: a fundamental construction kit (FCK) on *left* gives rise to new evolved "derived" construction kits, such as the DCK on the *right*, from which new trajectories can begin, rapidly producing new more complex designs, e.g. organisms with new morphologies and new information-processing mechanisms. The *shapes* and *colours* (crudely) indicate qualitative differences between components of old and new construction kits, and related trajectories. A DCK trajectory uses larger components and is therefore much shorter than the equivalent FCK trajectory

Figure 1 shows how a common initial construction kit can generate many possible trajectories, in which components of the kit are assembled to produce new instances (living or non-living). The space of possible trajectories for combining basic constituents is enormous, but routes can be shortened and search spaces shrunk by building derived construction kits (DCKs) that assemble larger structures in fewer steps,[6] as indicated in Fig. 2.

The history of technology, science and engineering includes many transitions in which new construction kits are derived from old ones. That includes the science and technology of digital computation, where new advances used an enormous variety of discoveries and inventions, including punched cards (used in Jacquard

[6]Assembly mechanisms are part of the organism, as illustrated in a video of grass growing itself from seed (https://www.youtube.com/watch?v=JbiQtfr6AYk). In mammals with a placenta, more of the assembly process is shared between mother and offspring.

looms), through many types of electronic device, many types of programming language, many types of external interface (not available on Turing machines!), many types of operating system, many types of network connection, and many types of virtual machine, in an enormous variety of applications. Particular inventions were generalised, using mathematical abstractions, to patterns that could be reused in new contexts. New applications frequently led to production of new more powerful tools.

Natural selection did all this on an even larger scale, with far more variety, probably discovering many obscure problems and solutions still unknown to us. (An educational moral: teaching only what has been found most useful can discard future routes to possible major new advances—like depleting a gene pool.)

Biological construction kits derived from the FCK can combine to form new Derived Construction Kits (DCKs), some specified in genomes, and (very much later) some discovered or designed by individuals (e.g. during epigenesis; Sect. 2.3) or by groups, for example new languages. Compared with derivation from the FCK, the rough calculations above show how DCKs can enormously speed up searching for new complex entities with new properties and behaviours. See Fig. 2.

DCKs that evolve in different species in different locations may have overlapping functionality, based on different mechanisms: a form of *convergent evolution*. E.g., mechanisms enabling elephants to learn to use trunk, eyes, and brain to manipulate food may share features with those enabling primates to learn to use hands, eyes, and brain to manipulate food. In both cases, competences evolve in response to structurally similar affordances in the environment. This extends Gibson's ideas in [17] to include affordances for a species or a collection of species.[7]

2.2 Construction Kit Ontologies

A construction kit (and its products) can exist without being described. However, scientists need to use various forms of language in order to describe the entities they observe or postulate in explanations, and to formulate new questions to be answered. So a physicist studying the FCK will need one or more construction kits for defining concepts, formulating questions, formulating theories and conjectures, constructing models, etc. Part of the process of science is extending construction kits for theory formation. Something similar must be done by natural selection: extending useful genetic information structures that store specifications for useful components.

This relates to claims that have been made about requirements for control systems and for scientific theories. For example, if a system is to be capable of distinguishing between N different situations and responding differently to them, it must be capable

[7]Implications for evolution of vision and language are discussed in http://www.cs.bham.ac.uk/research/projects/cogaff/talks/#talk111.

of being in at least N different states (recognition+control states). This is a variant of Ashby's "Law of Requisite Variety" [3].

Many thinkers have discussed representational requirements for scientific theories, or for specifications of designs. Chomsky [9] identified requirements for theories of language, which he labelled *observational adequacy* (covering the variety of observed uses of a particular language), *descriptive adequacy* (covering the intuitively understood principles that account for the scope of a particular language) and *explanatory adequacy* (providing a basis for explaining how any language can be acquired on the basis of data available to the learner). These labels were vaguely echoed by McCarthy and Hayes [29], who described a form of representation as *metaphysically* adequate if it can express anything that can be the case, *epistemologically* adequate if it can express anything that can be known by humans and future robots, and *heuristically* adequate if it supports efficient modes of reasoning and problem solving. (I have simplified all these proposals.)

Requirements can also be specified for powers of biological construction kits. The fundamental construction kit (FCK) must have the power to make any form of life that ever existed or will exist possible, using huge search spaces if necessary. DCKs may meet different requirements, e.g. each supporting fewer types of life form, but enabling those life forms to be "discovered" in a shorter time by natural selection, and replicated (relatively) rapidly. Early DCKs may support the simplest organisms that reproduce by making copies of themselves perhaps as Ganti [15] described.

At later stages of evolution, DCKs are needed that can construct organisms that change their properties during development and change their control mechanisms appropriately as they grow [60]. This requires the ability to produce individuals whose features are *parametrised*, with parameters that change over time. More sophisticated DCKs must be able to produce species with epigenetic mechanisms that modify their knowledge and their behaviours not merely as required to accommodate their own growth but also to cope with changing physical environments, new predators, new prey and new shared knowledge. A special case of this is having genetic mechanisms able to support development of a wide enough range of linguistic competences to match any type of human language developed in any social or geographical context. However, the phenomenon is far more general than language development, as discussed in the next section.

2.3 Construction Kits Built During Development (Epigenesis)

Some new construction kits are products of evolution of a species and are initially shared among only a few members of the species (barring genetic abnormalities), alongside cross-species construction kits shared between species, such as those used in mechanisms of reproduction and growth in related species. Evolution also discovered the benefits of "meta-construction-kits": mechanisms that allow members of a species to build new construction kits during their own development.

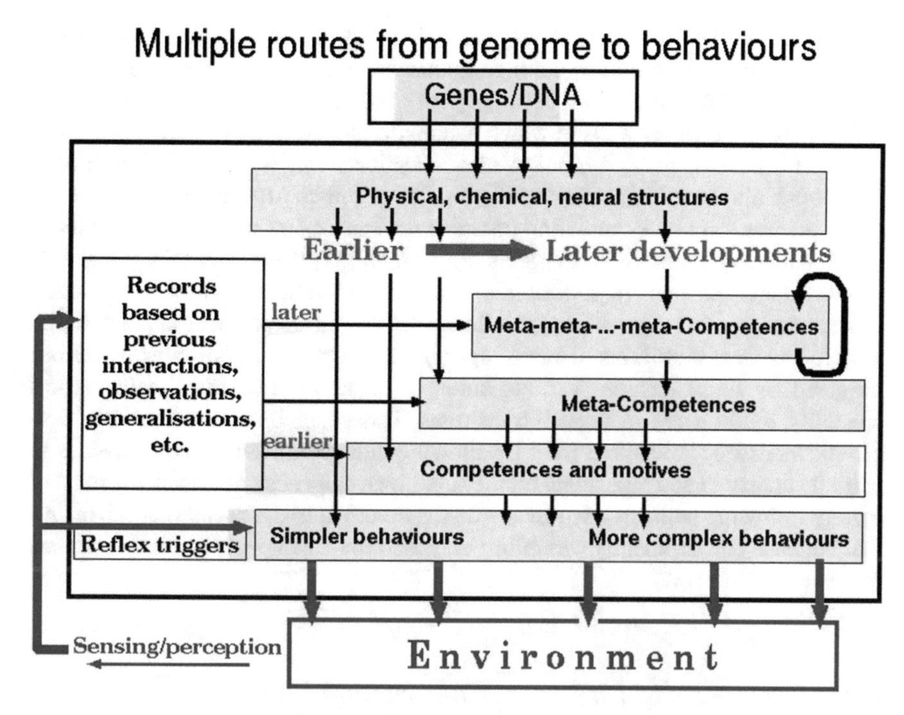

Fig. 3 A construction kit gives rise to very different individuals if the genome interacts with the environment in increasingly complex ways during development. Precocial species use only the downward routes on the *left*, producing preconfigured competences. Competences of altricial species, using staggered development, may be far more varied. Results of using earlier competences interact with the genome, producing meta-configured competences on the *right*

Examples include mechanisms for learning that are initially generic mechanisms shared across individuals, and developed by individuals on the basis of their own previously encountered learning experiences, which may be different in different environments for members of the same species. Human language learning is a striking example: things learnt at earlier stages make new things learnable that might not be learnable by an individual transferred from a different environment partway through learning a different language. This contrast between genetically specified and individually built capabilities for learning and development was labelled a difference between "pre-configured" and "meta-configured" competences in [8], summarised in Fig. 3. The meta-configured competences are partly specified in the genome, but that specification is combined with information abstracted from individual experiences. Mathematical development and language development in humans both seem to be special cases of growth of meta-configured competences. Karmiloff-Smith presented closely related ideas in [19].

Construction kits used for assembly of new organisms that start as seeds or eggs make possible many different processes in which components are assembled in parallel, using abilities of the different sub-processes to constrain one another.

Nobody knows the full variety of ways in which parallel construction processes can exercise mutual control in developing organisms. One implication of Fig. 3 is that there are no simple correlations between genes and organism features.

Explaining the many ways in which a genome can orchestrate parallel processes of growth, development, formation of connections, etc. is a huge challenge. A framework allowing abstract specifications in a genome to interact with details of the environment in instantiating complex designs is illustrated schematically in Fig. 3. An example might be Popper's proposal in [34] that newly evolved desires of individual organisms (e.g. desires to reach fruit in taller trees) could indirectly and gradually, across generations, influence selection of physical characteristics (e.g. longer necks, abilities to jump higher) that improve success-rates of actions triggered by those desires. Various kinds of creativity, including mathematical creativity, might result from such transitions. This generalises Waddington's "epigenetic landscape" metaphor [65], by allowing individual members of a species to partially construct and repeatedly modify their own epigenetic landscapes instead of merely following paths in a landscape that is common to the species. Mechanisms that increase developmental variability may also make new developmental defects possible (e.g. autism?).

2.4 The Variety of Biological Construction Kits

As products of physical construction kits become more complex, with more ways of contributing to needs of organisms, and directly or indirectly to reproductive fitness, they require increasingly sophisticated control mechanisms. New sorts of control often use new types of information. Processing that information may require new mechanisms. That may require new construction kits for building new types of information-processing mechanism. The simplest organisms use only a few types of (mainly chemical) sensor, providing information about internal physical and chemical states and the immediate external physical environment. They have very few behavioural options. They acquire, use and replace fragments of information, using the same forms of information throughout their life, to control deployment of a fixed repertoire of capabilities.

More complex organisms acquire information about enduring spatial locations in extended terrain, including static and changing routes between static and changing resources and dangers. They need to construct and use far more complex (internal or external) information stores about their environment, and, in some cases, "meta-semantic" information about information processing, in themselves and in others, e.g. conspecifics, predators and prey.

What forms can such information take? Many controlled systems have states that can be represented by a fixed set of physical measures, often referred to as "variables", representing states of sensors, output signals, and internal states of various sorts. Relationships between such state-components are often represented mathematically by equations, including differential equations, and constraints (e.g.

inequalities) specifying restricted, possibly time-varying, ranges of values for the variables, or magnitude relations between the variables. A system with N variables (including derivatives) has a state of a fixed dimension, N. The only way to record new information in such a system is in static or dynamic values for numeric variables—changing "state vectors"—and possibly alterations in the equations. A typical example is Powers [36], inspired by Wiener [68] and Ashby [2]. There are many well-understood special cases, such as simple forms of homeostatic control using negative feedback. Neural net-based controllers often use large numbers of variables clustered into strongly interacting sub-groups, groups of groups, etc.

For many structures and processes, a set of numerical values and rates of change linked by equations (including differential equations) expressing their changing relationships is an adequate form of representation, but not for all, as implied by the discussion of types of *adequacy* in Sect. 2.2. That's why chemists use *structural* formulae, e.g. diagrams showing different sorts of bonds between atoms and collections of diagrams showing how bonds change in chemical reactions. Linguists, programmers, computer scientists, architects, structural engineers, map makers and users, mathematicians studying geometry and topology, composers, and many others, work in domains where structural diagrams, logical expressions, grammars, programming languages, plan formalisms, and other *non-numerical* notations express information about structures and processes that is not usefully expressed in terms of collections of numbers and equations linking numbers.[8]

Of course, any information that can be expressed in 2-D written or printed notation, such as grammatical rules, parse trees, logical proofs, and computer programs, can also be converted into a large array of numbers by taking a photograph and digitising it. Although such processes are useful for storing or transmitting documents, they add so much irrelevant numerical detail that the original functions are obstructed, such as checking whether an inference is valid, manipulating a grammatical structure by transforming an active sentence to a passive one, determining whether two sentences have the same grammatical subject, removing a bug from a program, or checking whether a geometric construction proves a theorem—unless the original non-numerical structures are extracted, often at high cost.

Similarly, collections of numerical values will not always adequately represent information that is biologically useful for animal decision making, problem solving, motive formation, learning, etc. Moreover, biological sensors are poor at acquiring or representing very precise information, and neural states often lack reliability and stability. (Such flaws can be partly compensated for by using many neurons per numerical value and averaging.) More importantly, the biological functions, e.g. of

---

[8]Examples include: https://en.wikipedia.org/wiki/Parse_tree
https://en.wikipedia.org/wiki/Structural_formula
https://en.wikipedia.org/wiki/Flowchart
https://en.wikipedia.org/wiki/Euclidean_geometry
https://en.wikipedia.org/wiki/Entity-relationship_model
https://en.wikipedia.org/wiki/Programming_language.

visual systems, may have little use for absolute measures if their functions are based on *relational* information, such as that A is closer to B than to C, A is biting B, A is keeping B and C apart, A can fit through the gap between B and C, the joint between A and B is non-rigid, A cannot enter B unless it is reoriented, and many more. As Schrödinger [37] pointed out, topological structures of molecules can reliably encode a wide variety of types of genetic information, and may also turn out to be useful for recording other forms of structural information. Do brains employ them? Chomsky [9] pointed out that using inappropriate structures in models can divert attention from important biological phenomena that need to be explained—see Sect. 2.2, above. Max Clowes, who introduced me to AI in 1969, made similar points about research in vision around that time.[9] So a subtask for this project is to identify types of non-numerical, e.g. relational, information content that are of biological importance, and the means by which such information can be stored, transmitted, manipulated, and used, and to explain how the mechanisms performing those tasks can be built from the FCK, using appropriate DCKs.

2.5 Increasingly Varied Mathematical Structures

Electronic computers made many new forms of control possible, including use of logic, linguistic formalisms, planning, learning, problem solving, vision, theorem proving, teaching, map-making, automated circuit design, program verification, and many more. The world wide web is an extreme case of a control system made up of millions of constantly changing simpler control systems, interacting in parallel with each other and with millions of display devices, sensors, mechanical controllers, humans, and many other things. The types of control mechanism in computer-based systems now extend far beyond the numerical sorts familiar to control engineers.[10]

Organisms also need multiple control systems, not all numerical. A partially constructed percept, thought, question, plan or terrain description has parts and relationships, to which new components and relationships can be added and others removed as construction proceeds and errors are corrected. So the structures change—unlike a fixed-size collection of variables assigned changing values. Non-numerical types of mathematics are needed for describing or explaining such systems, including topology, geometry, graph theory, set theory, logic, formal grammars, and theory of computation. A full understanding of mechanisms and processes of evolution and development may need new branches of mathematics, including mathematics of non-numerical structural processes, such as chemical change, or changing "grammars" for internal records of complex structured information. The importance of non-numerical information structures has been understood by many

[9] http://www.cs.bham.ac.uk/research/projects/cogaff/81-95.html#61.

[10] Often misleadingly labelled "non-linear"—like calling apples, apes and avalanches non-bananas! http://en.wikipedia.org/wiki/Control_theory http://en.wikipedia.org/wiki/Nonlinear_control.

mathematicians, logicians, linguists, computer scientists and engineers, but many scientists still focus only on numerical structures and processes. They sometimes seek to remedy failures by using statistical methods, which can be spectacularly successful in restricted contexts, as shown by recent AI successes, whose limitations I have commented on elsewhere.[11]

The FCK need not be able to produce all biological structures and processes *directly*, in situations without life, but it must be rich enough to support successive generations of increasingly powerful DCKs that together suffice to generate all possible biological organisms evolved so far, and their behavioural and information-processing abilities. Moreover, the FCK, or DCKs derived from it, must include abilities to acquire, manipulate, store, and use information structures in DCKs that can build increasingly complex machines that encode information, including non-numerical information. Since the 1950s we have also increasingly discovered the need for new *virtual* machines as well as *physical* machines [54, 56].

Large-scale physical processes usually involve a great deal of variability and unpredictability (e.g. weather patterns), and sub-microscopic indeterminacy is a key feature of quantum physics; yet, as Schrödinger pointed out in [37], life depends on very complex objects built from very large numbers of small-scale structures (molecules) that can preserve their *precise* chemical structure despite continual thermal buffetting and other disturbances. Unlike non-living natural structures, important molecules involved in reproduction and other biological functions are copied repeatedly, predictably transformed with great precision, and used to create very large numbers of new molecules required for life, with great, but not absolute, precision. This is *non-statistical* structure preservation, which would have been incomprehensible without quantum mechanics, as explained by Schrödinger. That feature of the FCK resembles "structure-constraining" properties of construction kits such as Meccano, TinkerToy and Lego[12] that support structures with more or less complex, discretely varied topologies, or kits built from digital electronic components, that also provide extremely reliable preservation and transformations of precise structures, in contrast with sand, water, mud, treacle, plasticine, and similar materials. Fortunate children learn how structure-based kits differ from more or less amorphous construction kits that produce relatively flexible or plastic structures with non-rigid behaviours—as do many large-scale natural phenomena, such as snowdrifts, oceans, and weather systems.

Schrödinger's 1944 book stressed that quantum mechanisms can explain the structural stability of individual molecules and explained how a set of atoms in different arrangements can form discrete stable structures with very different properties (e.g. in propane and isopropane, only the location of the single oxygen atom differs, but that alters both the topology and the chemical properties of the

[11]E.g. http://www.cs.bham.ac.uk/research/projects/cogaff/misc/impossible.html.

[12]https://en.wikipedia.org/wiki/Meccano, https://en.wikipedia.org/wiki/Tinkertoy and https://en.wikipedia.org/wiki/Lego.

molecule).[13] He also pointed out the relationship between the number of discrete changeable elements and information capacity, anticipating [39]. Some complex molecules with quantum-based structural stability are simultaneously capable of *continuous* deformations, e.g. folding, twisting, coming together, moving apart, etc., all essential for the role of DNA and other molecules in reproduction, and many other biochemical processes. This combination of discrete topological structure (forms of connectivity), used for storing very precise information for extended periods, and non-discrete spatial flexibility, used in assembling, replicating and extracting information from large structures, is unlike anything found in digital computers, although it can to some extent be approximated in digital computer models of molecular processes.

Highly deterministic, very small-scale, discrete interactions between very complex, multi-stable, enduring molecular structures, combined with continuous deformations (folding, etc.) that alter opportunities for the discrete interactions, may have hitherto unnoticed roles in brain functions, in addition to their profound importance for reproduction and growth. Much recent AI and neuroscience uses statistical properties of complex systems with many continuous scalar quantities changing randomly in parallel, unlike symbolic mechanisms used in logical and symbolic AI, though the latter are still far too restricted to model animal minds. The Meta-Morphogenesis project has extended a set of examples studied four decades earlier (e.g. in [45]) of types of mathematical discovery and reasoning that use perceived *possibilities* and *impossibilities* for change in geometrical and topological structures. Further work along these lines may help to reveal biological mechanisms that enabled the great discoveries by Euclid and his predecessors that are still unmatched by AI theorem provers (discussed in Sect. 5).

2.6 Thermodynamic Issues

The question sometimes arises whether formation of life from non-living matter violates the second law of thermodynamics, because life increases the amount of order or structure in the physical matter on the planet, reducing entropy. The standard answer is that the law is applicable only to closed systems, and the earth is not a closed system, since it is constantly affected by solar and other forms of radiation, asteroid impacts, and other external influences. The law implies only that our planet could not have generated life forms without energy from non-living sources, e.g. the sun (though future technologies may reduce or remove such dependence). Some of the ways in which pre-existing dispositions can harness external sources of energy to increase local structure are discussed in a collection

[13] E.g. see James Ashenhurst's tutorial: http://www.masterorganicchemistry.com/2011/11/10/dont-be-futyl-learn-the-butyls/.

of thoughts on entropy, evolution, and construction kits: http://www.cs.bham.ac.uk/research/projects/cogaff/misc/entropy-evolution.html.[14]

Our discussion so far suggests that the FCK has two sorts of components: (a) a generic framework including space-time and generic constraints on what can happen in that framework, and (b) components that can be non-uniformly and dynamically distributed in the framework. The combination makes possible formation of galaxies, stars, clouds of dust, planets, asteroids, and many other lifeless entities, as well as supporting forms of life based on derived construction kits (DCKs) that exist only in special conditions. Some local conditions, e.g. extremely high pressures, temperatures, and gravitational fields (among others), can mask some parts of the FCK, i.e. prevent them from functioning. So, even if all sub-atomic particles required for earthly life exist at the centre of the sun, local factors can rule out earth-like life forms. Moreover, if the earth had been formed from a cloud of particles containing no carbon, oxygen, nitrogen, iron, etc., then no DCK able to support life as we know it could have emerged, since that requires a region of space-time with a specific manifestation of the FCK, embedded in a larger region that can contribute additional energy (e.g. solar radiation), and possibly other resources.

As the earth formed, new physical conditions created new DCKs that made the earliest life forms possible. Ganti [15], usefully summarised in [23] and [14], presents an analysis of requirements for a minimal life form, the "chemoton", with self-maintenance and reproductive capabilities. Perhaps still unknown DCKs made possible formation of pre-biotic chemical structures, and also the *environments* in which a chemoton-like entity could survive and reproduce. Later, conditions changed in ways that supported more complex life forms, e.g. oxygen-breathing forms. Perhaps attempts to identify the first life form in order to show how it could be produced by the FCK are misguided, because several important pre-life construction kits were necessary: i.e. several DCKs made possible by conditions on earth were necessary for precursors. Some of the components of the DCKs may have been more complex than their living products, including components providing *scaffolding* for constructing life forms, rather than materials.

2.7 *Scaffolding in Construction Kits*

An important feature of some construction kits is that they contain parts that are used during assembly of products of the kit, but are not included in the products. For example, Meccano kits include spanners and screwdrivers, used for manipulating screws and nuts during assembly and disassembly, though they are not normally

[14]Partly inspired by memories of a talk by Lionel Penrose in Oxford around 1960 about devices he called *droguli*—singular *drogulus*. Such naturally occurring multi-stable physical structures seem to me to render redundant the apparatus proposed by Deacon in [12] to explain how life apparently goes against the second law of thermodynamics. See https://en.wikipedia.org/wiki/Incomplete_Nature.

included in the models constructed. Similarly, kits for making paper dolls and their clothing[15] may include pencils and scissors, used for preparing patterns and cutting them out. But the pencils and scissors are not parts of the dolls or their clothing. When houses are built, many items are used that are not part of the completed house, including tools and scaffolding frameworks to support incomplete structures. A loose analogy can be made with the structures used by climbing plants, e.g. rock faces, trees, or frames provided by humans: these are essential for the plants to grow to the heights they need but are not parts of the plant. More subtly, rooted plants that grow vertically make considerable use of the soil penetrated by their roots to provide not only nutrients but also the stability that makes tall stalks or trunks possible, including in some cases the ability to resist strong winds most of the time. The soil forms part of the scaffolding. A mammal uses parts of its mother as temporary scaffolding while developing in the womb, and continues to use the mother during suckling and later when fed portions of prey caught by parents. Other species use eggs with protective shells and food stores. Plants that depend on insects for fertilization can be thought of as using scaffolding in a general sense.

This concept of scaffolding may be crucial for research into origins of life. As far as I know, nobody has found candidate non-living chemical substances made available by the FCK that have the ability spontaneously to assemble themselves into primitive life forms. It is possible that the search is doomed to fail because there never were such substances. Perhaps the earliest life forms required not only materials but also scaffolding—e.g. in the form of complex molecules that did not form parts of the earliest organisms but played an essential causal role in assembly processes, bringing together the chemicals needed by the simplest organisms. Evolution might then have produced new organisms without that reliance on the original scaffolding. The scaffolding mechanisms might later have ceased to exist on earth, e.g. because they were consumed and wiped out by the new life forms, or because physical conditions changed that prevented their formation but did not destroy the newly independent organisms. A similar suggestion was recently made by Mathis et al. [27]. So it is quite possible that many evolutionary transitions, including transitions in information processing, our main concern, depended on forms of scaffolding that later did not survive and were no longer needed to maintain what they had helped to produce. So research into evolution of information processing, our main goal, is inherently partly speculative.

2.8 Biological Construction Kits

How did the FCK generate complex life forms? Is the Darwin-Wallace theory of natural selection the whole answer, as suggested in [6]? Bell writes: *"Living complexity cannot be explained except through selection and does not require*

---

[15]https://en.wikipedia.org/wiki/Paper_doll.

any other category of explanation whatsoever." No: the explanation must include both *selection* mechanisms and *generative* mechanisms, without which selection processes will not have a supply of new viable options. Moreover, insofar as environments providing opportunities, challenges and threats are part of the selection process, the construction kits used by evolution include mechanisms not intrinsically concerned with life, e.g. volcanoes, earthquakes, asteroid impacts, lunar and solar tides, and many more.

The idea of evolution producing construction kits is not new, though they are often referred to as "toolkits". Coates et al. [10] ask whether there is "a genetic toolkit for multicellularity" used by complex life-forms. Toolkits and construction kits normally have *users* (e.g. humans or other animals), whereas the construction kits we have been discussing (FCKs and DCKs) do not all need separate users.

Both generative mechanisms and selection mechanisms change during evolution. Natural selection (blindly) uses the initial enabling mechanisms provided by physics and chemistry not only to produce new organisms, but also to produce new richer DCKs, including increasingly complex information-processing mechanisms. Since the mid 1900s, spectacular changes have also occurred in human-designed computing mechanisms, including new forms of hardware, new forms of virtual machinery, and networked social systems all unimagined by early hardware designers. Similar changes during evolution produced new biological construction kits, e.g. grammars, planners, and geometrical constructors, not well understood by thinkers familiar only with physics, chemistry and numerical mathematics.

Biological DCKs produce not only a huge variety of physical forms and physical behaviours, but also forms of *information processing* required for increasingly complex control problems, as organisms become more complex and more intelligent in coping with their environments, including interacting with predators, prey, mates, offspring, conspecifics, etc. In humans, that includes abilities to form scientific theories and discover and prove theorems in topology and geometry, some of which are also used unwittingly in practical activities.[16] I suspect many animals come close to this in their *systematic* but unconscious abilities to perform complex actions that use mathematical features of environments. Abilities used unconsciously in building nests or in hunting and consuming prey may overlap with topological and geometrical competences of human mathematicians. (See Sect. 6.2.) For example, search for videos of weaver birds building nests.

[16]Such as putting a shirt on a child: http://www.cs.bham.ac.uk/research/projects/cogaff/misc/shirt.html. I think Piaget noticed some of the requirements.

3 Concrete (Physical), Abstract and Hybrid Construction Kits

Products of a construction kit may be concrete, i.e. physical; or abstract, like a theorem, a sentence, or a symphony; or hybrid, e.g. a written presentation of a theorem or poem.

Concrete Kits Construction kits for children include physical parts that can be combined in various ways to produce new physical objects that are not only larger than the initial components but also have new shapes and new behaviours. Those are *concrete* construction kits. The FCK is (arguably?) a concrete construction kit. Lego, Meccano, twigs, mud, and stones, can all be used in construction kits whose constructs are physical objects occupying space and time: *concrete* construction kits.

Abstract Kits There are also non-spatial *abstract* construction kits, for example components of languages, such as vocabulary and grammar, or methods of construction of arguments or proofs. Physical *representations* of such things, however, can occupy space and/or time, e.g. a spoken or written sentence, a diagram, or a proof presented on paper. Using an abstract construction kit, e.g. doing mental arithmetic or composing poetry in your head, requires use of one or more physical construction kits, directly or indirectly implementing features of the abstract kit.

There are (deeply confused) fashions emphasising "embodied cognition" and "symbol grounding" (previously known as "concept empiricism" and demolished by Immanuel Kant and twentieth century philosophers of science). These fashions disregard many examples of thinking, perceiving, reasoning and planning that require abstract construction kits. For example, planning a journey to a conference does not require physically trying possible actions, like water finding a route to the sea. Instead, you may use an abstract construction kit able to *represent* possible options and ways of combining them. Being able to talk requires use of a grammar specifying abstract structures that can be assembled using a collection of grammatical relationships, in order to form new abstract structures with new properties relevant to various tasks involving information. Sentences allowed by a grammar for English are abstract objects that can be instantiated physically in written text, printed text, spoken sounds, Morse code, etc.; so a grammar is an abstract construction kit whose constructs can have concrete (physical) instances. The idea of a grammar is not restricted to verbal forms: it can be extended to many complex structures, e.g. grammars for sign languages, circuit diagrams, maps, proofs, architectural layouts and even molecules.

A grammar does not fully specify a language: a structurally related *semantic* construction kit is required for building possible *meanings*. Use of a language depends on language users, for which more complex construction kits are required, including products of evolution, development and learning. Evolution of various types of language, including languages used only *internally*, is discussed in [52].

In computers, digital circuitry implements abstract construction kits via intermediate abstract kits—virtual machines—presumably also required in brains.

Hybrid Abstract+Concrete Kits These are combinations, e.g. physical chess board and chess pieces combined with the rules of chess, lines and circular arcs on a physical surface instantiating Euclidean geometry, puzzles like the mutilated chess-board puzzle, and many more. A particularly interesting hybrid case is the use of physical objects (e.g. blocks) to instantiate arithmetic, which may lead to the discovery of prime numbers when certain attempts at rearrangement fail—and an explanation of the impossibility is found.[17]

In some hybrid construction kits such as games like chess, the concrete (physical) component may be redundant for some players, e.g. chess experts who can play without physical pieces on a board. But communication of moves needs physical mechanisms, as does the expert's brain (in ways that are not yet understood). Related abstract structures, states and processes can also be implemented in computers, which can now play chess better than most humans, without replicating human brain mechanisms. In contrast, physical components are indispensable in hybrid construction kits for outdoor games, such as cricket [69]. (I don't expect to see good robot cricketers soon.)

Physical computers, programming languages, operating systems and virtual machines form hybrid construction kits that can make things happen when they run. A logical system with axioms and inference rules can be thought of as an abstract kit supporting construction of logical proof sequences, usually combined with a physical notation for written proofs. A purely logical system cannot have physical causal powers, whereas its concrete instances can, e.g. teaching a student to distinguish valid from invalid proofs. Natural selection "discovered" the power of hybrid construction kits using virtual machinery long before human engineers did. In particular, biological virtual machines used by animal minds outperform current engineering designs in some ways, but they also generate much confusion in the minds of philosophical individuals who are aware that something more than purely physical machinery is at work, but don't yet understand how to implement virtual machines in physical machines [54, 56, 58].

Animal perception, learning, reasoning, and intelligent behaviour require *hybrid* construction kits. Scientific study of such kits is still in its infancy. Work done so far on the Meta-Morphogenesis project suggests that natural selection "discovered" and used a staggering variety of types of hybrid construction kit that were essential for reproduction, for developmental processes (including physical development and learning), for performing complex behaviours, and for social/cultural phenomena.

[17]A possibility discussed in http://www.cs.bham.ac.uk/research/projects/cogaff/misc/toddler-theorems.html#primes. Contributions from observant parents and child-minders are welcome. Much deeper insights come from extended individual developmental trajectories than from statistics of snapshots of many individuals.

3.1 Kits Providing External Sensors and Motors

Some construction kits can be used to make toys with moving parts, e.g. wheels or grippers, that interact with the environment. A toy car may include a spring, whose potential energy can be transformed into mechanical energy via gears, axles and wheels in contact with external surfaces. Further interactions, altering the direction of motion, may result from collisions with fixed or mobile objects in the environment or from influence of some control device.

As noted in [45, Chap. 6], the distinction between internal and external components is often arbitrary. For example, a music box may play a tune under the control of a rotating disc with holes or spikes. The disc can be thought of either as part of the music box or as part of a changing environment.

If a toy train set has rails or tracks used to guide the motion of the train, then the wheels can be thought of as sensing the environment and causing changes of direction. This is partly like and partly unlike a toy vehicle that uses an optical sensor linked to a steering mechanism, so that the vehicle can follow a line painted on a surface. The railway track provides both information and the forces required to change direction. A painted line, however, provides only the information, and other parts of the vehicle have to supply the energy to change direction, e.g. an internal battery that powers sensors and motors. Evolution uses both sorts, e.g. wind blowing seeds away from parent plants and a wolf following a scent trail left by its prey. An unseen wall uses force to stop your forward motion in a dark room, whereas a visible, or lightly touched, wall provides only information [55]. More sophisticated kits use sensors, e.g. optical, auditory, tactile, inertial, or chemical sensors, providing information that internal mechanisms can use to evaluate and select goals, control actions, interact with conspecifics, predict events in the environment, evaluate hypotheses, and other functions.

3.2 Mechanisms for Storing, Transforming and Using Information

Often information is acquired, used, and then lost because it is overwritten, e.g. sensor information in simple servo-control systems with "online intelligence", where only the latest sensed state is used for deciding whether to speed something up, change direction, etc. In more complex control systems, with "offline intelligence", sensor information is saved, possibly combined with previously stored information, and remains available for use on different occasions for different purposes.[18]

[18]Trehub [61] proposed an architecture for vision that allows snapshots from visual saccades to be integrated in a multi-layer fixation-independent visual memory.

In the "offline" case, the underlying construction kit needs to be able to support stores of information that grow with time and can be used for different purposes at different times. A control decision at one time may need items of information obtained at several different times and places, for example information about properties of a material, where it can be found, and how to transport it to where it is needed. Sensors used online may become faulty or require adjustment. Evolution may provide mechanisms for testing and adjusting. When used offline, stored information may need to be checked for falsity caused by the environment changing, as opposed to sensor faults. The offline/online use of visual information has caused much confusion among researchers, including muddled attempts to interpret the difference in terms of "what" and "where" information.[19] Contrast Sloman [46].

Ways of acquiring and using information have been discovered and modelled by AI researchers, psychologists, neuroscientists, biologists and others. However, evolution has produced many more. Some of them require not additional storage space but very different sorts of information-processing architectures. A range of possible architectures is discussed in [45–47, 50, 51], whereas AI engineers typically seek one architecture for a project. A complex biological architecture may use sub-architectures that evolved at different times, meeting different needs in different niches.

This raises the question whether evolution produced "architecture kits" able to combine evolved information-processing mechanisms in different ways long before software engineers discovered the need. Such a kit could be particularly important for species that produce new subsystems, or modify old ones, during individual development, e.g. during different phases of learning by apes, elephants, and humans, as described in Sect. 2.3, contradicting the common assumption that a computational architecture must remain fixed.[20]

3.3 Mechanisms for Controlling Position, Motion and Timing

All concrete construction kits (and some hybrid kits) share a deep common feature insofar as their components, their constructs and their construction processes involve space and time, both during assembly and while working. Those behaviours include both relative motion of parts, e.g. wheels rotating, joints changing angles, and also motion of the whole object relative to other objects, e.g. an ape grasping a berry. *A consequence of spatiality is that objects built from different construction kits can interact, by changing their spatial relationships (e.g. if one object enters, encircles or grasps another), applying forces transmitted through space, and using spatial sensors to gain information used in control.* Products of different kits can interact in

[19]http://en.wikipedia.org/wiki/Two-streams_hypothesis.

[20]The BICA society aims to bring together researchers on biologically inspired cognitive architectures. Some examples are here: http://bicasociety.org/cogarch/.

varied ways, e.g. one being used to assemble or manipulate another, or one providing energy or information for the other. Contrast the problems of getting software components available on a computer to interact sensibly: merely locating them in the same virtual or physical machine will not suffice. Some rule-based systems are composed of condition-action rules, managed by an interpreter that constantly checks for satisfaction of conditions. Newly added rules may then be invoked simply because their conditions become satisfied, though "conflict resolution" mechanisms may be required if the conditions of more than one rule are satisfied.[21]

New concrete kits can be formed by combining two or more kits. In some cases this will require modification of a kit, e.g. combining Lego and Meccano by adding pieces with Lego studs or holes alongside Meccano-sized screw holes. In other cases, mere spatial proximity and contact suffices, e.g. when one construction kit is used to build a platform and others are used to assemble a house on it. Products of different biological construction kits may also use complex mixtures of juxtaposition and adaptation.

Objects that exist in space and/or time often need timing mechanisms. Organisms use "biological clocks" operating on different time scales controlling repetitive processes, including daily cycles, heartbeats, breathing, and wing or limb movements required for locomotion. More subtly, there are adjustable speeds and adjustable rates of change: e.g. a bird in flight approaching a perch; an animal running to escape a predator and having to decelerate as it approaches a tree it needs to climb; a hand moving to grasp a stationary or moving object, with motion controlled by varying coordinated changes of joint angles at waist, shoulder, elbow and finger joints so as to bring the grasping points on the hand into suitable locations relative to the intended grasping points on the object. (This can be very difficult for robots, when grasping novel objects in novel situations, if they use ontologies that are too simple.) There are also biological mechanisms for controlling or varying rates of production of chemicals (e.g. hormones).

So biological construction kits need many mechanisms able to measure time intervals and to control rates of repetition or rates of change of parts of the organism. These kits may be combined with other sorts of construction kit that combine temporal and spatial control, e.g. changing speed and direction.

3.4 Combining Construction Kits

At the molecular level there is now a vast, and rapidly growing, amount of biological research on interacting construction kits, for example interactions between different parts of the reproductive mechanism during development of a fertilised egg, interactions between invasive viral or bacterial structures and a host organism,

[21]Our SimAgent toolkit is an example: http://www.cs.bham.ac.uk/research/projects/poplog/packages/simagent.html [49].

and interactions with chemicals produced in medical research laboratories. In computers, the ways of combining different toolkits include the application of functions to arguments, although both functions and their arguments can be far more complex than the cases most people encounter when learning arithmetic. A function could be a compiler, its arguments could be arbitrarily complex programs in a high-level programming language, and the output of the function might be either a report on syntactic errors in the input program, or a machine code program ready to run.

Applying functions to arguments is very different from assembling structures in space-time, where inputs to the process form parts of the output. If computers are connected via digital-to-analog interfaces linking them to surrounding matter, or if they are mounted on machines that allow them to move around in space and interact, that adds a kind of richness that goes beyond application of functions to arguments.

The additional richness is present in the modes of interaction of chemical structures that include both digital (on/off chemical bonds) and continuous changes in relationships, as discussed by Turing [64], in the paper on chemistry-based morphogenesis that inspired this Meta-Morphogenesis project [57].

3.5 Combining Abstract Construction Kits

Section 2.1 showed how a new DCK using combinations of old components can make some new developments very much quicker to reach—fewer steps are required, and the total search space for a sequence of steps to a solution may be dramatically reduced. Combining *concrete* construction kits uses space-time occupancy. Combining *abstract* construction kits is less straightforward. Sets of letters and numerals are combined to form labels for chess board squares, e.g. "a2", "c5", etc. A human language and a musical notation can form a hybrid system for writing songs. A computer operating system (e.g. Linux) can be combined with programming languages (e.g. Lisp, Java). In organisms, as in computers, products of different kits may share *information*, e.g. information for sensing, predicting, explaining or controlling, including information about information [55]. Engineers combining different kinds of functionality find it useful to design re-usable information-processing *architectures* that provide frameworks for combining different mechanisms and information stores (see footnote 20), especially in large projects where different teams work on sensors, learning, motor systems, reasoning systems, motivational systems, various kinds of metacognition, etc., using specialised tools. The toolkit mentioned in footnote 21 is an example framework. It is often necessary to support different sorts of **virtual** machinery interacting simultaneously with one another and with internal and external physical environments during perception and motion. This may require new general frameworks for assembling complex *information-processing architectures*, accommodating multiple interacting virtual machines, with different modifications developed at different times [30, 31, 50]. Self-extension is a topic for further research—see footnote 17.

Creation of new construction kits may start by simply recording parts of successful assemblies, or, better still, parametrized parts, so that they can easily be reproduced in modified forms—e.g. as required for organisms that change size and shape while developing. Eventually, parametrized stored designs may be combined to form a *"meta-construction kit"* able to extend, modify or combine previously created construction kits, as human engineers have recently learnt to do in software development environments. Evolution needs to be able to create new meta-construction kits using natural selection. Natural selection, the great creator/meta-creator, is now spectacularly aided and abetted by its products, especially humans!

4 Construction Kits Generate Possibilities and Impossibilities

Explanations of how things are possible (Sect. 1) can refer to construction kits either manufactured, e.g. Meccano and Lego, or composed of naturally occurring components, e.g. boulders, mud, or sand. (Not all construction kits have sharp boundaries.) Each kit makes possible certain types of construct, instances of which can be built by assembling parts from the kit. Some construction kits use *products of products of* biological evolution, e.g. birds' nests assembled from twigs or leaves.

In some kits, features of components, such as shape, are inherited by constructed objects. E.g. objects composed only of Lego bricks joined in the "standard" way have external surfaces that are divisible into faces parallel to the surfaces of the first brick used. However, if two Lego bricks are joined at a corner only, using only one stud and one socket, it is possible to have continuous relative rotation (because studs and sockets are circular), violating that constraint, as Ron Chrisley pointed out in a conversation. This illustrates the fact that constructed objects can have "emergent" features none of the components have, e.g. a hinge is a non-rigid object that can be made from two rigid objects with aligned holes through which a screw is passed.

So, a construction kit that makes some things possible and others impossible can be extended so as to remove some of the impossibilities, e.g. by adding a hinge to Lego, or adding new parts from which hinges can be assembled.

4.1 Construction Kits for Making Information Users

Not everything that can play a role in acquisition, storage or transfer of information has information-processing capabilities. Consider a lump of plasticine or damp clay that can be deformed under pressure, then retains the deformation. If a coin is pressed against it the lump will change its shape. Entities with information-processing capabilities (e.g. archaeologists) can use the depression as a source of information about the coin. But the deformed lump of material is not an information user. If the depression is used to control a process, e.g. making copies of the coin, or to help a historian years later, then the deformed material is used as a source of information about the coin. The fact that some part of a brain is changed by perceptual processes in an organism does not imply that that portion of the brain is an information user. It may play a role analogous to the lump of clay, or a footprint in soil. Additional mechanisms are required if the information is to be *used*: different mechanisms for different types of use. A photocopier acquires information from a sheet of paper, but all it can do with the information is produce a replica (possibly after slight modifications such as changes in contrast, intensity or magnification). Different mechanisms are required for recognising text, correcting spelling, analysing the structure of an image, interpreting it as a picture of a 3-D scene, using information about the scene to guide a robot, building a copy of the scene, or answering a question about which changes are possible. Thinking up ways of using the impression as a source of information about the coin is left as an exercise for the reader.

Biological construction kits for producing information-processing mechanisms evolved at different times. Sloman [47] discusses the diversity of uses of information from sensors, including sharing of sensor information between different uses, concurrently or sequentially. Subsystems can compete for sensors (e.g. concentrating on the road or admiring the scenery). Information vehicles such as sound or light provide multi-purpose information about the source or reflector of the sound or light, e.g. used for deciding whether to flee, or for controlling actions such as grasping or avoiding the information-source.

Some information-using mechanisms are direct products of biological evolution, e.g. reflex protective blinking mechanisms. Others are grown by epigenetic mechanisms influenced by context. For example, humans in different cultures start with a generic language construction kit (sometimes misleadingly labelled a "universal grammar") which is extended and modified to produce locally useful linguistic mechanisms. Language-specific mechanisms, such as mechanisms for acquiring, producing, understanding and correcting textual information, must have evolved long after mechanisms shared between many species that can use visual information for avoiding obstacles or grasping objects. In some species, diversity in the construction kits produced by individual genomes, can lead to even greater diversity in adults, especially if they develop in different physical and cultural environments using the epigenetic mechanisms suggested in Sect. 2.3 and Fig. 3.

4.2 *Different Roles for Information*

Despite huge diversity in biological construction kits and the mechanisms in individual organisms, some themes recur, such as functions of different sorts of information in control: e.g. information about how things actually are or might be ("belief-like" information contents), information about how things need to be or might need to be for the individual information user ("desire-like" information contents), and information about how to achieve or avoid certain states ("procedural" information contents). Each type has different subtypes: across species, across members of a species and across developmental stages in an individual. How a biological construction kit supports all those requirements depends on the environment, the animal's sensors, its needs, the local opportunities, and the individual's history. Different mechanisms performing such functions may share a common evolutionary precursor after which they diverged. Moreover, mechanisms with similar functions can evolve independently: convergent evolution.

Information relating to targets and how to achieve or maintain them is *control* information: the most basic type of biological information, from which all others are derived. A simple case is a thermostatic control, discussed by McCarthy [28]. It has (at least) two sorts of information: (a) a *target* temperature ("desire-like" information), (b) *current temperature* ("belief-like" information). A discrepancy between them causes the thermostat to select between turning a heater on, or off, or doing nothing. This very simple homeostatic mechanism uses information and a source of energy to achieve or maintain a target state. There are very many variants on this schema, based on the type of target (e.g. a measured state or some complex relationship), the type of control (on, off, or variable, with single or multiple effectors), and the mechanisms by which targets and control actions are selected, which may be modified by learning, and may use simple actions or complex plans.

As Gibson [16] pointed out, acquisition of information often requires cooperation between processes of sensing and acting. Saccades are visual actions that constantly select new information samples from the environment (or the optic cone). Uses of the information vary widely according to context, e.g. controlling grasping, controlling preparation for a jump, controlling avoidance actions, or sampling text to be read. A particular sensor can therefore be shared between many control subsystems [47], and the significance of the sensor state will depend partly on which subsystems are connected to the sensor at the time and partly on which other mechanisms receive information from the sensor (which may change dynamically—a possible cause of some types of "change blindness").

The study of varieties of use of information in organisms is exploding, and includes many mechanisms on molecular scales as well as many intermediate levels of informed control, including sub-cellular levels (e.g. metabolism), physiological processes of breathing, temperature maintenance, digestion, blood circulation, control of locomotion, feeding and mating of large animals and coordination in communities, such as collaborative foraging in insects and trading systems of humans. Slime moulds include spectacular examples in which modes of acquisition and use of information change dramatically.[22]

The earliest organisms must have acquired and used information about things inside themselves and in their immediate vicinity, e.g. using chemical detectors in

[22]http://www.theguardian.com/cities/2014/feb/18/slime-mould-rail-road-transport-routes.

Fig. 4 Between the simplest and most sophisticated organisms there are many intermediate forms with very different information processing requirements and capabilities

an enclosing membrane. Later, evolution extended those capabilities in dramatic ways (crudely indicated in Fig. 4). In the simplest cases, local information is used immediately to select between alternative possible actions, as in a heating control or a trail-following mechanism. Uses of motion in haptic and tactile sensing and the use of saccades, changing vergence, and other movements in visual perception all use the interplay between sensing and doing characteristic of "online intelligence". But there are cases ignored by Gibson and anti-cognitivists, namely organisms that exhibit "offline intelligence", using perceptual information for tasks other than controlling immediate reactions, for example, reasoning about remote future possibilities or attempting to explain something observed or working out that bending a straight piece of wire will enable a basket of food to be lifted out of a tube as in Fig. 4 [67] or correcting a stored generalisation. Doing that requires use of previously acquired information about the environment, including particular information about individual objects and their locations or states, general information about learnt laws or correlations and information about what is and is not possible. (Compare footnote 11.)

An information-bearing structure (e.g. the impression of a foot, the shape of a rock) can provide very different information to different information users, or to the same individual at different times, depending on (a) what kinds of sensors they have, (b) what sorts of information-processing (storing, analysing, comparing, combining, synthesizing, retrieving, deriving, using, etc.) mechanisms they have, (c) what sorts of needs or goals they can serve by using various sorts of information (knowingly or not), and (d) what information they already have. So, from the fact that changes in some portion of a brain correlate with changes in some aspect of the environment, we cannot conclude much about what information about the environment the brain acquires and uses or how it does that, since typically that will depend on context.

4.3 Motivational Mechanisms

It is often assumed that every information user, U, constantly tries to achieve rewards or avoid punishments (negative rewards), and that each new item of information, I, will make some actions more likely for U, and others less likely, on the basis of what

U has previously learnt about which actions increase positive rewards or decrease negative rewards under conditions indicated by I. But animals are not restricted to acting on motives selected *by them* on the basis of expected rewards. They may also have motive generators that are simply triggered as "internal reflexes" just as evolution produces phototropic reactions in plants without giving plants any ability to anticipate benefits to be gained from light. Some reflexes, instead of directly triggering *behaviour*, trigger *new motives*, which may or may not lead to behaviour, depending on the importance of other competing motives. For example, a kind person watching someone fall may acquire a motive to rush to help, which is not acted on if competing motives are too strong. It is widely believed that all motivation is reward-based. But a new motive triggered by an internal reflex need not be associated with some reward. It may be "architecture-based motivation" rather than "reward-based motivation" [53]. Triggering of architecture-based motives in playful, intelligent young animals can produce kinds of delayed learning that the individuals could not possibly anticipate, and therefore cannot be motivated by [19].

Unforeseeable biological benefits of automatically triggered motives include acquisition of new information by sampling properties of the environment. The new information may not be immediately usable, but in combination with information acquired later and genetic tendencies activated later, as indicated in Fig. 3, it may turn out to be important during hunting, caring for young, or learning a language. A toddler may have no conception of the later potential uses of information gained in play, though the ancestors of that individual may have benefited from the presence of the information-gathering reflexes. In humans this seems to be crucial for mathematical development.

During evolution, and also during individual development, the sensor mechanisms, the types of information processing, and the uses to which various types of information are put, become more diverse and more complex, while the information-processing architectures allow more of the processes to occur in parallel (e.g. competing, collaborating, invoking, extending, recording, controlling, redirecting, enriching, training, abstracting, refuting, or terminating). Without understanding how the architecture grows, which information-processing functions it supports, and how they diversify and interact, we are likely to reach wrong conclusions about biological functions of the parts: e.g. over-simplifying the functions of sensory subsystems, or over-simplifying the variety of concurrent control mechanisms involved in producing behaviours. Moreover, the architectural knowledge about how such a system works, like information about the architecture of a computer operating system, may not be expressible in sets of equations, or statistical learning mechanisms and relationships. (Ideas about architectures for human information processing can be found in [24, 30, 31, 40, 50, 59], among many others.)

Construction kits for building information-processing architectures with multiple sensors and motor subsystems in complex and varied environments differ widely in the designs they can produce. Understanding that variety is not helped by disputes about which architecture is best. A more complete discussion would need to survey the design options and relate them to actual choices made by evolution or by individuals interacting with their environments.

5 Mathematics: Some Constructions Exclude or Necessitate Others

Physical construction kits (e.g. Lego, plasticine, or a combination of paper, scissors and paste) have parts and materials with physical properties (e.g. rigidity, strength, flexibility, elasticity, adhesion, etc.), possible relationships between parts and possible processes that can occur when the parts are in those relationships (e.g. rotation, bending, twisting and elastic or inelastic resistance to deformation).

Features of a physical construction kit—including the shapes and materials of the basic components, ways in which the parts can be assembled into larger wholes, kinds of relationships between parts and the processes that can occur involving them—explain the possibility of *entities* that can be constructed and the possibility of *processes*, including processes of construction and behaviours of constructs.

Construction kits can also explain necessity and impossibility. A construction kit with a large initial set of generative powers can be used to build a structure realising some of the kit's possibilities, in which some further possibilities are excluded, namely all extensions that do not include what has so far been constructed. If a Meccano construction has two parts in a substructure that fixes them a certain distance apart, then no extension can include a new part that is wider than that distance in all dimensions and is in the gap. Some extensions to the part-built structure that were previously possible become impossible unless something is undone. That example involves a limit produced by a gap size. There are many more examples of impossibilities that arise from features of the construction kit.

Euclidean geometry includes a construction kit that enables construction of closed planar polygons (triangles, quadrilaterals, pentagons, etc.), with interior angles whose sizes can be summed. If the polygon has three sides, i.e. it is a triangle, then the interior angles must add up to exactly half a rotation. Why? In this case, no physical properties of a structure (e.g. rigidity or impenetrability of materials) are involved, only spatial relationships. Figure 5 provides one way to answer the question, unlike the standard proofs, which use parallel lines. It presents a proof, found by Mary Pardoe, that internal angles of a planar triangle sum to a straight line, or 180°. (I am ignoring the question how to verify that the surface is planar.)

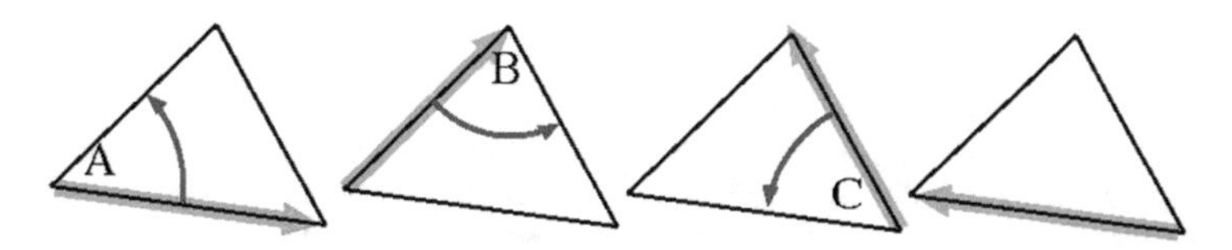

Fig. 5 The sequence demonstrates how the three-cornered shape has the consequence that summing the three angles necessarily produces half a rotation (180°). Since the position, size, orientation, and precise shape of the triangle can be varied without affecting the possibility of constructing the sequence, this is a proof that generalises to any planar triangle. It nowhere mentions Euclid's parallel axiom, used by "standard" proofs. This unpublished proof was reported to me by Mary Pardoe, a former student who became a mathematics teacher, in the early 1970s

Most humans are able to look at a physical situation, or a diagram representing a class of physical situations, and reason about constraints on a class of possibilities sharing a common feature. This may have evolved from earlier abilities to reason about changing affordances in the environment (Gibson [17]). Current AI perceptual and reasoning systems still lack most of these abilities, and neuroscience cannot yet explain what's going on (as opposed to where it's going on?). (See footnote 11.)

These illustrate mathematical properties of construction kits (partly analogous to mathematical properties of formal deductive systems and AI problem solving systems). As parts (or instances of parts) of the FCK are combined, structural relations between components of the kit have two opposed sorts of consequences: they make some further structures *possible* (e.g. constructing a circle that passes through all the vertices of the triangle), and other structures *impossible* (e.g. relocating the corners of the triangle so that the angles add up to 370°). These possibilities and impossibilities are *necessary* consequences of previous selection steps. The examples illustrate how a construction kit with mathematical relationships can provide the basis for necessary truths and necessary falsehoods in some constructions (as argued in Sloman [41, Chap. 7]).[23] Being able to think about and reason about alterations in some limited portion of the environment is a very common requirement for intelligent action [48]. It seems to be partly shared with other intelligent species, e.g. squirrels, nest builders, elephants, apes, etc. Since our examples of making things possible or impossible, or changing ranges of possibilities, are examples of causation (mathematical causation), this also provides the basis for a Kantian notion of causation based on mathematical necessity [18], so that not all uses of the notion of "cause" are Humean (i.e. based on empirical correlations), even if some are. Compare Sect. 5.3.[24]

Neuroscientific theories about information processing in brains currently omit the processes involved in such mathematical discoveries, so AI researchers influenced too much by neuroscience may fail to replicate important brain functions. Progress may require major conceptual advances regarding what the problems are and what sorts of answers are relevant.

We now consider ways in which evolution itself can be understood as discovering mathematical proofs—proofs of possibilities.

[23]Such relationships between possibilities provide a deeper, more natural basis for understanding modality (necessity, possibility, impossibility) than so-called "possible world semantics". I doubt that most normal humans who can think about possibilities and impossibilities base that ability on thinking about truth in the whole world, past, present and future, and in the set of alternative worlds.

[24]For more on Kantian vs. Humean causation, see the presentations on different sorts of causal reasoning in humans and other animals by Chappell and Sloman at the Workshop on Natural and Artificial Cognition (WONAC, Oxford, 2007): http://www.cs.bham.ac.uk/research/projects/cogaff/talks/wonac. Varieties of causation that do not involve mathematical necessity, but only probabilities (Hume?) or propensities (Popper), will not be discussed here.

5.1 Proof-Like Features of Evolution

A subset of the FCK produced fortuitously as a side effect of formation of the earth supported (a) primitive life forms and (b) processes of evolution that produced more and more complex forms of life, including new, more complex, derived, DCKs. New products of natural selection can make more complex products more reachable, as with toy construction kits and mathematical proofs. However, starting from those parts will make some designs unreachable except by disassembling some parts.

Moreover, there is not just one sequence: different evolutionary lineages evolving in parallel can produce different DCKs. According to the "Symbiogenesis" theory, different DCKs produced independently can sometimes merge to support new forms of life combining different evolutionary strands.[25] Creation of new DCKs in parallel evolutionary streams with combinable products can hugely reduce part of the search space for complex designs, at the cost of excluding parts of the search space reachable from the FCK. For example, use of DCKs in the human genome may speed up development of language and typical human cognitive competences, while excluding the possibility of "evolving back" to microbe forms that might be the only survivors after a cataclysm.

5.2 Euclid's Construction Kit

An old example, of great significance for science, mathematics, and philosophy, is the construction kit specified in Euclidean geometry, starting with points, lines, surfaces and volumes, and methods of constructing new more complex geometrical configurations using a straight edge for drawing straight lines in a plane surface, and a pair of compasses for drawing circular arcs. This construction kit makes it possible to bisect, but not trisect, an arbitrary planar angle. A slight extension, the "Neusis construction", known to Archimedes, allows line segments to be translated and rotated in a plane while preserving their length, and certain incidence relations. This allows arbitrary angles to be trisected! (See http://www.cs.bham.ac.uk/research/projects/cogaff/misc/trisect.html.)

The ability of (at least some) humans to discover such things must depend on evolved information-processing capabilities of brains that are as yet unknown and not yet replicated in AI reasoning systems. The idea of a space of possibilities generated by a physical construction kit may be easier for most people to understand than the comparison with generative powers of grammars, formal systems, or geometric constructions, though the two must be connected, since grammars and mathematical systems are abstract construction kits that can be parts of hybrid construction kits.

[25] http://en.wikipedia.org/wiki/Symbiogenesis.

Concrete construction kits corresponding to grammars can be built out of physical structures. For example, a collection of small squares with letters and punctuation marks, and some blanks, can be used to form sequences that correspond to the words in a lexicon. A cursive ("joined up") script requires a more complex physical construction kit. Human sign-languages are far more demanding, since they involve multiple body parts moving concurrently.

Some challenges for construction kits used by evolution, and also challenges for artificial intelligence and philosophy, arise from the need to explain both how natural selection makes use of mathematical properties of construction kits related to geometry and topology, in producing organisms with spatial structures and spatial competences, and also how various subsets of those organisms (e.g. nest-building birds) developed specific topological and geometrical reasoning abilities used in controlling actions or solving problems; and finally how at least one species developed abilities to reflect on the nature of those competences and eventually, through unknown processes of individual development and social interaction, using unknown representational and reasoning mechanisms, managed to produce the rich, deep and highly organised body of knowledge published as Euclid's *Elements* (see footnote 1).

There are important aspects of those mathematical competences that, as far as I know, have not yet been replicated in Artificial Intelligence or Robotics.[26] Is it possible that currently understood forms of digital computation are inadequate for the tasks, whereas chemistry-based information-processing systems used in brains are richer, because they combine both discrete and continuous operations, as discussed in Sect. 2.5? (That's not a rhetorical question: I don't know the answer.)

5.3 Mathematical Discoveries Based on Exploring Construction Kits

Some mathematical discoveries result from observation of naturally occurring physical construction kits and noticing how constraints on modes of composition of components generate constraints on resulting constructs. E.g. straight line segments on a surface can be joined end to end to enclose a finite region, but that is impossible with only two lines, as noted by Kant [18]. Likewise, flat surfaces can be combined to enclose a volume, such as a tetrahedron or cube, but it is impossible for only three flat surfaces to enclose a finite space. It is not clear how humans detect such impossibilities: no amount of trying and failing can establish impossibility. Kant had no access to a twentieth-century formal axiomatisation of Euclidean geometry. What he, and before him Euclid, Archimedes and others had were products of evolution. What products?

[26] Several are listed at http://www.cs.bham.ac.uk/research/projects/cogaff/misc/mathstuff.html.

Many mathematical domains (perhaps all of them) can be thought of as sets of possibilities generated by construction kits. Physicists and engineers deal with hybrid concrete and abstract construction kits. The space of possible construction kits is also an example. As far as I know, this domain has not been explored systematically by mathematicians, though many special cases have.

In order to understand biological evolution on this planet, we need to understand the sorts of construction kits made possible by the existence of the physical universe, and in particular the variety of construction kits inherent in the physics and chemistry of the materials of which our planet was formed, along with the influences of its environment (e.g. solar radiation, asteroid impacts). An open research question is whether a construction kit capable of producing all the non-living structures on the planet would also suffice for evolution of all the forms of life on this planet, or whether life and evolution have additional requirements, e.g. cosmic radiation.

5.4 *Evolution's (Blind) Mathematical Discoveries*

Insofar as construction kits have mathematical properties, life and mathematics are closely interconnected, as we have already seen. More complex relationships arise after evolution of mathematical metacognitive mechanisms. On the way to achieving those results, natural selection often works as "a blind theorem prover". Many of the "theorems" are about new *possible* structures, processes, organisms, ecosystems, etc. The proofs that they are possible are implicit in the evolutionary trajectories that lead to occurrences. Proofs are often thought of as abstract entities that can be represented physically in different ways (using different formalisms) for communication, persuasion (including self-persuasion), predicting, explaining and planning. A physical sequence produced unintentionally, e.g. by natural selection or by plant growth, that leads to a new sort of entity is a proof that some construction kit makes that sort of entity possible. The evolutionary or developmental trail answers the question: how is that sort of thing possible? So biological evolution can be construed as a "blind theorem prover", despite there being no intention behind the proof. Proofs of *impossibility* (or *necessity*) raise more complex issues, to be discussed elsewhere.

These observations seem to support a new kind of "Biological-evolutionary" foundation for mathematics that is closely related to Immanuel Kant's philosophy of mathematics in his *Critique of Pure Reason* (1781). I attempted to defend his ideas in Sloman [41]. This answers questions like "How is it possible for things that make mathematical discoveries to exist?", an example of explaining a possibility (see footnote 3). Attempting to go too directly from hypothesized properties of the primordial construction kit (or the physical universe) to explaining advanced capabilities such as human self-awareness, without specifying all the relevant construction kits, including required temporary scaffolding, will fail, because shortcuts omit essential details of both the problems and the solutions, like mathematical proofs with gaps.

Many of the "mathematical discoveries" (or inventions?) produced (blindly) by evolution depend on mathematical properties of physical structures or processes or problem types, whether they are specific solutions to particular problems (e.g. use of negative feedback control loops) or new construction kit components that are usable across a very wide range of different species (e.g. the use of a powerful "genetic code", the use of various kinds of learning from experience, the use of new forms of representation for information, the use of new physical morphologies to support sensing, or locomotion, or consumption of nutrients, etc.).

These mathematical "discoveries" started happening long before there were any humans doing mathematics (refuting claims that humans create mathematics). Many of the discoveries were concerned with what is possible, either absolutely or under certain conditions, or for a particular sort of construction kit. Other discoveries, closer to what are conventionally thought of as mathematical discoveries, are concerned with limitations on what is possible, i.e. necessary truths. Some discoveries are concerned with probabilities derived from statistical learning, but I think the relative importance of statistical learning in biology has been vastly overrated because of misinterpretations of evidence (to be discussed elsewhere). In particular, the discovery that something important is possible does not require statistical evidence: a single instance suffices. No amount of statistical evidence can show that something is impossible: structural constraints need to be analysed. For human evolution, a particularly important subtype of mathematical discovery was the unwitting discovery and use of mathematical (e.g. topological) structures in the environment, a discovery process that starts in human children before they are aware of what they are doing, and in some species without any use of language for communication. Examples are discussed in the "Toddler Theorems" document referenced in footnote 17.

6 Varieties of Derived Construction Kit

DCKs may differ (a) at different evolutionary stages within a lineage, (b) across lineages (e.g. in different coexisting organisms), and (c) during development of an individual that starts as a single cell and produces mechanisms that support different kinds of growth, development and information processing at different stages (Sect. 2.3). New construction kits can also be produced by cultures or ecosystems (e.g. human languages) and applied sciences (e.g. bioengineering, computer systems engineering). New cases build on what was previously available. Sometimes separately evolved DCKs are combined, for instance in symbiosis, sexual reproduction, and individual creative learning.

What sort of kit makes it possible for a child to acquire competence in any one of the thousands of different human languages (spoken or signed) in the first few years of life? Children do not merely learn pre-existing languages: they *construct* languages that are new for them, constrained by the need to communicate with conspecifics, as shown dramatically by Nicaraguan deaf children who developed a sign language going beyond what their teachers understood [38]. There are also languages that might have developed but have not (yet). Evolution of human spoken language may have gone from purely internal languages needed for perception, intention, etc., through collaborative actions, then (later) signed communication, then spoken communication, as argued in Sloman [52].

If language acquisition were mainly a matter of learning from expert users, human languages could not have existed, since initially there were no expert users to learn from, and learning could not get started. This argument applies to any competence thought to be based entirely on learning from experts, including mathematical expertise. So data mining in samples of expert behaviours will never produce AI systems with human competences—only inferior subsets at best.

The history of computing since the earliest calculators illustrates changes that can occur when new construction kits are developed. There have not only been changes in size, speed and memory capacity: there have also been profound qualitative changes, e.g. in new layers of virtual machinery, such as new sorts of mutually interacting causal loops linking virtual machine control states with portions of external environments, as in the use of GPS-based navigation. Long before that, evolved virtual machines provided semantic contents referring to non-physical structures and processes, e.g. mathematical problems, rules of games, and mental contents referring to possible future mental contents (e.g. "What will I see if...?"), including contents of other minds.

I claim, but will not argue here, that some new machines cannot be *fully described* in the language of the FCK even though they are *fully implemented* in physical mechanisms. (See Sect. 2.2 on ontologies.) We now understand many key components and many modes of composition that provide platforms on which *human-designed* layers of computation can be constructed, including subsystems closely but not rigidly coupled to the environment (e.g. a hand-held video camera).

Several different "basic" abstract construction kits have been proposed as sufficient for the forms of (discrete) computation required by mathematicians: namely Turing machines, Post's production systems, Church's Lambda Calculus, and several more, each capable of generating the others. The Church-Turing thesis claims that each is sufficient for all forms of computation.[27] There has been an enormous amount of research in computer science and computer systems engineering on forms of computation that can be built from such components. One interpretation of the Church-Turing thesis is that these construction kits generate all *possible* forms of information processing. But it is not at all obvious that those discrete mechanisms suffice for all biological forms of information processing. For example, chemistry-based forms of computation include both discrete mechanisms (e.g. forming or releasing chemical bonds) of the sort Schrödinger discussed, and continuous process, e.g. folding, twisting, etc. used in reproduction and other processes. Ganti [15] shows how a chemical construction kit can support forms of biological information processing that don't depend only on external energy sources

[27]For more on this, see http://en.wikipedia.org/wiki/Church-Turing_thesis.

(a fact that's also true of battery-powered computers), and can also support growth and reproduction using internal mechanisms, which human-made computers cannot do (yet).

There seem to be many different sorts of construction kit that allow different sorts of information processing to be supported, including some that we don't yet understand. In particular, the physical/chemical mechanisms that support the construction of both physical structures and information-processing mechanisms in living organisms may have abilities not available in digital computers.[28]

6.1 A New Type of Research Project

Very many biological processes and associated materials and mechanisms are not well understood, though knowledge about them is increasing rapidly. It is hard to know how many of the derived construction kits have not yet been identified and studied. I am not aware of any systematic attempt to identify features of the FCK that suffice to explain the possibility of all known evolved biological DCKs. Researchers in fundamental physics and cosmology do not normally attempt to ensure that their theories explain the many materials and process types that have been explored by natural selection and its products, in addition to known facts about physics and chemistry. Schrödinger [37] pointed out that a theory of the physical basis of life should explain such phenomena, though he could not have appreciated some of the requirements for sophisticated forms of information processing, because, at the time he wrote, scientists and engineers had not learnt what we now know. Curiously, although he mentioned the need to explain the occurrence of metamorphosis in organisms, the example he gave was the transformation from a tadpole to a frog. He could have given more spectacular examples, such as the transformation of a caterpillar to a butterfly via an intermediate stage as a chemical soup in an outer case, from which the butterfly later emerges.[29]

Penrose [32] attempted to show how features of quantum physics explain obscure features of human consciousness, especially mathematical consciousness, but he ignored all the intermediate products of biological evolution from which animal mental functions build. Human mathematics, at least the ancient mathematics done before the advent of modern algebra and logic, seems to build on animal abilities, for instance abilities to see various types of affordance. The use of diagrams and spatial models by Penrose could be an example of that. It is unlikely that there are very abstract human mathematical abilities that somehow grow directly out of quantum

[28]Examples of human mathematical reasoning in geometry and topology that have, until now, resisted replication on computers are presented in http://www.cs.bham.ac.uk/research/projects/cogaff/misc/torus.html and http://www.cs.bham.ac.uk/research/projects/cogaff/misc/triangle-sum.html.

[29]http://en.wikipedia.org/wiki/Pupa
http://en.wikipedia.org/wiki/Holometabolism

mechanical aspects of the FCK, without depending on the mostly unknown layers of perceptual, learning, motivational, planning, and reasoning competences produced by billions of years of evolution.

Twentieth-century biologists understood some of the achievements of the FCK in meeting physical and chemical requirements of various forms of life, though they used different terminology from mine, e.g. Haldane.[30] However, the task can never be finished, since the process of construction of new derived biological construction kits may continue indefinitely, producing new kits with components and modes of composition that allow production of increasingly complex types of *structure* and *behaviour* in organisms. That idea is familiar to computer scientists and engineers since very many new sorts of computational construction kit (new programming languages, new operating systems, new virtual machines, new development toolkits) have been developed from old ones in the last half century, making possible new kinds of computing system that could not previously be built from the original computing machinery without introducing new intermediate layers, including new virtual machines that are able to detect and record their own operations, a capability that is often essential for debugging and extending computing systems. Sloman [56] discusses the importance of virtual machinery in extending what information-processing systems can do and the properties they can have, including radical self-modification while running.

6.2 Construction Kits for Biological Information Processing

Each newly evolved mechanism provides opportunities for yet more control at higher levels: a recurring process that can repeatedly generate opportunities for *additional* mechanisms for (information-based) control of recently evolved mechanisms, for example choosing between competences or "tuning" them dynamically, often on the basis of their mathematical properties.

Implicit mathematical discovery processes enable production of competences used in interpretation of sensory information—e.g. locating perceived objects, events and processes in 2-D or 3-D space and time, deriving coherent wholes from separate information fragments. Further enhancements may include: new mechanisms for prediction; for motive generation and selection; for construction, comparison, selection and control of new plans, with resulting new behaviours.

Many of evolution's new mathematical discoveries, e.g. use of negative feedback control loops, were used in new designs producing useful behaviours, e.g. controlling temperature, osmotic pressure and other states; use of geometric constraints by bees whose cooperative behaviours produce hexagonal cells in honeycombs; and use of new ontologies for separating situations requiring different behaviours, e.g. manipulating different materials or hunting different kinds of prey.

[30]http://en.wikipedia.org/wiki/J._B._S._Haldane

As a result, construction kits used by evolution produced metacognitive mechanisms enabling individuals to notice and reflect on their own discoveries (enabling some of them to notice and remove flaws in their reasoning). Such metacognitive capabilities are required for abilities to communicate discoveries to others, discuss them, use them in shared practical tasks (e.g. making tools, clothes or weapons, building shelters, planning routes, discussing what will or will not work, and why), then later organising them into complex, highly structured bodies of shared knowledge, such as Euclid's *Elements* (footnote 1). I don't think anyone knows how long all of this took, what the detailed evolutionary changes were, or how the required mechanisms of perception, motivation, intention formation, reasoning, planning and retrospective reflection actually evolved. The brain mechanisms involved are also mostly unknown.

Explaining how all that could happen, and what it tells us about the nature of mathematics and biological/evolutionary foundations for mathematical knowledge, is a long-term goal of the Meta-Morphogenesis project. That includes seeking unnoticed overlaps between the human competences discovered by metacognitive mechanisms, and similar competences in animals that lack the metacognition, such as young humans making and using mathematical discoveries, on which they are unable to reflect because the required architecture has not yet developed. Other intelligent species make and use similar "proto-mathematical" discoveries without the meta-cognitive abilities required to notice what they are doing.

This could stimulate new research in robotics attempting to replicate such competences. Most of the naturally occurring mathematical abilities have not yet been replicated in Artificial Intelligence systems or robots, unlike logical, arithmetical, and algebraic competences that are relatively new to humans and (paradoxically?) easier to replicate on computers. Examples of topological reasoning about equivalence classes of closed curves not yet modelled in computers (as far as I know) are referenced in footnote 28. Even the ability to reason about alternative ways of putting a shirt on a child (footnote 16) is still lacking. It is not clear whether the difficulty of replicating such mathematical reasoning processes is due to the need for a kind of construction kit that digital computers (e.g. Turing machines) cannot support, or due to our lack of imagination in using computers to replicate some of the products of biological evolution, or both! Perhaps there are important forms of representation or types of information-processing architecture still waiting to be discovered by AI researchers. Alternatively, the gaps may be connected with properties of chemistry-based information-processing mechanisms combining discrete and continuous interactions, or other physical properties that cannot be replicated exactly (or even approximately) in familiar forms of computation. (This topic requires more detailed mathematical analysis.)

6.3 Representational Blind Spots of Many Scientists

Although I cannot follow all the details of writings of physicists, I think it is clear that most debates regarding what should go into a fundamental theory of matter ignore most of the biological demands on such a theory. For example, presentations on dynamics of physical systems make deep use of branches of mathematics concerned with sets of numerical values, and the ways in which different measurable or hypothesized physical values do or do not covary, as expressed in (probabilistic or non-probabilistic) equations of various sorts. But the biological functions of complex physiological structures, especially structures that change in complexity as they develop, don't necessarily have those forms.

Biological mechanisms include: digestive mechanisms; mechanisms for transporting chemicals; mechanisms for detecting and repairing damage or infection; mechanisms for storing reusable information about an extended structured environment; mechanisms for creating, storing and using complex percepts, thoughts, questions, values, preferences, desires, intentions and plans, including plans for cooperative behaviours; and mechanisms that transform themselves into new mechanisms with new structures and functions.

Forms of mathematics used by physicists are not necessarily useful for studying such biological mechanisms. Logic, grammars and map-like representations are sometimes more appropriate, though I think little is actually known about the variety of forms of representation (i.e. encodings of information) used in human and animal minds and brains. We may need entirely new forms of mathematics for biology, and therefore for specifying what physicists need to explain.

Many physicists, engineers and mathematicians who move into neuroscience assume that states and processes in brains need to be expressed as collections of numerical measures and their derivatives plus equations linking them, a form of representation that is well supported by widely used tools such as Matlab but not necessarily best suited for the majority of types of mental content (e.g. grammatical and semantic structures of thoughts like those expressed here). Related challenges are posed by attempts to model chemical processes, where complex molecules form and interact with multiple changing chemical bonds along with changing geometrical and topological relationships—one of the reasons for the original invention of symbolic chemical notations now being extended in computer models of changing interacting molecular structures. (There are many online videos of computer simulations of chemical reactions including protein folding processes.)

6.4 Representing Rewards, Preferences, Values

It is often assumed that all intelligent decision making uses positive or negative scalar rewards or utility values that are comparable across options [26]. But careful attention to consumer magazines, political debates, and the varieties of indecision in normal human life shows that reality is far more complex. For example, many preferences are expressed in rules about how to choose between certain options. Furthermore, preferences can be highly sensitive to changes in context. A crude example is the change in preference for type of car after having children. Analysis of examples in consumer reports led to the conclusion that "better" is a complex, polymorphic, logical concept with a rich structure that cannot be reduced to simple comparisons of numerical values [42, 43]. Instead of a linear reward or utility metric, choices for intelligent individuals or for natural selection involve a complex network of partial orderings, with "annotated" links between nodes (e.g. "better" qualified by conditions: "better for", "better if", "better in respect of"). In the Birmingham CogAff project [50], those ideas informed computational models of simple agents with complex choices to be made under varying conditions, but the project merely scratched the surface, as reported in [4, 5, 70, 71]. Most AI/Cognitive Science models use much shallower notions of motivation.

Despite all the sophistication of modern psychology and neuroscience, I believe they currently lack the conceptual resources required to describe either functions of brains in dealing with these matters, including forms of development and learning required, or the mechanisms implementing those functions. In particular, we lack deep explanatory theories about mechanisms that led to mathematical discoveries over thousands of years, including brain mechanisms producing mathematical conjectures, proofs, counter-examples, proof-revisions, new scientific theories, new works of art and new styles of art. In part that's because models considered so far lack both sufficiently rich forms of information processing (computation) and sufficiently deep methodologies for identifying what needs to be explained. There are other unexplained phenomena concerned with artistic creation and enjoyment, and the mechanisms involved in finding something funny.

7 Computational/Information-Processing Construction Kits

Since the mid-twentieth century, we have been learning about abstract construction kits whose products are machines that can be used for increasingly complex tasks. Such construction kits include programming languages, operating systems, software development tools and environments, and network technology that allows ever more complex information-processing machines to be constructed by combining simpler ones. A crucial, but poorly understood, feature of that history is the growing use of construction kits based on virtual machinery, mentioned in Sect. 2. A complete account of the role of construction kits in biological evolution would need to include an explanation of how the fundamental construction kit (FCK) provided by the physical universe could be used by evolution to produce an increasing variety of types of *virtual* machinery as well as increasingly varied *physical* structures and mechanisms.

7.1 *Infinite, or Potentially Infinite, Generative Power*

A construction kit implicitly specifies a large, in some cases infinite, set of possibilities, though as an instance of the kit is constructed, each addition of a new component or feature changes the set of possibilities accessible in later steps of that construction process. For example, as you construct a sentence or phrase in a language, at each state in the construction there are alternative possible additions (not necessarily at the end) and each of those additions will alter the set of possible further additions consistent with the vocabulary and grammar of the language. When use of language is embedded in a larger activity, such as composing a poem, that context can modify the constraints that are relevant. Chemistry does something like that for types of molecule, types of process involving molecular changes, and types of structure made of multiple molecules. Quantum mechanics added important constraints to nineteenth-century chemistry, including both the possibility of highly stable structures (resistant to thermal buffetting) and also locks and keys as in catalysis. All of that is essential for life as we know it, and also for forms of information processing produced by evolution (mostly not yet charted).

Research in fundamental physics is a search for the construction kit that has the generative power to accommodate all the varieties of matter, structure, process, and causation that can exist in our universe. However, physicists generally seek only to ensure that their construction kits are capable of accounting for phenomena observed in the physical sciences, most of which do not include production of living matter, or processes of evolution, development, learning, and mathematical discovery found in living organisms. Most do not try to ensure that their fundamental theories can account for those features also. There are notable exceptions, including Schrödinger's 1944 book, but most physicists (understandably) ignore most of the details of life, including the variety of forms it can take, the variety of environments coped with, the different ways in which individual organisms cope and change, the ways in which products of evolution become more complex and more diverse over time, and the many kinds of information processing and control both in individuals and in colonies (e.g. ant colonies), societies, and ecosystems.

If cosmologists and other theoretical physicists attempted to account for a wider range of biological phenomena, including the phenomena discussed here in connection with the Meta-Morphogenesis project, they would find considerable explanatory gaps between current physical theories and the diversity of phenomena of life, not because there is something about life that goes beyond what science can explain, but because we do not yet have a sufficiently rich theory of the constitution of the universe, including the Fundamental Construct Kit. In part that seems to be a consequence of the forms of mathematics known to physicists. The challenge presented by Anderson [1] discussed in Sect. 10, below, supports this.

It may take many years of research to find out what is missing from current physics. Collecting phenomena that need to be explained, and trying as hard as possible to construct *detailed* explanations of those phenomena, including working models, is one way to make progress. That may pinpoint gaps in our theories and stimulate development of new, more powerful, theories. Compare the profound ways in which our understanding of possible forms of computation has been extended by unending attempts to put computation to new uses. Collecting examples of such challenges helps us assemble tests to be passed by future proposed theories: samples of possibilities that a deep physical theory needs to be able to explain.

Perhaps the most tendentious proposal here is that an expanded physical theory, instead of being expressed mainly in terms of equations relating measures, may need formalisms better suited to specification of a construction kit, perhaps sharing features of grammars, programming languages, partial orderings, topological relationships, architectural specifications, and the structural descriptions in chemistry. The theory will need to use appropriate kinds of mathematics for drawing out implications of the theories, including explanations of possibilities, both observed and unobserved, including possible future forms of intelligence. Theories of utility measures may need to be replaced, or enhanced with new theories of how benefits, evaluations, comparisons and preferences can be expressed (attempted in [42]). We must also avoid assuming optimality. Evolution produces designs as diverse as microbes, cockroaches, elephants and orchids, none of which is optimal or rational in any simple sense, yet many of them survive and sometimes proliferate, because they are lucky, at least for a while, as with human decisions, policies, preferences, cultures, etc.

8 Types and Levels of Explanation of Possibilities

Suppose someone uses a Meccano kit to construct a toy crane, with a jib that can be moved up and down by turning a handle, and a rotating platform on a fixed base that allows the direction of the jib to be changed. What's the difference between explaining how that is possible and how it was done? First of all, if nobody actually builds such a crane then there is no actual crane-building to be explained. Yet, insofar as the Meccano kit makes such cranes possible it makes sense to ask *how* it is possible. This has several types of answer, including answers at different levels of abstraction, with varying generality and economy of specification.

More generally, the question "How is it possible to create X using construction kit Y?", or, simply, "How is X possible?", has several types of answer, including answers at different levels of abstraction, with varying generality. I'll assume that a particular construction kit is referred to either explicitly or implicitly. The following is not intended to be an exhaustive survey of the possible types of answer. It is merely a first experimental foray, preparing the ground for future work:

1. **Structural conformity:** The first type of answer, structural conformity (grammaticality), merely identifies the parts and relationships between parts that are supported by the kit, showing that X (e.g. a crane of the sort in question) could be composed of such parts arranged in such relationships. An architect's drawings for a building, specifying materials, components, and their spatial and functional

relations, would provide such an explanation of how a proposed building is possible, including, perhaps, answering questions about how the construction would make the building resistant to very high winds, or to earthquakes up to a specified strength. This can be compared with showing that a sentence is acceptable in a language with a well-defined grammar by showing how the sentence would be parsed (analysed) in accordance with the grammar of that language. A parse tree (or graph) also shows how the sentence can be built up piecemeal from words and other grammatical units by assembling various substructures and using them to build larger structures. Compare this with using a chemical diagram to show how a collection of atoms can make up a particular molecule, e.g. the ring structure of C_6H_6 (Benzene).

Some structures are specified in terms of piecewise relations, where the whole structure cannot possibly exist, because the relations cannot hold simultaneously, e.g. X is above Y, Y is above Z, Z is above X. It is possible to depict such objects, e.g. in pictures of impossible objects by Reutersvard, Escher, Penrose, and others.[31] Some logicians and computer scientists have attempted to design languages in which specifications of impossible entities are necessarily syntactically ill-formed. This leads to impoverished languages with restricted practical uses, e.g. strongly typed programming languages. For some purposes less restricted languages, needing greater care in use, are preferable, including human languages, as I have tried to show in [44].

2. **Process possibility:** The second type of answer demonstrates constructability by describing a sequence of spatial trajectories by which such a collection of parts could be assembled. This may include processes of assembly of temporary scaffolding (Sect. 2.7) to hold parts in place before the connections have been made that make them self-supporting or before the final supporting structures have been built (as often happens in large engineering projects, such as bridge construction). Many different possible trajectories can lead to the same result. Describing (or demonstrating) any such trajectory explains both how that construction process is possible and how the end result is possible. There may be several different routes to the same end result.

   In some cases, a complex object has type 1 possibility although not type 2. For example, from a construction kit containing several rings it is possible to assemble a *pile* of three rings, but not possible to assemble a *chain* of three rings even though each of the parts of the chain is exactly like the parts of the pile.

3. **Process abstraction:** Some possibilities are described at a level of abstraction that ignores detailed routes through space, and covers *many* possible alternatives. For example, instead of specifying precise trajectories for parts as they are assembled, an explanation can specify the initial and final state of each trajectory, where each state-pair may be shared by a vast, or even infinite, collection of different possible trajectories producing the same end state, e.g. in a continuous space.

[31] http://www.cs.bham.ac.uk/research/projects/cogaff/misc/impossible.html.

In some cases, the possible trajectories for a moved component are all continuously deformable into one another (i.e. they are topologically equivalent); for example the many spatial routes by which a cup could be moved from a location where it rests on a table to a location where it rests on a saucer on the table, without leaving the volume of space above the table. Those trajectories form a continuum of possibilities that is too rich to be captured by a parametrized equation for a line with a number of variables. If trajectories include passing through holes, or leaving and entering the room via different doors or windows, then the different possible trajectories will not all be continuously deformable into one another: there are different equivalence classes of trajectories sharing common start and end states, for example, the different ways of threading a shoe lace with the same end result.

The ability to abstract away from detailed differences between trajectories sharing start and end points, thereby implicitly recognizing invariant features of an infinite collection of possibilities, is an important aspect of animal intelligence that I don't think has been generally understood. Many researchers assume that intelligence involves finding *optimal* solutions. So they design mechanisms that search using an optimisation process, ignoring the possibility of mechanisms that can find sets of possible solutions (e.g. routes) initially considered as a class of *equivalent* options, leaving questions about optimal assembly to be settled later, if needed. These remarks are closely related to the origins of abilities to reason about geometry and topology.[32]

4. **Grouping:** Another form of abstraction is related to the difference between **1** and **2**. If there is a sub-sequence of assembly processes whose order makes no difference to the end result, they can be grouped to form an unordered "composite" move containing an unordered set of moves. If N components are moved from initial to final states in a sequence of N moves, and it makes no difference in what order they are moved, merely specifying the set of N possibilities without regard for order collapses N factorial sets of possible sequences into one composite move. If N is 15, that will collapse 1,307,674,368,000 different sequences into one.

   Sometimes a subset of moves can be made in parallel. For example someone with two hands can move two or more objects at a time while transferring a collection of items from one place to another. Parallelism is particularly important in many biological processes where different processes occurring in parallel constrain one another so as to ensure that instead of all the possible states that could occur by moving or assembling components separately, only those end states occur that are consistent with parallel constructions. In more complex cases, the end state may depend on the relative speeds of sub-processes

[32]Illustrated in these discussion notes:
http://www.cs.bham.ac.uk/research/projects/cogaff/misc/changing-affordances.html
http://www.cs.bham.ac.uk/research/projects/cogaff/misc/triangle-theorem.html
http://www.cs.bham.ac.uk/research/projects/cogaff/misc/torus.html.

and also on continuously changing spatial relationships. This is important in epigenesis, since all forms of development from a single cell to a multi-celled structure depend on many mutually constraining processes occurring in parallel.

For some construction kits, certain constructs made of a collection of subassemblies may require different subassemblies to be constructed in parallel if completing some too soon could make the required final configuration unachievable. For example, rings being completed before being joined could prevent formation of a chain.

5. **Iterative or recursive abstraction:** Some process types involve unspecified numbers of parts or steps, although each instance of the type has a definite number, for example a process of moving chairs by repeatedly carrying a chair to the next room until there are no chairs left to be carried, or building a tower from a collection of bricks, where the number of bricks can be varied. A specification that abstracts from the number can use a notion like "repeat until", or a recursive specification: a very old idea in mathematics, such as Euclid's algorithm for finding the highest common factor of two numbers. Production of such a generic specification can demonstrate a large variety of possibilities inherent in a construction kit in an extremely powerful and economical way. Many new forms of abstraction of this type have been discovered by computer scientists developing programming languages, for operating not only on numbers but many other structures, e.g. trees and graphs.

   Evolution may also have "discovered" many cases long before humans existed by taking advantage of mathematical structures inherent in the construction kits available and the trajectories by which parts can be assembled into larger wholes. This may be one of the ways in which evolution produced powerful new genomes, and reusable genome components that allowed many different biological assembly processes to result from a single discovery, or a few discoveries, at a high enough level of abstraction.

   Some related abstractions may have resulted from parametrisation: processes by which details are removed from specifications in genomes and left to be provided by the context of development of individual organisms, including the physical or social environment. (See Sect. 2.3 on epigenesis.)

6. **Self-assembly:** If, unlike construction of a toy Meccano crane or a sentence or a sorting process, the process to be explained is a self-assembly process, like many biological processes, then the explanation of how the assembly is possible will not merely have to specify trajectories through space by which the parts become assembled, but also:

   - what causes each of the movements (e.g. what manipulators are required);
   - where the energy required comes from (an internal store, or external supply?);
   - whether the process involves pre-specified information about required steps or required end states, and, if so, what mechanisms can use that information to control the assembly process;
   - how that prior information structure (e.g. specification of a goal state to be achieved, or plan specifying actions to be taken) came to exist, e.g. whether it

was in the genome as a result of previous evolutionary transitions, or whether it was constructed by some planning or problem-solving mechanism in an individual, or whether it was provided by a communication from an external source;
- how these abilities can be acquired or improved by learning or reasoning processes or by random variation (if they can).

7. Use of explicit intentions and plans: None of the explanation types above presupposes that the possibility being explained has ever been represented explicitly by the machines or organisms involved. Explaining the possibility of some structure or process that results from intentions or plans would require specifying pre-existing information about the end state and in some cases also intermediate states, namely information that existed before the process began—information that can be used to control the process (e.g. intentions, instructions, or sub-goals, and preferences that help with selections between options). It seems that some of the reproductive mechanisms that depend on parental care make use of mechanisms that generate intentions and possibly also plans in carers, for instance intentions to bring food to an infant, intentions to build nests, intentions to carry an infant to a new nest, intention to migrate to another continent when the temperature drops, and many more. Use of intentions that can be carried out in multiple ways selected according to circumstances rather than automatically triggered reflexes could cover a far wider variety of cases, but would require provision of greater intelligence in individuals.

Sometimes an explanation of possibility prior to construction is important for engineering projects where something new is proposed and critics believe that the object in question could not exist, or could not be brought into existence using available known materials and techniques. The designer might answer sceptical critics by combining answers of any of the above types, depending on the reasons for the scepticism.

Concluding Comment on Explanations of Possibilities Those are all examples of components of explanations of assembly processes, including self-assembly. In biological reproduction, growth, repair, development, and learning there are far more subdivisions to be considered, some of them already studied piecemeal in a variety of disciplines. In the case of human development, and to a lesser extent development in other species, there are many additional sub-cases involving construction kits both for creating information structures and for creating information-processing mechanisms of many kinds, including perception, learning, motive formation, motive comparison, intention formation, plan construction, plan execution, language use, and many more. A subset of cases with further references can be found in [51].

The different answers to "How is it possible to construct this type of object?" may be correct as far as they go, though some provide more detail than others. More subtle cases of explanations of possibility include differences between reproduction via egg-laying and reproduction via parturition, especially when followed by caring for offspring. The latter allows a parent's influence to continue during development, as does teaching of younger individuals by older ones. This also allows development of cultures suited to different environments.

To conclude this rather messy section: the investigation of different types of generality in modes of explanation for possibilities supported by a construction kit is also relevant to modes of specification of new designs based on the kit. Finding economical forms of abstraction may have many benefits, including both reducing search spaces when trying to find a new design and also providing a generic design that covers a broad range of applications tailored to detailed requirements. Of particular relevance in a biological context is the need for designs that can be adjusted over time, e.g. during growth of an organism, or shared across species with slightly different physical features or environments. Many of the points made here are also related to structural changes in both computer programming languages and software design specification languages. Evolution may have beaten us to important ideas. That these levels of abstraction are possible is a metaphysical feature of the universe, implied by the generality of the FCK.

9 Alan Turing's Construction Kits

Turing [62] showed that a rather simple sort of machine, now known as a Turing machine, could be used to specify an infinite set of constructions with surprisingly rich mathematical features. The set of possibilities was infinite because a Turing machine is defined to have an infinite (or indefinitely extendable) linear "tape" divided into discrete locations in which symbols can be inserted. A feature of a Turing machine that is not in most other construction kits is that it can be set up and then started, after which it will modify initial structures and build new ones, possibly indefinitely, though in some cases the machine will eventually halt.

Another type of construction kit with related properties is Conway's Game of Life,[33] a construction kit that creates changing patterns in 2D regular arrays. Stephen Wolfram has written a great deal about the diversity of constructions that can be explored using such cellular automata. Neither a Turing machine nor a Conway game has any external sensors: once started they run according to their stored rules and the current (changing) state of the tape or grid cells. In principle, either of them could be attached to external sensors able to produce changes to the tape of a Turing machine or the states of some of the cells in the Life array. However, any such extension would significantly alter the powers of the machine, and theorems about what such a machine could or could not do would change.

Modern computers use a variant of the Turing machine idea, where each computer has a finite memory but with the advantage of much more direct access between the central computer mechanism and the locations in the memory (a von Neumann architecture). Increasingly, computers have also been provided with

[33]http://en.wikipedia.org/wiki/Conway.27s.Game.of.Life.

a variety of external interfaces connected to sensors or motors so that while running they can acquire information (e.g. from keyboards, buttons, joysticks, mice, electronic piano keyboards, network connections, and many more) and can also send signals to external devices. Theorems about disconnected Turing machines may not apply to machines with rich two-way interfaces connected to the environment.

Turing machines and Game of Life machines can be described as "self-propelling" because, once set up, they can be left to run according to the general instructions they have and the initial configuration on the tape or in the array. But they are not really self-propelling: they have to be implemented in physical machines with an external power supply. In contrast, Ganti [15] shows how the use of chemistry as a construction kit provides "self-propulsion" for living things, though every now and again the chemicals need to be replenished. A battery-driven computer is a bit like that, but someone else has to make the battery.

Living things make and maintain themselves, at least after being given a kick-start by their parent or parents. They do need constant, or at least frequent, external inputs, but, for the simplest organisms, those are only chemicals in the environment and energy from chemicals or heat energy via radiation, conduction or convection. John McCarthy pointed out in a conversation that some animals also use externally supplied mechanical energy, e.g. rising air currents used by birds that soar. Unlike pollen grains, spores, etc. propagated by wind or water, the birds use internal information-processing mechanisms to control how the wind energy is used, as does a human piloting a glider.

9.1 Beyond Turing Machines: Chemistry

Turing also explored other sorts of construction kits, including types of neural nets and extended versions of Turing machines with "oracles" added. Shortly before his death (in 1954), he published [64], in which he explored a type of pattern-forming construction kit in which two chemical substances can diffuse through the body of an expanding organism and interact strongly wherever they meet. He showed that that sort of construction kit could generate many of the types of surface pattern observed on plants and animals. I have been trying to show how that can be seen as a very simple example of something far more general.

One of the important differences between types of construction kit mentioned above is the difference between kits supporting only discrete changes, e.g. to a first approximation Lego and Meccano (ignoring variable length strings and variable angle joints), and kits supporting continuous variation, e.g. plasticine and mud (ignoring, for now, the discreteness at the molecular level).

One of the implications of such differences is how they affect abilities to search for solutions to problems. If only big changes in design are possible, the precise change needed to solve a problem may be inaccessible (as many who have played with construction kits will have noticed). On the other hand, if the kit allows arbitrarily small changes, it will, in principle, permit exhaustive searches in some sub-spaces. The exhaustiveness comes at the cost of a very much larger (infinite, or potentially infinite!) search-space. That feature could be useless, unless the space of requirements has a structure that allows approximate solutions to be useful. In that case, a mixture of big jumps to get close to a good solution, followed by small jumps to home in on a (locally) optimal solution can be very fruitful—a technique that has been used by Artificial Intelligence researchers, called "simulated annealing".[34]

Wagner [66] claims that the structure of the search space generated by the molecules making up the genome increases the chance of useful approximate solutions to important problems to be found with *relatively* little searching (compared with other search spaces), after which small random changes allow improvements to be found. I have not yet read the book, but it seems to illustrate the importance for evolution of the types of construction kit available.[35] I have not yet had time to check whether the book discusses uses of abstraction and the evolution of mathematical and meta-mathematical competences discussed here. Nevertheless, it seems to be an (unwitting) contribution to the Meta-Morphogenesis project. Recent work by Jeremy England at MIT[36] may turn out also to be relevant.

9.2 Using Properties of a Construction Kit to Explain Possibilities

A formal axiomatic system can be seen as an abstract construction kit with axioms and rules that support construction of proofs that end in theorems. The theorems are formulae that can occur at the end of a proof using only axioms and inference rules in the system. The kit explains the possibility of some theorems based on the axioms and rules. The non-theorems of an axiomatic system are formulae for which no such proof exists. Proving that something is a non-theorem can be difficult, and requires a proof in a meta-system.

Likewise, a physical construction kit can be used to demonstrate that some complex physical objects can occur at the end of a construction process. In some cases there are objects that are describable but cannot occur in a construction using that kit: e.g. an object whose outer boundary is a surface that is everywhere curved cannot be produced in a construction based on Lego bricks or a Meccano set, though one could occur in a construction based on plasticine or soap film.

[34]One of many online explanations is at http://www.theprojectspot.com/tutorial-post/simulated-annealing-algorithm-for-beginners/6.

[35]An interview with the author is online at https://www.youtube.com/watch?v=wyQgCMZdv6E.

[36]https://www.quantamagazine.org/20140122-a-new-physics-theory-of-life/

9.3 Bounded and Unbounded Construction Kits

A rectangular grid of squares combined with the single digit numbers, 0,1,...,9 (strictly numerals representing numbers), allows construction of a set of configurations in which numbers are inserted into the squares subject to various constraints, e.g. whether some squares can be left blank, whether certain pairs of numbers can be adjacent, or whether the same number can occur in more than one square. For a given grid and a given set of constraints, there will be a finite set of possible configurations (although it may be a very large set). If, in addition to insertion of a number, the "construction kit" allows extra empty rows or columns to be added to the grid no matter how large it is, then the set of possible configurations becomes infinite. Many types of infinite construction kit have been investigated by mathematicians, logicians, linguists, computer scientists, musicians and other artists.

Analysis of chemistry-based construction kits for information-processing systems would range over a far larger class of possible systems than Turing machines (or digital computers), because of the mixture of discrete and continuous changes possible when molecules interact, e.g. moving together, moving apart, folding, and twisting, but also locking and unlocking, using catalysts [20]. I don't know whether anyone has a deep theory of the scope and limits of chemistry-based information processing.

Recent discoveries indicate that some biological mechanisms use quantum-mechanical features of the FCK that we do not yet fully understand, providing forms of information processing that are very different from what current computers do. For example a presentation by Seth Lloyd summarises quantum phenomena used in deep-sea photosynthesis, avian navigation, and odour classification.[37] This may turn out to be the tip of the iceberg of quantum-based information-processing mechanisms.

There are some unsolved, very hard, partly ill-defined, problems about the variety of functions of biological vision, e.g. simultaneously interpreting a very large, varied and changing collection of visual fragments, perceived from constantly varying viewpoints as you walk through a garden with many unfamiliar flowers, shrubs, bushes, etc. moving irregularly in a changing breeze. Could some combination of quantum entanglement and non-local interaction play a role in rapidly and simultaneously processing a large collection of mutual constraints between multiple visual fragments? The ideas are not yet ready for publication, but work in progress is recorded here: http://www.cs.bham.ac.uk/research/projects/cogaff/misc/quantum-evolution.html.

Some related questions about perception of videos of fairly complex moving plant structures are raised here: http://www.cs.bham.ac.uk/research/projects/cogaff/misc/vision/plants/.

[37]https://www.youtube.com/watch?v=wcXSpXyZVuY

10 Conclusion: Construction Kits for Meta-Morphogenesis

A useful survey by Keller of previous attempts to show how life and its products relate to the physical world [21, 22] concluded that attempts so far have not been successful. Keller ends with the suggestion that the traditional theory of dynamical systems is inadequate for dealing with constructive processes and needs to be expanded to include "objects, their internal properties, their construction, and their dynamics", i.e. a theory of *"Constructive dynamical systems"*. This chapter outlines a project to do that and more, including giving an account of branching layers of new *derived* construction kits produced by evolution, development and other processes. The physical world clearly provides a very powerful (chemistry-based) fundamental construction kit that, together with natural selection processes and processes within individuals as they develop, produced an enormous variety of organisms on this planet, based on additional derived construction kits (DCKs), including concrete, abstract and hybrid construction kits and, most recently, new sorts of construction kit used as a toy or an engineering resource.

The idea of a construction kit is offered as a new unifying concept for Philosophy of mathematics, Philosophy of science, Philosophy of biology, Philosophy of mind and Metaphysics. The aim is to explain how it is possible for minds to exist in a material world and to be produced by natural selection and its products. Related questions have been raised about the nature of mathematics and its role in life. The ideas are still at an early stage of development and there are probably many more distinctions to be made, and a need for a more formal, mathematical presentation of properties of and relationships between construction kits, including the ways in which new derived construction kits can be related to their predecessors and their successors. The many new types of computer-based *virtual* machinery produced by human engineers since around 1950 provide examples of non-reductive supervenience (as explained in [56]). They are also useful as relatively simple examples to be compared with far more complex products of evolution.

In [13], a distinction is made between two "principled" options for the relationship between the basic constituents of the world and their consequences. In the "Humean" option there is nothing but the distribution of structures and processes over space and time, though there may be some empirically discernible patterns in that distribution. The second option is "modal realism", or "dispositionalism", according to which there is something about the primitive stuff and its role in space-time that constrains what can and cannot exist, and what types of process can and cannot occur.

I am arguing for a "multi-layer" version of the modal realist option (developing ideas in [41, 48, 56]).

I suspect that a more complete development of this form of modal realism can contribute to answering the problem posed in Anderson's famous paper [1], namely how we should understand the relationships between different levels of complexity in the universe (or in scientific theories). The reductionist alternative claims that when the physics of elementary particles (or some other fundamental physical level) has been fully understood, everything else in the universe can be explained in terms of mathematically derivable consequences of the basic physics. Anderson contrasts this with the anti-reductionist view that different levels of complexity in the universe require "entirely new laws, concepts and generalisations" so that, for example, biology is not applied chemistry and psychology is not applied biology. He writes: "Surely there are more levels of organization between human ethology and DNA than there are between DNA and quantum electrodynamics, and each level can require a whole new conceptual structure". However, the structural levels are not merely in the concepts used by scientists, but actually in the world.

We still have much to learn about the powers of the fundamental construction kit (FCK), including: (1) the details of how those powers came to be used for life on earth, (2) what sorts of derived construction kit (DCK) were required in order to make more complex life forms possible, (3) how those construction kits support "blind" mathematical discovery by evolution, mathematical competences in humans and other animals and, eventually, meta-mathematical competences, then meta-meta-mathematical competences, at least in humans, (4) what possibilities the FCK has that have not yet been realised, (5) whether and how some version of the FCK could be used to extend the intelligence of current robots, (6) whether currently used Turing-equivalent forms of computation have at least the same information-processing potentialities (e.g. abilities to support all the biological information-processing mechanisms and architectures), and (7) if those forms of computation lack the potential, then how are biological forms of information processing different? Don't expect complete answers soon.

In future, physicists wishing to show the superiority of their theories should attempt to demonstrate mathematically and experimentally that they can explain more of the potential of the FCK to support varied construction kits required for, and produced by, biological evolution than rival theories can. Will that be cheaper than building bigger, better colliders? Will it be harder?[38]

End Note

As I was finishing off this paper, I came across a letter Turing wrote to W. Ross Ashby in 1946 urging Ashby to use Turing's ACE computer to implement his ideas about modelling brains. Turing expressed a view that seems to be unfashionable among AI researchers at present (2015), but accords with the aims of this paper:

> In working on the ACE I am more interested in the possibility of producing models of the actions of the brain than in the practical applications to computing.
> http://www.rossashby.info/letters/turing.html

[38]Here's a cartoon teasing particle physicists: http://www.smbc-comics.com/?id=3554.

It would be very interesting to know whether he had ever considered the question whether digital computers might be incapable of accurately modelling brains making deep use of chemical processes. He also wrote in [63], *"In the nervous system chemical phenomena are at least as important as electrical."* But he did not elaborate on the implications of that claim.[39]

Acknowledgements My work on the Meta-Morphogenesis project, including this paper, owes a great dept to Barry Cooper, who unfortunately died in October 2015. I had never met him until we both contributed chapters to a book published in 2011 on Information and Computation. Barry and I first met, by email, when we reviewed each others' chapters. Later, out of the blue, he invited me to contribute to the Turing centenary volume he was co-editing [11]. I contributed three papers. He then asked me for a contribution to Part 4 (on Emergence and Morphogenesis) based on Turing's paper on morphogenesis published in 1952, 2 years before he died. That got me wondering what Turing might have done if he had lived another 30–40 years. So I offered Barry a paper proposing "The Meta-Morphogenesis Project" as an answer. He accepted it (as the final commentary paper in the book), and ever since then I have been working full-time on the project. He later encouraged me further by inviting me to give talks and to contribute a chapter to this book. As a result we had several very enjoyable conversations. He changed my life by giving me a new research direction, which does not often happen to 75-year old retired academics! (Now 5 years older.) I wish we could continue our conversations. I also owe much to the highly intelligent squirrels and magpies in our garden, who have humbled me. Finally, my sincere thanks to Ronan Nugent (Senior Editor, Springer Computer Science) who took on the enormous workload required for dealing with the final corrections.

References

1. P.W. Anderson, More is different. Sci. New Ser. **177**(4047), 393–396 (1972)
2. W.R. Ashby, *Design for a Brain* (Chapman and Hall, London, 1952)
3. W.R. Ashby, *An Introduction to Cybernetics* (Chapman and Hall, London, 1956)
4. L.P. Beaudoin, Goal processing in autonomous agents. PhD thesis, School of Computer Science, The University of Birmingham, Birmingham, 1994
5. L.P. Beaudoin, A. Sloman, A study of motive processing and attention, in *Prospects for Artificial Intelligence*, ed. by A. Sloman, D. Hogg, G. Humphreys, D. Partridge, A. Ramsay (IOS Press, Amsterdam, 1993), pp. 229–238
6. G. Bell, *Selection The Mechanism of Evolution*, 2nd edn. (Oxford University Press, Oxford, 2008)
7. K. Bennett, Construction zone: no hard hat required. Philos. Stud. **154**, 79–104 (2011)
8. J.M. Chappell, A. Sloman, Natural and artificial meta-configured altricial information-processing systems. Int. J. Unconv. Comput. **3**(3), 211–239 (2007)
9. N. Chomsky, *Aspects of the Theory of Syntax* (MIT Press, Cambridge, MA, 1965)
10. J. Coates, Umm-E-Aiman, B. Charrier, Understanding "green" multicellularity: do seaweeds hold the key? Front. Plant Sci. (2014). doi: 10.3389/fpls.2014.00737
11. S.B. Cooper, J. van Leeuwen (eds.) *Alan Turing: His Work and Impact* (Elsevier, Amsterdam, 2013)

[39]I think it will turn out that the ideas about "making possible" used here are closely related to Alastair Wilson's ideas about grounding as "metaphysical causation" [69].

12. T.W. Deacon, *Incomplete Nature: How Mind Emerged from Matter* (W. W. Norton and Company, New York, 2011)
13. M. Esfeld, D. Lazarovici, V. Lam, M. Hubert, The physics and metaphysics of primitive stuff. Br. J. Philos. Sci. **67**(4), 1–29 (2015)
14. C. Fernando, Review of "*The Principles of Life*" by Tibor Ganti. (2003, Oxford University Press.). Artif. Life **14**(4), 467–470 (2008)
15. T. Ganti, in *The Principles of Life*, ed. by E. Szathmáry, J. Griesemer. (Oxford University Press, New York, 2003). Translation of the 1971 Hungarian edition
16. J.J. Gibson, *The Senses Considered as Perceptual Systems* (Houghton Mifflin, Boston, 1966)
17. J.J. Gibson, *The Ecological Approach to Visual Perception* (Houghton Mifflin, Boston, 1979)
18. I. Kant, *Critique of Pure Reason* (Macmillan, London, 1781). Translated (1929) by Norman Kemp Smith
19. A. Karmiloff-Smith, *Beyond Modularity: A Developmental Perspective on Cognitive Science* (MIT Press, Cambridge, MA, 1992)
20. S. Kauffman, *At Home in the Universe: The Search for Laws of Complexity* (Penguin Books, London, 1995)
21. E.F. Keller, Organisms, machines, and thunderstorms: a history of self-organization, part one. Hist. Stud. Nat. Sci. **38**(1 (Winter)), 45–75 (2008)
22. E.F. Keller, Organisms, machines, and thunderstorms: a history of self-organization, part two. Complexity, emergence, and stable attractors. Hist. Stud. Nat. Sci. **39**(1 (Winter)), 1–31 (2009)
23. G. Korthof, Review of *The Principles of Life* by Tibor Ganti (2003). Updated 6 Oct 2014
24. J.E. Laird, A. Newell, P.S. Rosenbloom, SOAR: an architecture for general intelligence. Artif. Intell. **33**, 1–64 (1987)
25. I. Lakatos, Falsification and the methodology of scientific research programmes, in *Philosophical Papers, Vol I*, ed. by J. Worrall, G. Currie (Cambridge University Press, Cambridge, 1980), pages 8–101
26. R.D. Luce, H. Raiffa, *Games and Decisions: Introduction and Critical Survey* (Wiley, New York; Chapman and Hall, London, 1957)
27. C. Mathis, T. Bhattacharya, S.I. Walker, The emergence of life as a first order phase transition. Tech. rep., Arizona State University (2015)
28. J. McCarthy, Ascribing mental qualities to machines, in *Philosophical Perspectives in Artificial Intelligence*, ed. by M. Ringle (Humanities Press, Atlantic Highlands, 1979), pp. 161–195, http://www-formal.stanford.edu/jmc/ascribing/ascribing.html
29. J. McCarthy, P.J. Hayes, Some philosophical problems from the standpoint of AI, in *Machine Intelligence 4*, ed. by B. Meltzer, D. Michie (Edinburgh University Press, Edinburgh, 1969), pp. 463–502, http://www-formal.stanford.edu/jmc/mcchay69/mcchay69.html
30. M.L. Minsky, *The Society of Mind* (William Heinemann, London, 1987)
31. M.L. Minsky, *The Emotion Machine* (Pantheon, New York, 2006)
32. R. Penrose, *Shadows of the Mind: A Search for the Missing Science of Consciousness* (Oxford University Press, Oxford, 1994)
33. K.R. Popper, *The Logic of Scientific Discovery* (Routledge, London, 1934)
34. K.R. Popper, *Unended Quest* (Fontana/Collins, Glasgow, 1976)
35. K.R. Popper, Natural Selection and the Emergence of Mind. Dialectica **32**(3–4), 339–355 (1978)
36. W.T. Powers, *Behavior, the Control of Perception* (Aldine de Gruyter, New York, 1973)
37. E. Schrödinger, *What Is Life?* (Cambridge University Press, Cambridge, 1944)
38. A. Senghas, Language emergence: clues from a new Bedouin sign language. Curr. Biol. **15**(12), R463–R465 (2005)
39. C. Shannon, A mathematical theory of communication. Bell Syst. Tech. J. **27**, 379–423, 623–656 (1948)
40. H.A. Simon, Motivational and emotional controls of cognition, in *Reprinted in Models of Thought*, ed. by H.A. Simon (Yale University Press, Newhaven, 1967), pp. 29–38
41. A. Sloman, Knowing and understanding: relations between meaning and truth, meaning and necessary truth, meaning and synthetic necessary truth (DPhil Thesis). PhD thesis, Oxford University, 1962. http://www.cs.bham.ac.uk/research/projects/cogaff/07.html#706

42. A. Sloman, How to derive "better" from "is". Am. Philos. Q. **6**, 43–52 (1969)
43. A. Sloman, "Ought" and "better". Mind **LXXIX**(315), 385–394 (1970)
44. A. Sloman, Tarski, Frege and the liar paradox. Philosophy **46**(176), 133–147 (1971). http://www.cs.bham.ac.uk/research/projects/cogaff/62-80.html#1971-03
45. A. Sloman, *The Computer Revolution in Philosophy* (Harvester Press/Humanities Press, Hassocks, Sussex/New Jersey, 1978). http://www.cs.bham.ac.uk/research/cogaff/62-80.html#crp,2015
46. A. Sloman, Image interpretation: the way ahead?, in *Physical and Biological Processing of Images (Proceedings of an International Symposium Organised by The Rank Prize Funds, London, 1982.)*, ed. by O.J. Braddick, A.C. Sleigh (Springer, Berlin, 1983), pp. 380–401
47. A. Sloman, The mind as a control system, in *Philosophy and the Cognitive Sciences*, ed. by C. Hookway, D. Peterson (Cambridge University Press, Cambridge, 1993), pp. 69–110
48. A. Sloman, Actual possibilities, in *Principles of Knowledge Representation and Reasoning: Proceedings of the 5th International Conference (KR '96)*, ed. by L.C. Aiello, S.C. Shapiro (Morgan Kaufmann Publishers, Boston, 1996), pp. 627–638
49. A. Sloman, The SimAgent TOOLKIT – for Philosophers and Engineers (And Some Biologists, Psychologists and Social Scientists) (1996), http://www.cs.bham.ac.uk/research/projects/poplog/packages/simagent.html
50. A. Sloman, The cognition and affect project: architectures, architecture-schemas, and the new science of mind. Tech. rep., School of Computer Science, University of Birmingham, Birmingham (2003), Revised August 2008
51. A. Sloman, Requirements for a fully deliberative architecture (Or component of an architecture). Research Note COSY-DP-0604, School of Computer Science, University of Birmingham, Birmingham (2006)
52. A. Sloman, Evolution of minds and languages. What evolved first and develops first in children: languages for communicating, or languages for thinking (Generalised Languages: GLs)? (2008)
53. A. Sloman, Architecture-based motivation vs reward-based motivation. Newsl. Philos. Comput. **09**(1), 10–13 (2009)
54. A. Sloman, How virtual machinery can bridge the "explanatory gap", in natural and artificial systems, in *Proceedings SAB 2010, LNAI 6226*, ed. by S. Doncieux et al. (Springer, Heidelberg, 2010), pp. 13–24
55. A. Sloman, What's information, for an organism or intelligent machine? How can a machine or organism mean?, in *Information and Computation*, ed. by G. Dodig-Crnkovic, M. Burgin (World Scientific, New Jersey, 2011), pp. 393–438
56. A. Sloman, Virtual machine functionalism (the only form of functionalism worth taking seriously in philosophy of mind and theories of consciousness). Research note, School of Computer Science, The University of Birmingham (2013)
57. A. Sloman, Virtual machinery and evolution of mind (part 3) meta-morphogenesis: evolution of information-processing machinery, in *Alan Turing - His Work and Impact*, ed. by S.B. Cooper, J. van Leeuwen (Elsevier, Amsterdam, 2013), pp. 849–856
58. A. Sloman, R.L. Chrisley, Virtual machines and consciousness. J. Conscious. Stud. **10**(4–5), 113–172 (2003)
59. R. Sun, The CLARION cognitive architecture: extending cognitive modeling to social simulation, in *Cognition and Multi-Agent Interaction*, ed. by R. Sun (Cambridge University Press, New York, 2006), pp. 79–99. http://www.cogsci.rpi.edu/~rsun/sun.clarion2005.pdf
60. D.W. Thompson, *On Growth and Form* (Cambridge University Press, Cambridge, 1917). Revised Edition 1948
61. A. Trehub, *The Cognitive Brain* (MIT Press, Cambridge, 1991)
62. A.M. Turing, On computable numbers, with an application to the Entscheidungsproblem. Proc. Lond. Math. Soc. **42**(2), 230–265 (1936)
63. A.M. Turing, Computing machinery and intelligence. Mind **59**, 433–460 (1950); Reprinted in E.A. Feigenbaum, J. Feldman (eds.) *Computers and Thought* (McGraw-Hill, New York, 1963), pp. 11–35

64. A.M. Turing, The chemical basis of morphogenesis. Philos. Trans. R. Soc. Lond. B **237**, 37–72 (1952)
65. C.H. Waddington, *The Strategy of the Genes* (Macmillan, New York, 1957)
66. A. Wagner, *Arrival of the Fittest: Solving Evolution's Greatest Puzzle* (Oneworld Publications, London, 2014)
67. A.A.S. Weir, J. Chappell, A. Kacelnik, Shaping of hooks in New Caledonian crows. Science **297**, 981 (2002)
68. N. Wiener, *Cybernetics: or Control and Communication in the Animal and the Machine*, 2nd edn. (The MIT Press, Cambridge, 1961)
69. A. Wilson, Metaphysical causation. Tech. rep., University of Birmingham, Department of Philosophy (2015). Video lecture: https://www.youtube.com/watch?v=j2l1wKrtlxs
70. I.P. Wright, Emotional agents. PhD thesis, School of Computer Science, The University of Birmingham, 1997
71. I.P. Wright, A. Sloman, L.P. Beaudoin, Towards a design-based analysis of emotional episodes. Philos. Psychiatry Psychol. **3**(2), 101–126 (1996)

Printed in the United States
By Bookmasters